ECOLOGY AND CONTROL OF THE NATURAL ENVIRONMENT

YU. A. IZRAEL'
USSR State Committee for Hydrometeorology and Control of the Natural Environment, Moscow, USSR

Ecology and Control of the Natural Environment

Springer Science+Business Media, B.V.

Izraėl', ÎU. A. (ÎUriĭ Antonievich)
[Ėkologiîa i kontrol' sostoîaniîa prirodnoĭ sredy. English]
Ecology and control of the natural environment / by Yu.A. Izrael.
p. cm.
Translation of: Ėkologiîa i kontrol' sostoîaniîa prirodnoĭ sredy. Izd. 2., dop.
Includes bibliographical references.
ISBN 978-94-010-5499-7 ISBN 978-94-011-3390-6 (eBook)
DOI 10.1007/978-94-011-3390-6
1. Ecology. 2. Environmental monitoring. 3. Man--Influence on nature. I. Title.
QH541.I9713 1991
363.7'063--dc20 91-2597

ISBN 978-94-010-5499-7

The first edition of this book received the 1983
V. N. Sukachev Gold Medal of the USSR Academy of Sciences,
awarded for outstanding work in the field of ecology.
This is a revised and extended translation of the
second edition (1984).

Translated from Russian by N. V. Kostyleva (from the Preface and up to and including Chapter 4) and N. Yu. Vail (from Chapter 5 and up to and including the Conclusion)

Translation edited by Ron Hardin

Printed on acid-free paper

Table of Contents

Preface

In recent decades it has become increasingly urgent to protect human health and wellbeing from the possible negative consequences of man's economic activities, both at the actual production sites and in areas where the impact is felt. These negative effects have gradually become more and more widespread, presenting a major hazard to the natural environment, taking on an international character, and assuming global proportions.

For the countries of Europe and North America, transport of pollutants and acid rain across boundaries is a serious problem. After the Chernobyl reactor accident, regular measurements of radioactive isotopes became imperative.

It is obvious that drastic measures, including steps taken on an international level, are required to limit the negative anthropogenic impact on the environment.

Under the conditions of this growing man-caused impact on nature, the existing ecological reserves of the biosphere should be husbanded especially carefully. We must determine the regimes of rational utilization of these reserves and of judicious management of the natural environment, thereby maintaining a high quality of the biosphere and preserving nature's regenerative capacity. Reliable methods should be developed to keep the environment from being overloaded and to safeguard the elements of the biosphere from injury.

Given such a situation, it is of particular importance to have objective information about the critical factors of the human impact and the actual state of the biosphere, as well as to obtain forecasts of its future state.

In this connection, it has become crucial to establish special systems of observations, so as to monitor the state of the natural environment, both at sites of major human impact and on a global scale.

It has also become necessary to determine the maximum permissible ecological loads, so as to limit (normalize) existing anthropogenic effects, allowing for the combined hazardous effect of many factors as well as their potential ecological, economic, and social consequences.

The problem of environmental monitoring has become ecological as well as geophysical. Its social significance is also evident.

In this book the basic approaches to, and the various aspects of, environmental monitoring are presented. The primary emphasis is placed on the geophysical and ecological aspects of monitoring man-induced changes in the natural environment.

During the last few years this system has been supplemented by observations of the background; a system of ecological monitoring is being realized.

Since the publication of the first edition of this book (1979) global ecological problems have come more and more into the limelight. In this connection, a large new section has been added, entitled 'Global ecological problems. Critical anthropogenic effects on the biosphere'.

This work discusses the basic ways of solving the problem of environmental protection, as well as means for practical realization of the contemplated solutions. The subject of international cooperation with regard to this is also dealt with.

Introduction

The renewable and nonrenewable natural resources of our planet, which until recently were considered to be inexhaustible, are being used intensively by man to satisfy his many needs.

The growing technology and power potential of human society are making possible the ever-increasing extraction of minerals and utilization of biological and other natural resources. Man gets from nature practically everything that is necessary for his life: energy, food, materials for building dwellings and constructing machines, etc. In addition, nature is necessary for human recreation, since it is a means of satisfying important esthetic requirements.

Man inevitably alters nature when he interacts with it. Intentional man-made changes in nature are mostly positive; they are necessary for human society: urban centers are constructed, agriculture and industry are developed, and the landscape is improved. Even some unintentional changes may be considered to be factors contributing to the development of the biosphere; sometimes such changes may lead to a favorable simplification of existing biocenoses, to a kind of rejuvenation of these [16].

The state of the natural environment, of the biosphere, is constantly changing. These changes differ in their nature, orientation, and extent, and they are nonuniformly distributed in space and time.

Natural changes in the state of the biosphere possess a very important characteristic, namely that, as a rule, they occur around some average level with respect to the constant level (here we are referring to a large-scale element of the biosphere and to averaging over a long time period). For instance, the climatic characteristics of large regions, the natural compositions of different environments, the cycles of substances in nature, the global biological productivity, and large ecological systems are all relatively constant.

Average values can change significantly only during long time periods (measured in millennia). Variations caused by natural disasters (volcanic eruptions, earthquakes, hurricanes) constitute exceptions; however, they are, as a rule, relatively local.

Anthropogenic changes in the state of the environment, which have become especially pronounced in recent decades, can be quite different in character. In some cases anthropogenic alterations can cause sharp, rapid changes in the state of the natural environment in a given region. Such effects may be subdivided into intentional (positive) changes and unintentional (negative) changes. Changes designed to meet the needs of human society (these were mentioned above) can be classified as intentional positive anthropogenic alterations in the state of the environment. On the other hand, the degradation of huge land areas, the destruction or significant transformation of the ecosystems of many lakes, the pollution of the World Ocean with petroleum products, the desertification of the Sahel region of Africa, and the substantial acidification of the surface water and soil in Europe and North America can be classified as unintentional negative anthropogenic changes in the state of the natural environment.

Some changes are dual in nature. On the one hand, they are positive, in that they achieve some objective (for instance, the construction of a cascade of reservoirs on a large river, to bring power to the region), while, on the other hand, they are negative, due to their ecological effects (for instance, expansion of the area under irrigation decreases the inflow of river water to lakes and seas).

However, the negative results of anthropogenic effects are not an inevitable consequence of the development of human society.

Man-induced deterioration of the natural environment is, as a rule, associated with errors in technological and ecological policies, with an inadequate level of technical development, with ignorance of the possible anthropogenic effects, and with a wide use of the environment to decrease the effects of harmful industrial wastes and of other active factors (e.g. the use of the environment to disperse harmful substances). Attempts at the maximum use of natural resources which are not supported by an adequate knowledge of the capabilities of nature, of its ability to regenerate and regulate itself, can also have serious negative ecological consequences.

The evolution of the biosphere under these new conditions leads us to pose the problem of the interrelations between man and nature in a new way. As stated in Ref. [16]: "Relations between man and nature should be based on trust; man should not assume the functions of the biosphere; rather, he should facilitate its operation. The joint efforts of man and nature can contribute to the formation of stable biogeocenoses in an environment altered by man."

Let us emphasize once again that, in spite of the enormous ecological reserves of the biosphere and the possibilities of evolution under new conditions, some anthropogenic modifications lead to extremely negative consequences which the biosphere cannot cope with (at least not rapidly). This refers, first of all, to pollution of the biosphere by chemical substances not usually found in nature (so-called 'xenobiotic' substances) and to a disturbance of the natural geochemical cycles.

Needless to say, this also pertains to an intensive, irrational use of natural resources, whereby nature's ability to reproduce renewable resources may be undermined, while nonrenewable resources will be depleted before human society manages to reform its economy and its relevant activities.

In such a situation it is extremely important to monitor the state of the natural environment, to observe its continual changes as well as to determine the trends of these changes (both for urgent practical action and for long-range planning). It is necessary to monitor not only the natural changes in the state of the environment, but also the anthropogenic alterations which are superimposed on the natural changes and which sometimes reinforce them.

Besides, we must also study the ecological reserves of the biosphere. A knowledge of these reserves, along with an assessment of the state of the environment, enables us to use properly the existing natural potential and resources and to manage nature and ensure environmental quality on a scientific basis. During such studies, particular attention should be focused on critical factors associated with the instability of natural processes, altering the direction of these processes as a result of relatively small primary changes.

Consequently, the study and detection of anthropogenic changes and fluctuations of the environment, against the background of natural changes and fluctuations, is a complex and urgent problem. This book will deal with precisely this problem.

Various geophysical agencies, such as meteorological, hydrological and agrometeorological services, services for observing the state of the seas and oceans, ionospheric, heliogeophysical, seismic, tsunami warning services, etc., have already been monitoring natural changes in the environment for many years. In the USSR the duties of a majority of these services are entrusted to the State Committee for Hydrometeorology.

The World Weather Watch (WWW) of the World Meteorological Organization coordinates observations of meteorological phenomena and data exchange on an international scale. Since meteorological and hydrological observations of natural changes in the biosphere, as well as methods of assessment and prediction of such changes, have been treated in a great number of works, therefore this monograph will practically not deal with these subjects at all.

In view of the need to distinguish anthropogenic changes in the environment, it was thought wise to establish a special information system for observing and analyzing the state of the environment, and in particular pollutants and their effects on the biosphere. In 1974 the author [8] suggested calling this system "a system for monitoring the state of the environment", or, more exactly, "a system for monitoring anthropogenic changes in the environment". The definition of a monitoring system presented in Ref. [19]:

'a system of repeated observations of the elements of the environment in space and time, for definite purposes and according to a prepared program', should be considered useful, but too general.

In accordance with our definition of a monitoring system in Ref. [8], its basic functions include observations of impact factors and the state of the environment, forecasts of its future state, and assessments of the actual and forecasted environmental states.

It should be emphasized that a monitoring system must not be considered to be a fundamentally new system of observations; rather it must be part of a universal system (service) for observing and monitoring the state of the natural environment. It must be based on the experience of already existing geophysical services and it must use the available required elements of these services [9], although it will also entail the introduction of some additional elements.

The term 'monitoring' is used here for two reasons: (1) the term 'control', which is more commonly used in Russian, includes the elements of management and regulation; 'monitoring', on the other hand, corresponds to an information system which does not include these elements (therefore, 'monitoring' is nearer in meaning to the term 'watch'; however, according to the author's definition, it also includes evaluating and forecasting the state of the environment); (2) the term 'monitoring' is widely used on an international level and is well known; major international projects have as their goal the establishment and development of regional and global monitoring systems.

For these reasons in this book, along with the term 'control', we will use the term 'monitoring', which is already widely used in Soviet scientific literature.*

'Control' of the environment, in a broader sense, will be understood to include monitoring the state of the environment and regulating the environmental quality (e.g., by establishing norms for pollutants), taking into consideration the geophysical, ecological, economic, and social aspects of the problem.

Monitoring includes observing the sources and factors of man-induced chemical, physical (radiation and mechanical), and biological modifications, as well as observing their effects on the environment, especially the responses

---

* The term 'monitoring' was borrowed from English relatively recently by Soviet scientists. In many cases it is used as a synonym for the tried-and-true Russian word 'kontrol' ', which has the sense of a *passive* observation (monitoring) of a quantity.

The English word 'control', on the other hand, implies an *active* regulation of something, except in such rare usages as 'quality control', where it approaches the more passive sense of the Russian 'kontrol' '.

In this work 'kontrol' ' has generally been translated as 'monitoring', except where an active regulation of the quantity is indicated or implied. Note that occasionally the author uses the word 'monitoring' to refer to a 'monitoring system' [Translation Editor].

of biological systems to these manmade intrusions. The observations can be carried out using various physical, chemical, and biological indices; integral indices of the state of natural systems seem to be especially promising. Monitoring should help to identify critical impact factors and to determine the biospheric elements which are most exposed to these effects. The scientific basis of environmental monitoring is described in Chapter 3.

In connection with the problem of the anthropogenic impact, it should be noted that the ability of any system to maintain a stable existence under the conditions of a given environment is achieved at a certain level of organization of this system [15]. Since the level of the organization of a system, expressed in the arrangement of the organisms and their way of functioning, is provided by the given environment, therefore the simplification of biocenoses, which Academician S. S. Shvarts regards as a positive factor for the environment [16], can present a certain hazard to individual structures.

Accordingly, one of the objectives of monitoring should be to find out whether the level of system organization corresponds to the environmental conditions necessary for this level. It is necessary to watch carefully, on the one hand, the state of living matter and its ability to restrain the increase in entropy (by lessening the conversion of free energy to inconvertible forms of energy [15]) and, on the other hand, the accumulation of waste and other effects of the anthropogenic impact, leading to an increase in entropy. It is interesting to note that the author of [11] suggested using the fact that entropy increases when life is maintained and develops to try to detect extraterrestrial civilizations on the basis of the presence of degraded energy (e.g., long-wave radiation). This is a striking example of the possibilities presented by the integral approach to the implementation of monitoring.

When monitoring, it is important to have data both on the abiotic component of the environment and on the state of the biota, the functioning of ecosystems, and their responses to the impact. This entire area is related to the geophysical and ecological aspects of environmental monitoring.

Assessment of the state of the environment is an extremely important element of monitoring.

The selection of the integral characteristics and indices, and the direct measurement of these, already constitute a stage of this assessment, since, on the basis of such measurements or an analysis of the measured quantities, it is possible to answer the question: what is the state of the environment? At the same time, it is impossible to evaluate the state of the environment only according to the intensity of a single factor or according to individual values of the function of the state of some element. Rather, these have to be compared to some specially worked-out criterion (e.g., compared with the concentration of a pollutant having the maximum permisssible concentration).

When determining the anthropogenic impact permissible for an ecosystem (against the background of the natural variability of its state), it is necessary to proceed from the ecological reserves of the given system (or its assimilation capacity) and the interval of permissible variations of its state (within the limits of the homeostatic plateau, and sometimes beyond these limits). It is important here to keep in mind the biological stability of the ecosystem and to take into account the relationships between the impact and the effects produced by this impact, as well as the threshold of these effects. When determining the ecological reserves of the environment (determined by the stability of the biosphere), it is, of course, necessary to know well (and to know how to distinguish) the critical factors of the man-induced impact and the critical elements of the biosphere, the effect on which can lead to sharp, sometimes avalanchelike, changes in the environment, which, on the contrary, demonstrates the instability of the biosphere at these critical points.

The ecological approach to the determination of the permissible loads differs from the health-and-hygiene approach, whose purpose is to determine the permissible environmental pollution (or some other load), first of all for an individual human organism but also for the population as a whole. This makes such an approach quite rigorous.

When establishing ecological norms, it is necessary to take into account the combined collective resistance of the ecosystem (biogeocenosis) to any impact, although in this case some individual species may prove to be a 'critical link'. The basic criterion for determining the permissible ecological loading is the absence of any decrease in the productivity, stability, and diversity of the system; the death of an individual organism does not seem critical in this case. This approach can be used to determine the permissible loading on a community or a biogeocenosis.

Of course, when developing the concept of the permissible ecological loading, we must specify some definite conditions (criteria) for preservation of the environment, of the system (or its individual elements). These conditions, the criteria of environmental quality, may be, and must be, different for unique (protected) and widespread ecosystems.

The concept of high environmental quality, providing for a stable existence and development of the given communities or ecosystems, as well as for the absence of unfavorable consequences for the populations forming the community or ecosystem, and the concept of a 'good' biogeocenosis, characterized by stability, high production and high productivity of all the links of the trophic chains, and heterogeneity of the trophic levels, are closely related to problems of establishing the permissible loads on the systems and problems of setting up ecological norms.

A comprehensive analysis of the environment is a universal tool for studying the most varied anthropogenic effects on the environment, enabling us to assess such effects. This approach was suggested by the author back in

1972.

This method, representing a systems analysis of heterogeneous anthropogenic impacts on the biosphere, assuming many, very different ways of impacting, was devised at the E. K. Fedorov Institute of Applied Geophysics of the USSR State Committee for Hydrometeorology and Environmental Monitoring.* It was later developed within the framework of Soviet-American cooperation as the project *Comprehensive Analysis of the Natural Environment.* From 1974 to 1979 four Soviet-American symposia on this problem were held. Specialists in very different fields attended: geophysicists, ecologists, mathematicians, biologists, hygienists, chemists, geneticists, climatologists, and economists. The proceedings of these symposia were published in the USSR and USA [3–6]. The method of this comprehensive analysis of the environment is described in Chapter 1.

As already noted, the main feature of the comprehensive analysis is a detailed consideration of all the basic aspects of the interaction. If the comprehensive analysis is considered step-by-step (taking into account the above-mentioned most important aspects), then we can say that the first stage is an analysis of the effects of the impacts of various factors on the natural environment, including an analysis of the reactions of the elements of the biosphere, a determination of the different kinds of damage, and an isolation of the critical impact factors and elements of the biosphere.

The objective of the second stage of the analysis is to determine the permissible ecological loads on natural systems.

The third stage of the comprehensive analysis comprises a calculation and determination of the permissible loads on an ecosystem or an entire region, taking into consideration the economic and social aspects. This stage to a great extent determines the 'decision making' process, i.e, the strategy of the regulation (management) of the environment.

The permissible loads on the elements of the biosphere and the principles of establishing ecological standards (norms) are discussed in Chapter 2.

It will always be important to determine the ecological reserves and the critical elements of the biosphere for a reasonable and effective interaction with the natural environment, and to determine the amount of natural resources that can (and should) be extracted without disturbing the basic features of the environment [21].

Regulation (and management) of the environmental quality is based on ecological forecasting, and it requires the development of ecological-economic models.

Environmental quality regulation should begin with a determination of the ecologically permissible human impact, and with the establishment of norms

---

* Now called the USSR State Committee for Hydrometeorology. On 8 January 1988 the USSR State Committee for Environmental Protection was established.

for this impact at each point in the environment and in large zones (e.g., norms for the maximum permissible concentrations (MPCs) and maximum permissible ecological loads (MPELs) for each organism and for the system as a whole, respectively). Moreover, criteria should be worked out to limit the amount of impact, for instance, norms for the maximum permissible discharges (MPDs) of pollutants into the environment.

When establishing these norms, it is necessary to take into consideration a great number of factors, for instance, to establish the MPDs we must take into account the hydrometeorological parameters, the relief, the distribution (in space and time) of organisms subjected to the impact, the sensitivity of these organisms to the impact, etc. In order to determine the norms of the impact, the methods of mathematical simulation and modeling under laboratory conditions and in nature are widely used.

To establish these norms, we must also determine the priorities of the impact factors and effects (from the standpoint of their critical character for a given situation). The priority (allowing for economic considerations and production efficiency) determines the nature and urgency of the measures to be taken to combat and prevent negative impacts. These measures are designed to regulate the environmental quality (setting up norms for different natural conditions, optimum distributions of possible impact sources in space and time, technical and technological measures to limit the effects). The character of these measures depends on the scientific-technical production level and, as already noted, on the ecological possibilities of the region. Specifying priorities for the measures taken is of particular importance when the material resources are limited.

If the achievement of some justified (from the health or ecological point of view) norm is impossible for economic or scientific-technical reasons, then temporarily coordinated values of the impacts may be introduced, with a gradual (step-by-step) transition to the norm. However, this option is only possible provided that the levels of the impact on man are not too high (not hazardous) and provided that the critical levels for ecosystems are not exceeded. Otherwise, 'prohibitive' decisions with respect to particular impact sources (factors) are required.

A comprehensive analysis requires that all kinds of damage (ecological, economic, and esthetic, as well as psychological and political) be taken into account.

When determining the degree of impact and possible damage, the probability approach to evaluating a given effect is promising, especially in planning various programs of economic development. This approach enables us to assess different possible versions of the impact for the implementation of these programs. It includes identification of a potential hazard, and also assessment of the risk and its social significance (allowing for economic, psychological, and other factors) when implementing individual components and the program as a whole. The calculation and evaluation of the risk consist in determining the probability of a consequence associated with one of the above kinds of damage.

In any general assessment of the natural environment, it is important to take into consideration forecasts of the future state of the biosphere. These make it possible to take into account effects which have not yet appeared; the trends in environmental changes can themselves serve as an evaluation of the total situation.

The complex of anthropogenic factors influencing the biosphere, and human health, is extremely diverse. It includes: environmental pollution by different substances, physical (including mechanical) modifications disturbing the natural cover of the planet, removal of renewable and nonrenewable resources, etc.

The results of these anthropogenic impacts are even more diverse. They can lead to changes in the basic elements of the biosphere and to numerous geophysical and ecological changes. Man-induced modifications affect human health and wellbeing directly, or via the above changes, and they lead to different social consequences.

The study and assessment of the negative consequences of anthropogenic impacts, with a view toward preventing them or reducing the damage, represent a task of primary importance; this task must be completed successfully in order to organize the economy and ensure public health.

Table 1 shows the effects of the basic anthropogenic factors on the biosphere, and on the health and wellbeing of the population. The various factors of the anthropogenic impact (divided into seven groups) are presented in Column I. The action of each factor (group) can lead to different consequences. The possible effects of these factors are given in Columns II–IV. Note that some of the consequences presented in Columns II–IV can themselves produce further effects and become, in turn, anthropogenic factors of the impact. Undoubtedly, these impacts may be accompanied by mutually intensifying effects. Perhaps the most serious problems are those associated with anthropogenic impacts characterized by: (1) large-scale appearance of the effects and signs of changes up to a global scale; (2) significant inertia of the consequences; and (3) acuteness of the negative consequences. Naturally, the division into these categories is somewhat arbitrary, since many major effects can be placed simultaneously in several groups.

The following global ecological problems associated with the anthropogenic impact, and especially with the entry of pollutants into the atmosphere and into other environments, should be mentioned in particular: possible climatic change, the risk of disturbing the ozone layer, acidification of natural environments due to the long-range transport of sulfur and nitrogen oxides and the fall of acid rain, pollution of the World Ocean, and possible changes in the electrical properties of the atmosphere. The new section of the book,

Table 1. Effects of anthropogenic factors on biosphere, and on human health and wellbeing.

I. Factors of anthropogenic impact	II. Change in properties of basic elements of biosphere
1. Emission of chemically and physically active substances into biosphere	1. Change in composition and properties of atmosphere (pollution, electrical conductivity, and radiation properties)
2. Emission of inert material (aerosol particles, etc.) into biosphere	2. Change in composition and properties of inland water (pollution, mineralization)
3. Direct heating of biosphere	3. Change in composition and properties of World Ocean water (pollution, etc.)
4. Physical (mechanical) impact, leading to change in land surface and vegetation cover (erosion, ploughing, urbanization, fires)	4. Change in state of biota as biogeophysical environment
5. Biological impact (development of agrocenoses, introduction of biological species, etc.)	5. Change in lithosphere (mechanical disturbances, waste accumulation, etc.)
6. Removal and destruction of nonrenewable and renewable resources	6. Change in cryosphere
7. Anthropogenic orderly transport flows of substances	7. Change in properties of land surface and soil (integrity, acidity, radiation characteristics)
	8. Change in geophysical properties of large systems: climatic system and biosphere as a whole

III. Geophysical and geochemical consequences and effects	IV. Ecological and biological consequences, disturbances of ecosystems
1. Large-scale changes in atmospheric and oceanic circulations	1. Change in terrestrial and water ecosystems, disturbance of their stability
2. Weather and climate change	2. Change in marine ecosystems (structural simplifications, etc.)
3. Redistribution and alteration of renewable (water, climate) resources	3. Genetic effects, degeneration
4. Disturbance of ozone layer and ionosphere (changes in UV and radiowave transmission)	4. Extinction of existing species, appearance of new species
5. Change in atmospheric transparency and solar radiation transmission	5. Decrease in bioproductivity, reproductive rate, and number of populations, degradation of forests, desertification (biological aspect)
6. Erosion of land surface, change in albedo of land surface	6. Soil degradation, desertification
7. Disturbance of natural geochemical cycles and of cycles of different elements	7. Change in biosphere's ability to regenerate renewable resources; exhaustion of nonrenewable resources
	8. Change in character of evolution of biosphere

Table 1. (continued) Effects of anthropogenic factors on biosphere, and on human health and wellbeing.

V. Effect on human health and wellbeing	VI. Social consequences
1. Decrease of working capacity	1. Change in food production, undernourishment, starvation
2. Esthetic damage, mood worsening	2. Change in structure of energy consumption
3. Disease, stress	3. Change in economics
4. Genetic effects	4. Social consequences on different scales; damage to wellbeing, possible disturbances of society's development
5. Change in life duration	
6. Decrease of rate of population growth	
7. Decrease of size of population on different scales	

Section III, including Chapters 8–11, describes and analyzes some global ecological problems.

Of the global problems associated not only with pollution, but also with an irrational, thoughtless use of renewable natural resources, the following should be noted: degradation of the soil and desertification; loss of biological diversity in a number of regions; deforestation, especially in tropical zones; and harming the living resources of the sea. The latter problems are only treated partially in different sections of the book; however, the considered scientific basis and structure of ecological monitoring largely cover these problems too.

A possible climatic change (on a global scale) should definitely be placed in the category of serious ecological problems.

There are many different man-induced modifications fostering climatic change; let us list some of these.

First of all, aerosols of anthropogenic origin may affect the radiation balance of the atmosphere. Different natural conversions associated with the construction of irrigation structures, urbanization, and a change in the area of planting can alter the albedo of the underlying surface.

Moreover, the accumulation of carbon dioxide and other substances absorbing long-wave radiation from the Earth's surface leads to warming. The appearance of oil films on the oceans disturbs the exchange of energy and gas between the ocean and the atmosphere. Finally, the diffusion of halogenated hydrocarbons and nitrogen monoxide into the atmosphere may affect the ozone layer of the atmosphere (the ozonosphere), which will alter the temperature distribution in the stratosphere.

The picture of the possible anthropogenic climatic changes is complicated by the fact that these changes are superimposed onto the natural variability and fluctuations of the climate. In this connection, it is difficult to forecast the climatic changes for the next few decades, although most scientists foresee a warming of the climate.

Apparently, the anthropogenic impact on the climatic system has achieved such dimensions that man-induced climatic changes are now highly probable. These will have a significant effect on human economic activity, on food production, on human wellbeing, and on the state of ecological systems.

That is why the Executive Council of the World Meteorological Organization decided to convene in February 1979 in Geneva a special World Conference on *Climate and Man.* At this conference, more than twenty experts presented survey papers on all aspects of the study of the global climate and its influence on mankind [2]. The author presented a paper dealing with the monitoring of climate-related data [10]. The gist of this report is given in Chapter 6.

The next global problem is pollution of the World Ocean.

As many as several million tons of oil and petroleum products already enter the World Ocean annually from shelf oil drilling, from ships (mainly due to accidents), from washing off the coasts, and through the atmosphere. This causes serious damage to marine ecosystems, and it disturbs the exchange of gas and energy between the ocean and the atmosphere (due to the formation of a thin oil film on the surface of the ocean). This effect manifests itself most strongly in the polar regions of the ocean, where the rate of degradation of petroleum products at low temperatures is extremely low.

The spreading of pollutants (radioactive substances, sulfur dioxide and its conversion products, and heavy metals) over long distances has already become an international problem. For instance, in Europe sulfur dioxide and its conversion products, transported over long distances, cause appreciable damage on the territories of many countries. The sources of the pollutants are often located far away from the country that suffers the damage.

The transport of pollutants across boundaries (and the monitoring of this pollution) was one of the basic topics at the All-Europe Conference on Cooperation in the Field of Environmental Protection (Geneva, November 1979), which was initiated by the Soviet Union. At a high level of the conference a *Convention on Transboundary Airborne Pollution over Long Distances* was adopted unanimously. A conference in June 1984 in Munich continued the discussion of the effects of transboundary air flows of toxic substances on forests and fresh-water bodies (in Europe), together with measures to prevent damage.

Particular attention should be given to large-scale high-inertia effects. These definitely include the genetic effects of pollution, as well as the slow accumulation of some substances in zones where substantial damage can be caused to a major element of the biosphere or to a community.

The significance of monitoring for the observation assessment, and forecasting of the state of the environment has already been noted above. Depending on the scale and the objects observed, monitoring can be subdivided into: (1) monitoring of factors related to human health (medical-ecological monitoring), (2) monitoring of the state of ecosystems (ecological monitoring), and (3) climate monitoring, including monitoring of the climatic system.

Other classifications of monitoring subsystems are also possible. For instance, Academician I. P. Gerasimov subdivides monitoring into biological, geoecological, and biosphere monitoring [7].

In this book much attention is given to ecological and climate monitoring (see Chapters 5 and 6). Problems related to medical-biological monitoring are only treated partially (for instance, pollution monitoring, which has received the main emphasis in this work, includes all the above categories; in addition, health criteria of environmental quality (MPCs) are also used).

In the author's opinion, ecological monitoring should have as its goals measuring, estimating and forecasting anthropogenic changes in the abiotic component of the biosphere (the main thing here being pollution) and the responses of biota to these changes, as well as subsequent changes in ecosystems resulting from the anthropogenic impact.

Apparently, the goal of modern ecology is becoming the study not of the manifestation of an individual organism, but rather of the mass impact of their aggregate on nature (which Vernadskii [1] considered within the framework of microbiology). This was also pointed out by Khil'mi [15].

Note that climate monitoring is, in its procedure, near to ecological monitoring, but, unlike the latter, it covers the climate system (atmosphere-ocean-land surface-cryosphere-biota), i.e. the system influencing the formation of climate and its changes.

The establishment of ecological and climate monitoring in the form of a permanent service (an ecological service and a service for the acquisition and representation of data on climate and the climatic system) represents an extremely important and urgent task. This service should become a component part of the production process, associated with the use of natural resources and with the impact on the environment.

As stated by Academicians I. P. Gerasimov and M. S. Gilyarov, a maximum 'ecologization' of both research and industry is required.

The ecological program which is being developed in the USSR at present should provide for an efficient use of ecological information and knowledge in the national economy, an increase of biological productivity on this basis, efficient nature protective measures, and a rational utilization of natural resources.

The ecological program is a scientific-technical program stimulating basic research in the field of ecology. It is designed to solve applied industrial problems, with wide applications to the interaction between humans and nature, and to develop ecological approaches to implementing the most diverse nature-protection measures, as well as to establish a unified Soviet ecological service.

Fundamental research and a complex of applied studies are envisaged within the framework of this program. The fundamental investigations include studies in the field of autecology, i.e., the ecology of organisms and their populations. It is necessary here to study the morphophysiological, behavioral, and other parameters of populations, and the relationships of the structure and properties of populations with the environmental conditions. In the field of synecology, i.e., the ecology of communities, it is necessary to study the conditions determining changes in the structure and functioning of natural and artificial communities and ecosystems, as well as the regularities of their formation and the conditions of stability.

One of the important components of the ecological program is the preservation of reference sections of the biosphere, in which rare (vanishing) species of animals and plants are preserved from extinction. Setting aside biosphere reserves will enable us to preserve pristine sections of the biosphere, that is, their functions will be even broader than just gene-pool preservation.

The fundamental research of the ecological program will include the following:

Investigations of human ecology and the genetic consequences of alterations of the environment.

Investigations of the ecology of ecosystems and global ecology, namely, estimating and forecasting the ecological consequences (for ecological systems and the population) of macroscale geophysical and geochemical changes, which have been discussed above.

The applied studies should in particular include the following.

A study of ecological problems associated with the development of agriculture, water resources, and power production (including nuclear power).

Ecological assessments related to regional problems and to the setup and functioning of large territorial-industrial complexes (partial diversion of river flows, construction of the Kamsk–Achinsk Power Complex, etc.).

Development and establishment of an optimum ecological monitoring system in the USSR.

A very important part of the investigations in accordance with the ecological program is the study of the ecological stability of biological systems with respect to an elevated level of the human impact on the environment.

The intensive study of different man-induced modifications of the biosphere, especially pollution, has yielded a number of new scientific trends during the last few years. For instance, the behavior and transformation of different pollutants in the atmosphere has become the subject matter of atmospheric chemistry, and the development of atomic meteorology and radioecology was connected with the problem of radioactive pollution. By analogy, we can speak about the ecological aspects of environmental pollution by heavy metals, pesticides, and other toxic substances. The study of the effects of such substances on ecosystems has led to the development of ecotoxicology (see, for example, [20]); its basic principles are presented in Chapter 2 of this book. Studies of the distribution in the atmosphere of anthropogenic pollutants affecting the state of the biota and climatic system pertain to atmospheric ecology.

Studies of man-induced changes in the state of the environment and biosphere (against the background of natural fluctuations and variations of this state) are being integrated more and more closely into a single entity, which could be considered as a new scientific field, namely biospherology or ecospherology.

The description, assessment, and interpretation of the state of the natural environment, as well as geoecological forecasting and the development of methods for the efficient exploitation of natural resources, will undoubtedly dovetail with modern geographical science.

This new trend in science has been developed intensively during recent years. Obviously, issues of the scientific justification of monitoring the state of the environment pertain to this trend.

The economic and social aspects of regulating environmental quality have already been mentioned above. It is obvious that high environmental quality must be achieved if the conditions for a harmonic development of society are to be satisfied. Therefore, when solving ecological-economic problems, man's requirements for material welfare on the one hand, and pure air, pure water, and an undisturbed natural environment on the other, have to be satisfied (since nature also serves to satisfy man's esthetic needs).

In this connection, when determining the priorities of measures designed to protect or regenerate the natural environment, it is necessary to take into account the economic cost of such works, as well as how possible damage from the anthropogenic impact is related to the production yield. This problem should obviously be solved under conditions of maximization of the yield, or, more exactly, of the ratio of the yield to the possible damage from the impact of a given process in various branches of the economy. This approach is described in Chapter 7.

The possibility that man, equipped with the latest technology, can affect nature is so great at present that there is an obvious urgent need to join international forces to study and prevent negative man-made consequences.

Recent scientific literature has included different, often pessimistic, forecasts of the consequences of the anthropogenic impact on the biosphere and warnings about the growing negative consequences of man's impact on nature (see, e.g., [14]) and about an impending ecological crisis [13, 17, 18]. Many scientists consider, however, that it is possible to avoid a crisis if a joint international plan of action aimed at preventing this crisis is acted upon. For instance, a document prepared recently by a number of specialists deals with this problem [21].

Ways of overcoming the possible ecological crisis and achieving social progress are discussed in detail in the monograph entitled *Ecological Crisis and Social Progress* by E.K. Fedorov [12].

In a declaration adapoted at the above-mentioned World Climate Conference (February 1979) it is said that "The nations of the world must work together to preserve the fertility of the soil, to avoid misuse of the world's resources, ... to lessen pollution of the atmosphere and the oceans. These actions will require great determination on the part of the countries with material resources, and they will be meaningful only in a world at peace".

Under these conditions, research on the possible geophysical and ecological consequences of a nuclear war has acquired particular importance. The results of this research indicate convincingly that nuclear wars (global or limited) are utterly senseless, since they lead to the destruction of such global systems as the climate system and can cause irreversible damage to the biosphere as a whole, and, consequently, to all mankind.

Therefore, this book is devoted mainly to a scientific justification of the establishment of a universal system of observation and monitoring of the natural environment, a system for evaluating its state and forecasting its future. Such a system is necessary for an efficient use of natural resources in the interests of human society.

The objective of this system is certainly not just to identify individual (sometimes particular) negative effects of the impact on nature, for instance, local pollution of the environment by some substances (which could probably be avoided in the future), but also to carry out a continuous monitoring of the general state of the biosphere, and of such integrated characteristics as the balance of convertible and inconvertible forms of energy and the accumulation of excessive entropy in systems.

It can be said with confidence that monitoring of the state of the physical environment, and of the organisms which inhabit it and the nature of their interactions, i.e., integrated monitoring of the biosphere during its harmonic development along with human society, the organizer of the biosphere, will always be necessary.

PART I

Comprehensive Analysis and Regulation of the Environment

Comprehensive analysis of the environment

Permissible loading of the biosphere and the establishment of ecological standards

Regulation of environmental quality

CHAPTER 1

Comprehensive Analysis of the Environment

1.1. The Role of Comprehensive Analysis in Optimizing the Interaction Between Man and Nature

The organization of the man/nature interaction calls for a careful study and complex analysis of the state of the environment and for the systematic approach that has recently come to be called a 'comprehensive analysis of the environment'.

The basic objective of a comprehensive environmental analysis is to study the diverse aspects of the impact of various factors, primarily anthropogenic factors, on the biospheric elements, and to carry out an overall analysis of the consequences of this impact, aimed at discovering which of its aspects have a primary importance and a high priority. The results of such an analysis are essential for optimization of the man/nature interaction.

The main distinguishing feature of a comprehensive analysis is the need to consider all the basic aspects of the interaction and relationships in the natural environment; its implementation involves the solution of a great number of problems relating to the competence of a variety of scientific trends. It is also characteristic of a comprehensive analysis that during such an analysis not a single interaction aspect of importance and priority is to be neglected.

Thus, a basic requirement of a comprehensive analysis is that all the important aspects of this interaction be included (if only tentatively at first) in the range of problems considered [15].

As an instrument for studying the many-sided anthropogenic impact on the biosphere and for the complex analysis of the consequences of such an impact, the comprehensive analysis of the environment has been developed over the years since 1972 and has formed a part of the program of Soviet-American cooperation in the field of environmental protection as an individual project [22].

When evaluating the role of the comprehensive analysis in planning and achieving the optimum interaction between man and the environment, consideration should be given to the ways and methods of its use for man/nature interactions of various types.

The first type of interaction appears during human economic activities; several kinds of such interactions could be mentioned here.

The first kind is a spontaneous anthropogenic impact, whereby, in the course of his economic activities, man uses the natural environment only as a sort of 'shock-absorber' between the sources of the impact and man, for instance, as a supposedly neutral disperser of industrial waste, and in so doing substantially affects the environment (producing mainly adverse effects) without any purposeful attempts to transform nature.

The environmental discharge of pollutants by numerous factories and plants is an example of such an interaction (man only utilizes some area, for instance, a site for construction or an environmental capacity for industrial waste discharge and dispersion, using practically no natural, except nonrenewable, resources of the given region).

The second kind is an impact using natural resources, accompanied by a substantial, but mainly unintentional, influence on nature and transformation of the biospheric elements (e.g., due to urbanization, land cultivation, open-pit mining, industrial construction, road building in tundra and highland forest tracts, etc.).

The third kind involves the use of natural potential by way of deliberate large-scale transformation, i.e. an intentional impact (e.g., reservoir construction, importation of water from large rivers, weather modification to enhance precipitation, etc.).

The described kinds of interaction have, along with their various positive effects, as a rule negative consequences in nature as well; in some cases the changes are only negative (viz. the first kind of interaction).

One more kind of man/nature interaction resulting in adverse effects should be mentioned, namely, preparation for war, and, what leads to even worse effects, war itself. In the latter case, man deliberately causes damage to animate nature by trying to ravage the enemy's territory for the sole purpose of defeating him. This case will not be considered below.

For optimal interaction between man and the environment (the last kind of interaction being certainly excluded), the negative consequences of the environmental impact will either cease completely or will be minimized with regard to ecological and economic considerations. In this case, human economic activities will not lead to adverse effects, and the first kind of interaction will actually disappear.

By this we mean that the organization of the optimal man/nature interaction will not affect in any way the rate and character of economic development, construction, or other kinds of human activities. It is obvious that, on the whole, industrialization and technological progress should not be considered only as factors potentially hindering the development of the biosphere. 'Deterioration' of the environment and disturbance of the ecological balance are not an inevitable consequence of the general strategy of an industrial society; rather, they are engendered by errors in the technological policy and by an inadequate level of technological progress [31].

One more type of interaction arises when optimizing the relationships between man and nature (the second type). This type can hardly be attributed directly to man's economic activities, since it is aimed at controlling the relations between man and nature. The following kinds of interactions belong to this type.

(1) Organizing observations, studying various factors of the environmental impact and the state of the biosphere, i.e. environmental (including monitoring of anthropogenic changes in the environment).
(2) Regulating the quality of the environment so as to achieve its optimal state, taking into account man's diverse interests (state of health, well-being, etc.).
(3) Making strategic decisions, regulating large-scale processes (both economic processes and natural processes). This kind of interaction can be united with the second kind, thereby expanding the concept of 'environmental quality control'.

The role of a comprehensive analysis of the natural environment in the strategy of optimizing the man/nature interaction is of primary importance with regard to regulating the environment.

The comprehensive analysis of the environment includes problems of environmental quality control and the establishment of monitoring systems. When studying the state of the environment and establishing a monitoring system, it is necessary to find answers to the following questions.

- What is the present state of the biosphere and what changes in the biosphere are expected in the future?
- What are the causes of the possible changes in the biosphere (including undesirable, and especially hazardous, critical changes) and what is the source of the harmful activity?
- Which loads, effects, and stresses on the natural environment are harmful (undesirable), from the point of view of the established and accepted criteria?
- What is the permissible level of impact for the biospheric elements and what ecological reserves does the biosphere (its elements) possess?

The answers should, of course, be such as would make it possible to use the data obtained to assess the investigated impacts on man, either directly or via the effects of changing conditions in the state of the natural environment.

When controlling the environmental quality and organizing the management of various processes so as to optimize the relationships between man and nature, it is necessary to develop a strategy for this control and management whereby a number of the following questions are answered.

- What environmental quality should be assumed to be high and acceptable (acceptable from the ecological point of view)?
- What environmental quality should be striven for, taking into consideration man's diverse purposes and interests (i.e., high or acceptable quality from both the ecological and economic points of view)?
- What criteria should be used to prepare an answer to the preceding question?
- Which actions are most appropriate or of highest priority from the point of view of reducing the ecological and esthetic damage?
- Which actions may be helpful for reducing the negative effects of the given anthropogenic impacts on nature? Will these actions not produce accompanying adverse effects?
- Where should such actions be more energetic and the environmental protection requirements more strict: in a polluted or unpolluted territory? Will pristine (clean) zones be specially set aside, and for what purposes?
- What actions are most appropriate with regard to the economic and social effects?
- What data should be available for determining the appropriate actions?
- How should the action priority be determined when ecological and economic interests clash?

Although the list of questions could be extended, any new questions would probably just be more detailed versions of the above.

To answer the majority of the questions posed, we have to carry out special investigations. Obviously, the optimization of the relations between man and nature should also be based on the answers to the above questions. The solutions of most problems can be based on a comprehensive analysis of the environment.

When stating the problem of comprehensive analysis as a whole, we had to formulate a number of premises stemming from the enumerated questions and select the basic trends of the investigations.

This work was accomplished within the framework of three Soviet-American symposia dedicated to the comprehensive analysis of the environment and published primarily in a number of papers [12–15, 28]. The specific features of the problems were already characterized and the most important trends in the investigations of those problems were already noted in the resolution (proceedings) of the first Soviet–American symposium [24].

It was stressed that a comprehensive analysis of the environment requires a knowledge of the pollution sources and objects, as well as of the kinds of impact and biological responses, taking into consideration regional peculiarities. The comprehensive analysis should be aimed at the integrated use of various methods to develop environmental protection programs, including an analysis for particular regions.

The development of approaches to a comprehensive environmental analysis should include a study of the 'dose–efffect' and 'dose–response' relationships in various experiments, and a study of the impact thresholds of various factors and of the effects of multimedia pollutants, as well as the development of methods for assessing the responses of complex ecological systems to changes in the state of the environment.

Here finding criteria for the permissible environmental loading and detecting the critical links in the biosphere responsible for such loads should become the most important element.

It was noted that a scientific justification of integrated environmental monitoring, including the development of a strategy for biological and genetic monitoring, will result from the work in this field. The comprehensive analysis should also include an assessment of the potential economic damage to the environment as a result of the anthropogenic impact. At the same time, much consideration should be given to developing ecological-economic models, so as to estimate the economic damage, on the one hand, and to indicate the economic effectiveness of the measures taken to prevent the environmental impact, primarily that of pollution, on the other.

At the first and subsequent Soviet–American symposia on comprehensive environmental analysis, the above-mentioned problems were formulated, answers to some of the questions were indicated in many papers, and scientific programs associated with comprehensive analysis gained ground [7–9].

Comprehensive environmental analysis comprises several stages (sections).

The first stage involves an analysis of the effects (Fig. 1.1, Block I) of various factors on the environment (at different levels).

This stage includes an analysis of the responses of organisms, populations, and ecological systems to plural impacts in various environments (Fig. 1.1, Block IA), an analysis of the consequences of the impacts (changes in state, diseases, and death of organisms, populations, and ecosystems), and a determination of the various kinds of damage caused by the impact (Block IB), revealing the critical factors of the impact (or combination of several factors) and the most sensitive (from the point of view of the consequences of the impact) elements of the biosphere (Block IC).

The objective of the second stage of the comprehensive analysis is to determine the permissible ecological impacts and loads (Block II).

The second stage includes a determination of the permissible loads on individual organisms (primarily on human beings) and different populations, taking into consideration the integrated and combined effect and the public-health requirements for environmental quality (Block IIA), a determination of the permissible loads on communities and ecosystems, taking into consideration the multiple ways of impacting on all the elements of a community or ecosystem simultaneously (Block IIB), as well as a determination of the permissible loads on a large system (the climate system and the biosphere) (Block IIC).

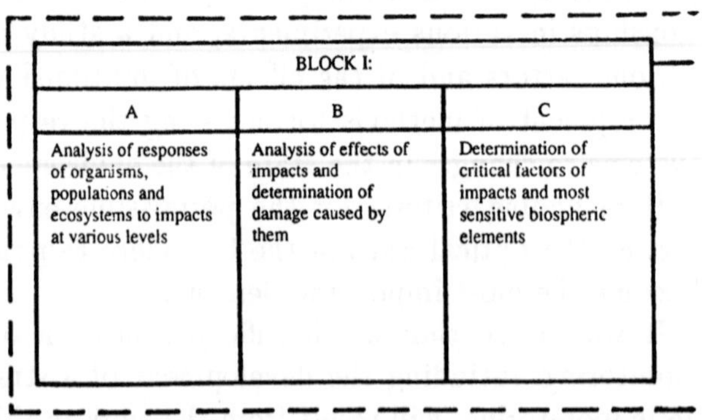

Figure 1.1. Block diagrams of comprehensive environmental analysis(1), establishment of ecological standards (2), and their connection with the block of strategy for environmental quality control (3).
Blocks: ***I*** **– analysis of effects of impact;** ***II – III*** **– determination of permissible impacts and loads;** ***IV – V*** **– working out and introduction of criteria and standards limiting impact, pollutant discharges, etc.;** ***VI*** **– implementation of technical measures aimed at limiting impacts (pollution, etc.).**

The concept of 'maximum permissible impacts' is in many cases identical to the concept of 'maximum permissible loads', especially for biological systems. However, these concepts may well be different for particular organisms (individuals); for example, the permissible load on an organism could signify the permissible content of a substance in the organism itself, in its tissues, while the permissible impact could signify the permissible amount of the substance in the environment. In any case, the permissible impacts and the permissible loads are supposed to be equivalent, according to the resulting permissible effect in the system within the limits of its normal state.

When determining the permissible load on ecosystems, we must find out what intensity (or integral of the intensity over time and space) of the impacting factor can be endured by the given ecosystem within the limits of the permissible state (or within the limits of the ecological reserve).

The third stage (Block III) of the comprehensive analysis comprises a determination of the permissible loads on an area or region from the ecological-economic point of view, taking into account different economical aspects, including the cost-gain ratios. This stage is rather diversified and has a pronounced effect on the strategy of environmental management and quality control.

Ecological regulation of the anthropogenic impact should start with a determination of the permissible impacts and loads on the ecosystems and natural environment at different levels. The next step in the process of establishing standards, therefore, involves developing criteria aimed at limiting the impact sources (for instance, developing standards for the maximum permissible environmental discharges) and reducing the impact effects

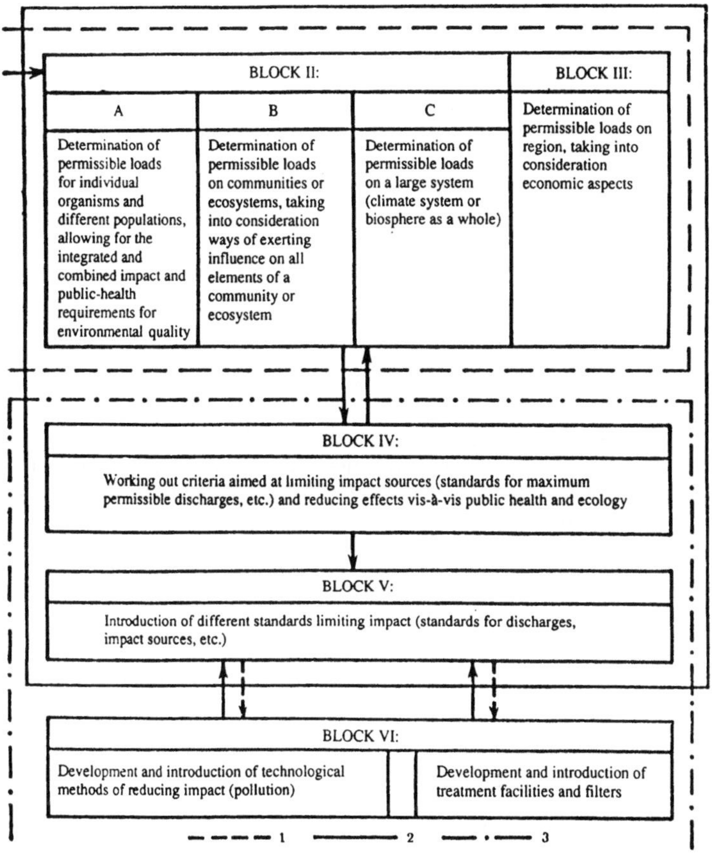

(Block IV). The establishment of public-health standards (establishment of maximum permissible concentrations in the environment for the population) also falls in this category, provided it is performed integrally, taking into consideration the effects of a great number of factors in different environments and from all existing sources (many of which have become 'natural', due to migration of the pollutants).

Thus, the above-mentioned criteria should be worked out with regard to both public health and ecology; a separate task is working out criteria from ecological-economic positions, taking into account the economic effects of using such criteria and the economic possibilities.

The last stage comprises establishment of standards for discharges and impact sources, based on the above criteria, and implementation of the described approach (Block V). In turn, the strategy of regulating the environmental quality comprises setting up criteria and introducing standards limiting the different anthropogenic impacts (taking into account economic aspects). The introduction of such standards is closely associated with developing different technical procedures aimed at the limitation of pollution

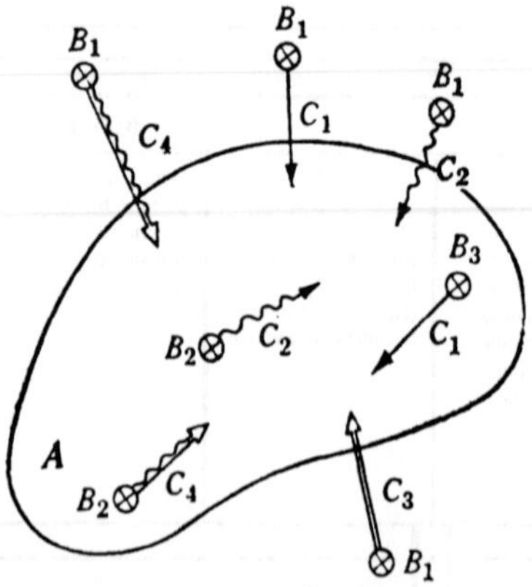

Figure 1.2. Impacts of various factors on element of biosphere A: B_i = impact sources (B_1 = external sources; B_2 = internal natural sources; B_3 = internal introduced sources); C_i = impact factors (C_1 = pollution; C_2 = ionizing radiations; C_3 = mechanical impact; C_4 = biological impact).

and other effects (Block VI).

A comprehensive environmental analysis, as applied to the establishment of monitoring and environmental quality control, in its first stage includes the analysis of factors affecting the environment and various consequences of the impact (ecological, economic, and other impacts). This part of the work, being in essence the first stage of the comprehensive analysis (Fig. 1.1, Block I), is based on the system of data obtained via the primary monitoring and via mathematical modeling, as well as from individual laboratory and field experiments.

The procedure of the first stage of the comprehensive environmental analysis will be considered below in more detail [15].

Several factors C_i stemming from different sources B_i affect a biospheric element A (Fig. 1.2). It is possible to assess separately the state of the abiotic component of the ecosystem, characterized by a magnitude A_a (Fig. 1.3), and that of the biotic component (individual biological elements), characterized by a magnitude A_b; the state of the ecosystem as a whole is characterized by a magnitude A_e. The state of the ecosystem after the impact will be characterized by the magnitudes A'_a, A'_b, and A'_e, respectively.

The sources of the impact can be external (with respect to the biospheric element A) and internal; the latter can be either natural components of the system or introduced from outside. The impact factors can be: pollution (chemical factor), radiation and mechanical impact (physical factor), and biological impact (competition, suppression, etc.).

Thus, state A'_e should synthesize the influence of all the dissimilar and diverse impacts. The analysis of state A'_e can be represented as (see Fig. 1.3):

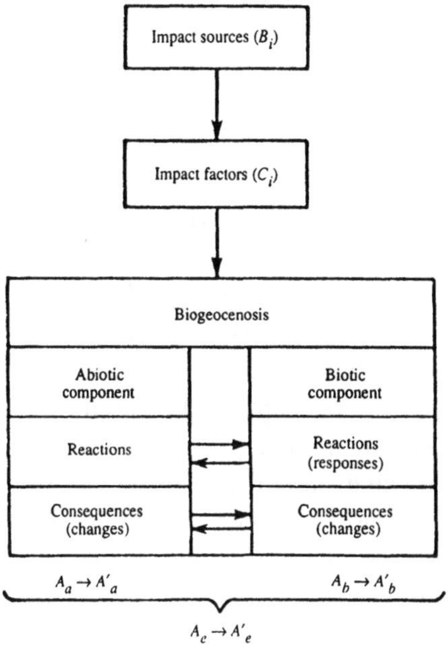

Figure 1.3. **Block diagram of sequence of effects of various factors on ecological system.**

(1) a change $A_e \to A'_e$, the amount of ecological (and esthetic) damage;
(2) the role played by each impact source B_i and factor C_i and their contribution to the above change.

To study the change $A_e \to A'_e$, it is necessary to know the initial natural state of the ecosystem A_e (the average and extreme values characterizing this state), and the background state (the state resulting from the background anthropogenic impacts in addition to the natural effects). It is also necessary to take into account the changes in the state of the ecosystem occurring not only under the influence of individual physical (chemical) factors, which could be easily associated with particular sources, but also under the influence of constant stresses of various kinds weakening individual organisms, the population, and the ecosystem as a whole (this could obviously be taken into account while describing state A_e, rather than by introducing extra sources into the calculation scheme). When evaluating the impact on a population, it is necessary to take into consideration the distribution of organisms not only in space and time $[n(R,t)]$, but also with respect to the degree of their susceptibility and sensitivity to the given impact, and, of course, to take into account the total number of organisms of different populations N_m subjected to the impact.

When determining the role of each source of the impact, it is necessary to know the laws governing the distribution of these factors in the homogeneous environment, as well as their potential effects and migration from other environments.

The next step in the given stage of the analysis is the determination of possible joint impacts of several different factors, their additivity, strengthening or weakening due to the joint impact, strengthening or weakening of the total effect on the system for a simultaneous effect at its different levels.

When determining the impact of different factors on individual organisms, on elements of the biosphere, and on whole ecosystems, it is important to know whether the given factor possesses a certain threshold of the impact above which the effect occurs and below which it is equal to zero, or if the threshold of the impact is determined as some acceptable probability of changes and/or adverse effects. For instance, for estimating the impact on individual organisms, the concept of thresholds is particularly significant and effective (e.g. for labour hygiene, when due to this concept rather strict specifications are established, aimed at limiting the impact and preventing occupational diseases). However, when estimating the permissible loads on communities or ecosystems, it seems more appropriate to use the concept of the probability approach to an impact (where possible, for any impact factors). It is precisely this kind of approach that would make possible an integrated, comprehensive assessment of the existing impact, or load on an ecosystem, taking into account the nonuniform distribution of the exposed organisms, including those at levels below the permissible load. Such an approach makes it possible to avoid the prohibited, from our point of view, scenario formally permitting, in spite of the established maximum permissible concentrations, the discharge of an enormous amount of pollutants into any natural body (this amount is determined using the formula for $\sigma_0 V$, where σ_0 is a concentration equal to or below the maximum permissible concentration, and V is the volume of the water or air body; in this case $\sigma_0 > 0$).

When determining the role played by each source of the impact, it seems appropriate to use:

- calculation of the concentrations, using certain laws governing the impact distribution, described in terms of known functions;
- mathematical modeling of the impact process as a whole or in part;
- laboratory and field experiments designed to reveal the effects of the impact of the given factor on individual populations, communities, and ecosystems;
- results of observations and field studies of the effects of the impact.

These tools will help us to achieve a synthesis of all the detected effects, in order to get a general integrated picture and then to solve, if necessary, the inverse problems.

The analysis of the state A'_e and the determination of the different kinds of damage take into account a great number of diverse factors and effects, but of most importance are the intensity of the given factor (e.g., the concentration), the toxicity or index of biological efficiency of the impact, and also the quantity N (absolute or relative) of organisms of the m-population subjected to the impact.

Strictly speaking, the analysis of a change $A_e \rightarrow A'_e$ also includes an analysis of $A'_e \rightarrow A''_e$, where A''_e corresponds to the state of the given ecosystem in the future. The point is that, for an integral consideration of the impact of different factors on organisms and populations distributed in space and time, integration over time is implied, i.e. A''_e is determined on the basis of available data. Thus, the analysis implies forecasting changes in the states of biospheric elements and determining the trends in these changes.

It is quite clear that a future manifestation of certain signs accumulated at present can change substantially the state of a population, of a biospheric element (due to the redistribution of certain toxicants, a large-scale appearance of negative hereditary effects, etc.). Experience has shown that the manifestation of some effects related to the impact is characterized by considerable inertia.

The number of factors and significant effects can be very great; it is necessary to select correctly the most essential ones, which are of highest priority from the point of view of the critical impact on the ecosystem as a whole.

To assess the ecological significance of such factors or effects, it is appropriate to use specially developed criteria, i.e. the maximum permissible concentrations (MPCs) for individual ingredients in various environments, the maximum permissible releases (MPRs) for the sources, and the maximum and permissible ecological loads (MPELs), for estimating the potential loads on communities or entire ecosystems.

Using the various criteria for assessment, we can list the following three methods (approaches) for determining the priority of one or another impact.

Method I. Determination from the point of view of sensitivity, i.e. the changes as compared with the initial state of the ecosystem (it shows how much A'_e differs from A_e).

Method II. Determination from the point of view of the criticality of the entire ecosystem or its most important parts (it shows whether A'_e is permissible from the point of view of the existence of the system or how close the state A'_e is to the critical one).

Method III. Determination from the point of view of the absolute change $A_e \rightarrow A'_e$, expressed as the product of the average change in individual organisms of the m-populations times the number of organisms exposed to the impact of all m (taking into account the 'weighting' factor).

Further, it is possible to determine (and isolate) the priority for a number of populations, integrated impacts, etc., on the basis of their group features.

The priorities determined with the aid of method I can be used to justify the organization or improvement of a monitoring system. The priorities determined by method II can be used to obtain a correct, justified organization of the measures to be taken (primarily urgent and short-term ones). The priorities determined by method III can be used to work out the general strategy of decision-making, while preparing a long-term plan of relevant actions. It is precisely this method that is essential for economic evaluations aimed at preventing substantial economic damage.

This point marks the end of the first stage of the comprehensive analysis of the state of the environment. Its results are used to improve the monitoring system, the calculation schemes, the modeling and the performance of experiments.

After this, the described stage of the analysis may be repeated (carried out again using new data). This will be the next approximation to the solution of the problem, i.e. the description of the actual state of the environment.

The performed analysis will make it possible to answer many of the questions stated above and to go on to the next stage of the comprehensive analysis, namely a determination of the permissible ecological impacts and loads, as well as economic estimates aiding in the development of a strategy for regulating environmental quality and for process management in the interests of nature protection against negative anthropogenic consequences.

The second stage of the comprehensive analysis is related to the determination of the permissible loads (impacts) on biospheric elements at different levels and to the development of appropriate ecological standards. This stage of the comprehensive analysis forms a basis for establishing ecological standards for the anthropogenic impact on the environment, including regional standards.

1.2. Analysis of the Ecological Loading on a Regional Scale

This section deals with the integrated comprehensive analysis of the environment on a regional scale, so as to find the permissible ecological loading on ecosystems on a large scale. Systems analysis and simulation mathematical modeling with respect to closed territorial-economic systems can be an appropriate instrument for these purposes. The application of the balance approach, and the inclusion of the so-called 'interbranch' balance technique in the methodology of simulation modeling, represent an important means of simplifying this methodology.

Since environmental pollution and other anthropogenic impacts are relatively 'confined' within the limits of territorial-economic regions, the establishment of regional standards is important. At the same time, a differential approach to various natural objects and biospheric elements (e.g. ecosystems) is also necessary. Natural objects may be divided into different categories, according to the scale of possible interference and the impact on these objects (ranging from unique, protected objects, where the interference is reduced to a minimum and sometimes should be reduced to zero, to objects where a considerable man-caused reconstruction of the ecosystem is possible and an intensive adaptation of the communities to man's requirements is necessary).

The above approaches to determining the permissible ecological loading on a large natural object or a region as a whole are being developed, e.g. for Lake Baikal [2, 18], a region with a developing industry and agriculture in which there is intensive use of natural resources. The uniqueness of Lake Baikal and its economic and esthetic values are characteristic features of this region.

Baikal contains about 23600 km^3 of exceptionally pure fresh water; its mineralization amounts to only about 100 mg/l. The water is very rich in oxygen, the content of which does not fall below 9 mg/l.

The flora and fauna of Lake Baikal are diverse; nearly one third of the plant organisms and two thirds of the animals are endemic, i.e. they are found only in this lake.

It is quite obvious that, on the one hand, further economic development of the Baikal region and the use of its riches are necessary, while, on the other hand, it is also necessary to preserve the lake in its unchanged state, to preserve its unqiue ecosystem whose functioning provides to a large extent the specific features of this lake, primarily the purity and transparency of its water.

The only industrial enterprise located directly on the shore of Lake Baikal is a pulp and paper mill. It was built in 1966 and is still monitored constantly [16, 17]. To protect the lake against pollution, powerful treatment facilities were built at the factory. The treatment includes three stages: biological, chemical, and mechanical. The zone influenced by wastes discharged into the lake after treatment (the continuous zone where the pollutant content differs from the background) amounts to 20–30 km^2, and the zone of partial pollution of the bottom in the form of separate patches is about 2 km^2 [16]. Changes in the microbiological indices and the biomass of zoobenthos are observed in the zone of a small patch of polluted ground (0.1 km^2) [17].

Although recently the size of the spot has increased and the area of the bottom pollution has reached 15–20 km^2. By the Governmental Decree of March 1987, the discharge of waste waters from the Baikal pulp-and-paper plant into Lake Baikal was prohibited and the plant was to go over (in 1993) to ecologically sound processes. It should be noted that during a decade of observations the zone influenced by the factory has remained practically at the same level [17].

The entire Baikal region undoubtedly constitutes a complex system having complex physiographical and economic peculiarities.

A peculiarity of a complex system is that such a system is characterized by a great number of variables and by the presence of feedback between the quantities; the system is distinguished by time lags, nonlinear relationships, and even breaks. The equations describing such a system do not have, as a rule, an analytical solution; here simulation mathematical modeling can be used successfully (see, for example, Forrester [30]).

It is precisely the systems analysis and the simulation mathematical modeling that make it possible to forecast potential states and changes of such systems and that make their management efficient.

The given approach can be described in the following way [2]:

- a specific statement of the given problem and a desirable solution to it are formulated;
- a verbal model with internal and external links is formulated;
- a graph of the verbal model is constructed and its mapping into a mathematical model is performed;
- the behavior of the system is simulated with the aid of a computer, and a comparison with real processes is carried out;
- in the case of unsatisfactory agreement, the above procedure is repeated with the required (possible) refinements.

Successive regression (adjustment) step-by-step leads to a model reflecting the real system.

Considering the approaches to determining the permissible ecological loads, it is necessary to dwell first of all on the problem of pollution standards for the region as a whole (both from the point of view of the permissible ecological load due to pollution and from the point of view of pollution 'production').

For a specific determination of the permissible ecological load on the given region, it is necessary to isolate (delineate) this region according to some feature corresponding to the problem posed. The 'openness' of the regional system for selecting many features (e.g. an economic feature) should be noted. To delineate the Lake Baikal region, the hydrological feature was selected, i.e. the lake basin was taken as the territory of the region (Fig. 1.4). In this region, certain types of landscape and systems of biogeocenoses are clearly defined.

Numerous studies carred out in this region can be drawn upon to synthesize the ecological-economic model of the region as a whole.

The modular principle of model construction used in [18] is very important, since every module has its own independent value, thus permitting reconstruction inside any model, without changing the model as a whole.

Forecasting changes in the ecological systems of the region, and, above all, identifying the sources of the anthropogenic impact and the causes of the changes occurring, as well as the potential impact of different parts of the region on adjacent parts, requires an analysis of separate territorial subsystems with relatively similar physical-geographical conditions (and similar ecosystems), the construction of models simulating economic activities, an analysis of the consequences of these activities, and an assessment of the consequences for each of the territorial subsystems. To do this, it is necessary to divide the region into territorial subsystems based on the above considerations. The isolation of subsystems is also necessary for determining the function describing the influence of one part of the region on any other part ('transfer' function).

It is very important to know, for example, how, for different ways of economic development, separate parts of the Baikal region can influence the arrival of any additional (to the natural fluxes) substances: mineral particles resulting from soil and rock erosion and various pollutants; it is also essential to know the relative contribution (and, consequently, the influence on the ecosystems) of each part of the region to the total flux of substances into the lake.

For instance, the portion (β_{ij}) of the flux of pollutants from any part (i) of the region to another part (j) will (averaged over some time interval) be

$$\beta_{ij} = \frac{\sum_l \sum_m f_{ijlm}\sigma_{il}}{\sum_i \sum_l \sum_m f_{ijlm}\sigma_{il} + \sum_l \sum_m f_{xijlm}\sigma_{xl} + \sum_l \sigma_{jl}} \tag{1.1}$$

where σ_{il} is the average pollution in the environment l of the ith part of the region; f_{ijlm} is the probability of pollutant transport from the environment l of the ith part of the region to the environment m of the jth part of the region (for the assumed time interval of averaging); f_{xjlm} is the same, but outside the region; σ_{xl} is the average pollution in environment l at the boundary of the region, entering from outside.

If the calculation of the transport is performed for environment l alone, then expression (1.1) becomes much simpler:

$$\beta_{ij} = \frac{f_{ij}\sigma_{ij}}{\sum_i f_{ij}\sigma_{il} + f_{xj}\sigma_{xl} + \sigma_{jl}} \tag{1.2}$$

An expression can also be written for the total balance of moving material which is potentially dangerous from the point of view of pollution. Here we mean the part of this material which is able to enter the environment; for example, for eroded soil areas, it may be the part of the soil subject to washout; for agricultural fertilizers, it may be the part subject to washout and further migration; for metals, it may be a small part which can be dissolved by acid precipitation, etc.

It seems appropriate to divide the ways of arrival of potential pollutants (various substances) into three categories: arrival via transport and migration through natural media (the atmosphere and water courses); arrival via the industrial or other transport of various materials; and arrival due to the movement (transfer) of food products. The last category deserves spe-

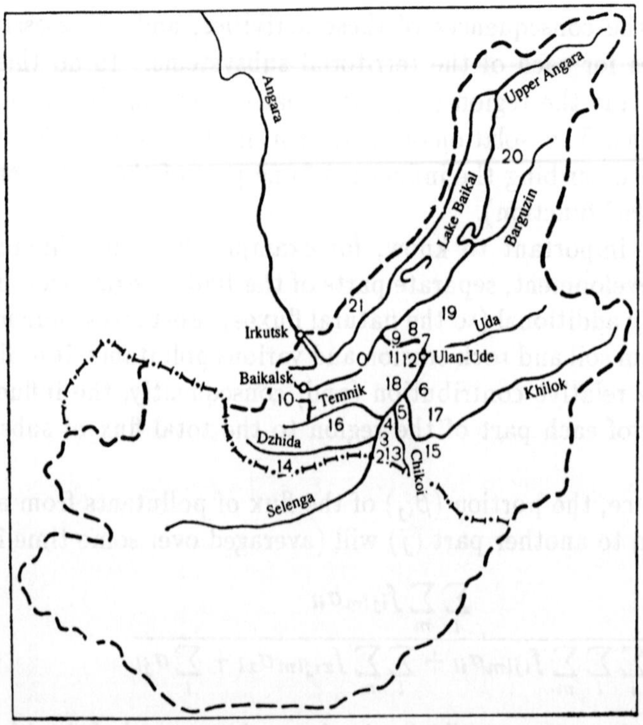

Figure 1.4. **Region of Lake Baikal. Dashed line indicates boundary of region. Numbers 2 to 21 denote territorial subsystems. 0 denotes Lake Baikal itself. 1 denotes external areas.**

cial study, since it is a way for pollutants to enter directly into the human organism.

Thus, the problem is obviously a complicated one. Let us consider the approaches to its solution in consecutive order, step-by-step, for the Lake Baikal region.

Fig. 1.4 shows the contour of the Baikal region and its subdivision into territorial subsystems (the subdivision principle here is relative homogeneity from the point of view of the possible influence on the Lake Baikal ecosystem, on the basis of the available description of the region). A total of 22 territorial subsystems were isolated, including the lake itself and an external (with respect to the region as a whole) subsystem (denoted in Fig. 1.5 by the number 1).

The interrelationship of the territorial subsystems is shown in Fig. 1.5. A brief description of the subsystems is presented below.

0. Lake Baikal, in the model an object of anthropogenic impacts.
1. External areas (with respect to the region as a whole), determining in essence the flow of different substances from outside into the Lake Baikal region, including the arrival of pollutants from the upper reaches of the Selenga River, from the Mongolian People's Republic.

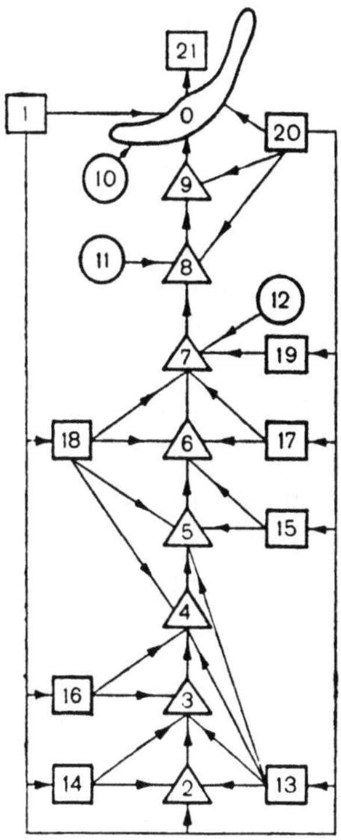

Figure 1.5. Territorial subsystems of Lake Baikal region, showing direction of pollutant transport.

2–9. The reaches of the Selenga, distinguished according to their contributions to the pollution.

10–12. Largest anthropogenic pollution sources, associated with urban centers and industrial enterprises (Ulan–Ude, Baikal pulp-and-paper works, Selenga pulp-and-carton works).

13–20. Parts of the Lake Baikal region corresponding to the basins of the largest rivers flowing into the lake and also characterized by relatively homogeneous land use.

21. The Angara River, flowing out of Lake Baikal.

Aside from the territorial subsystems, the most important functional subsystems were also distinguished: settlements, industry, transport; utilization of natural resources (ecosystems of Lake Baikal, forest, land, and water resources); recreation. The hydrological, ecological, and economic aspects of the state and development of the region were also considered.

To construct a model and to describe the functional relations between separate modules, the methods of system dynamics were used. At first, point models were investigated, and the dimensions of areas in which a system could be considered homogeneous were determined; then, the interaction between homogeneous areas was examined (subdivision of the region into separate territories represents a first, rough, approximation in solving the problem using this approach).

The inventory module and the functional module served as a basis for the model. Let us describe these separately.

The *inventory module* includes a list of all possible pollution sources in different media, together with their intensities (powers).

The data in this list are taken either from observations – detected sources and the intensities of their discharges (amount of pollutants discharged per unit time) – or from the technical documentation of the enterprises involved; alternatively, the amount of discharge is calculated tentatively on the basis of the production output, proceeding from the technological level of the industrial enterprises and the structure of the economic activities. Taking inventory was reduced to a certain standard format, including the subdivision of pollution sources into different categories, according to:

- types of sources (point, distributed over the are, stationary, mobile, etc.). The largest or potentially most hazardous sources are considered specially (e.g. the Baikal pulp-and-paper works and the Selenga pulp-and-carton works);
- natural environments into which waste is discharged and where pollutants spread (atmosphere, hydrosphere, lithosphere, biota);
- types of economic activities (industry, agriculture, municipal economy), as well as branches of the national economy, for connection with the system of interbranch balance;
- individual ingredients discharged into the environment;
- enlarged pollution categories, consolidating the ingredients into groups (heavy metals, petroleum products, pesticides, etc.);
- toxicity levels.

Pollutant spreading module (functional module). The basic problem of the present section is the determination of the functions $f_{ijlm}(t)$ (see Equation (1.1)) or, if we speak about transport only between different environments, the functions f_{lm}, where f_{ijlm} is the probability of pollutant transport during time Δt from area i to area j and from environment l to environment m. The quantities f_{11}, f_{12}, f_{13} and f_{14} correspond to the spreading of pollutants in the atmosphere (index 1), deposition on the soil surface (index 2), deposition on the water surface (index 3) and direct incidence into the biota (index 4) (resulting from dry depositions and washing out with precipitation).

From data on pollution sources and meteorological data, calculations are carried out of the spreading of pollutants in the atmosphere, the deposition (dry, as well as 'wet' with precipitation) onto the underlying surface, and the distribution of pollutants over the underlying surface.

Considerable literature has been devoted to calculations of pollutant spreading in the atmosphere and deposition onto the underlying surface. Some works present calculations of the short-range and long-range spreading of pollutants, discharged from transient and continuous sources located at different heights, having different rates of deposition onto an ideal surface and onto a surface with complex relief.

If $q_i(t_k)$ is taken to denote the integral pollutant concentration (integrated over the height $q_i(t_k) = \int_0^\infty C_i(t_k, h)\,dh$, where C_i is the volume concentration and h is the height) above an area i, then we have (for $l = m$)

$$\sum f_{ij} q_i(t_k) e^{-\frac{\Delta t}{T}} + Q_i(t_k)\Delta t = q(t_{k+1}) \tag{1.3}$$

where $\Delta t = t_{k+1} - t_k$; $Q_j(t_k)$ is the intensity of pollution sources in area j at moment t_k; T is the residence time of the pollutant in the atmosphere.

The concentration σ_i on the underlying surface is calculated using the formula

$$\sigma_i(t) = q_i(t)/T \tag{1.4}$$

and the probability f_{ij} is found from data on the recurrence of winds T in direction ω_{ij}:

$$f_{ij} = F(\omega_{ij}, r_{ij}/\Delta t) \tag{1.5}$$

where r_{ij} is the distance between areas i and j.

Equation (1.4) is written in a simplified form; strictly speaking, it is also necessary to take into account the pollution decrease in area j during the time Δt.

In the given calculations it is necessary to take into account possible sizable transfers of pollutants in the atmosphere from the adjacent regions and even from far away.

Some pollutants enter the atmosphere (index 1) from the land surface (index 3), e.g. dust and pesticides; mercury can evaporate and enter the atmosphere from the land (index 3) and water surface (index 2). In the two latter cases, it is necessary to determine functions f_{31} and f_{21}.

Pollutants can pass from the land surface into water courses (surface washout); their penetration into the biota (via roots) is also possible. The different forms of pollutant transfer from one environment to another are presented in Table 1.1.

The dynamics of the pollutant transfer (washing away) from the underlying surface into the water courses is described by the expression

$$\begin{aligned} \dot{\sigma} &= \frac{q}{T}S - \sigma_w, \\ \dot{\sigma}_w &= \frac{\sigma}{T_S} \end{aligned} \tag{1.6}$$

where S is the area over which the pollutant concentrations are integrated (averaged); σ is the pollution of the underlying surface; σ_w is the amount

Table 1.1 Possible forms of transfer (migration) of pollutants between different natural environments.

Probability f_{lm} of pollutant transfer from environment l to environment m	Natural environment	Possible form of pollutant transfer (migration)
1	2	3
f_{11}	Atmosphere–atmosphere	Atmospheric transport, characteristic for most pollutants
f_{12}	Atmosphere–hydrosphere	Deposition (washing out) of atmospheric pollution onto water surface
f_{13}	Atmosphere–land surface	Deposition (washing out) of atmospheric pollution onto land surface
f_{14}	Atmosphere–biota	Deposition of pollutants onto plant surface and their subsequent assimilation (nonroot arrival of pollutants into biota)
f_{21}	Hydrosphere–atmosphere	Evaporation of pollutants (e.g. petroleum products and mercury compounds) from water into atmosphere
f_{22}	Hydrosphere–hydrosphere	Transport (spreading) of pollutants in water systems
f_{23}	Hydrosphere–land surface (river and lake bottom)	Transfer of pollutants from water to soil (seepage, 'self-purification', deposition onto bottom of water body)
f_{24}	Hydrosphere–biota	Transfer of pollutants from surface water to biota (land and water ecosystems; arrival in human and animal organisms with drinking water)
f_{31}	Land surface–atmosphere	Transfer of pollutants from land surface to atmosphere (weathering, evaporation, dust transfer)
f_{32}	Land surface–hydrosphere	Washing out of pollutants from land during snowmelt, with precipitation, and by transient water courses
f_{33}	Land surface–land surface	Migration within soil, glaciers, snow cover (penetration of pollutants to various depths)
f_{34}	Land surface–biota	Arrival of pollutants to plants via roots
f_{41}	Biota–atmosphere	Evaporation from biota (unimportant)

Table 1.1 (continued)

1	2	3
f_{42}	Biota–hydrosphere	Arrival of pollutants from biota to water after death of organisms
f_{43}	Biota–land surface	Arrival of pollutants from biota to soil, primarily after death of organisms
f_{44}	Biota–biota	Migration along food chains

of pollutants washed into a water course; T_S is the residence time (lifetime) of the pollutant on the underlying surface; q/T is the rate of pollutant deposition from the atmosphere onto the underlying surface (assuming that $f_{31} \approx 0$).

When calculating the pollutant transfer by water courses, it is necessary to take into account the rate of pollutant arrival with sewage as well as the self purification processes occurring in water courses (the residence time of the pollutants in the water courses).

In the above modules establishing a link between pollutants in different environments (in connection with pollutant transfer and migration) it is necessary to take into account all the possible transfers listed in Table 1.1 (although one form, characterized by the function f_{41}, is unimportant for the calculation of pollutant transfers).

The module of assessing the influence zones of large sources, such as the Baikal pulp-and-paper works and the Selenga pulp-and-carton works, was distinguished as an individual module.

The results of special tracer experiments (with short-lived radioactive isotopes) were used to obtain relevant data on the dilution, dispersal, and spreading, in the Baikal water area adjacent to the paper mill, of substances from deep-water sites where purified wastes are discharged into the lake.

The next module is associated with forecasting the state of natural ecosystems. The influence of pollutants on the lake ecosystem and the effects of the pollution were determined using the calculated fields of the pollutants.

The principle of optimality was used as a basis for the forecasting method. The 'basic niches' were identified in the ecosystem, and the species systems occupying these niches were considered as a kind of 'generalized species'. A system of relations, the solution of which yields the parameters of an ecological system, depending on the anthropogenic impact, was obtained using the principle of optimality with the aid of the biological principle of interrelationship between the characteristics of communities. To investigate the ecosystem of Lake Baikal itself, a mathematical simulation submodel was developed reproducing the lake pelagial, where most endemics dwell and where the bulk of the production occurs, to a great extent determining the matter and energy turnover in the lake [18]. The given submodel is sufficiently stable (this was investigated by varying the parameters over a wide range), it describes the dynamics of the basic trophic levels of the ecosystem, and it comprises a system of ten ordinary differential equations. The basic parameters of the described submodel of the Baikal pelagial are presented in [18].

It is interesting that the intensity of the anthropogenic input of organic substances and biogenic elements to the lake varied in the calculations over a wide range, although at present such a problem does not exist on a wide scale for the lake. However, it might be important for protection of the lake against eutrophication in the future.

The last module of the model is the one assessing the damage from pollution and optimizing the use of the regional resources of the ecosystem. Different kinds of damage are considered:

- ecological damage, gene-pool impoverishment;
- esthetic damage, recreation damage (to tourism, resorts, etc.);
- damage to fisheries;
- damage from the pollution of water as a resource of clean water for industry and drinking.

Generally speaking, it is also necessary to consider the psychological damage associated with the impact on Lake Baikal; however, no quantitative approaches have been developed as yet to take account of this kind of damage. Considerable difficulties are encountered when determining the esthetic and ecological damage, particularly the damage associated with gene-pool impoverishment.

As far as the other kinds of damage are concerned, some studies have been made of this problem, for instance, a determination of the relation of the usefulness of the region from the point of view of fisheries to the damage caused by the anthropogenic impact on this trade.

Let us try to evaluate the damage caused by pollution of clean water and the damage to human recreation.

The damage from the pollution of water considered as an industrial resource can be estimated considering the functional associated with the usefulness of the lake water:

$$C_w = g(t,S)V(t,S) + q \tag{1.7}$$

where $g(t,S)$ is the regional cost of water as a quality function, which is also a function of the economic policy S and the time t; $V(t,S)$ is the industrial water consumption, q is the usefulness during the course of consumption.

The recreation damage can be found by considering the decrease, the loss, in the usefulness of the region C_R. The quantity C_R is determined from the expression

$$C_R = ar_1(t)T(t,S) + ar_2(t,S) \tag{1.8}$$

where $T(t,S)$ is the recreational usefulness of the lake in man/days (taking into account the time needed for the journey); r_1 is the average wages of a tourist (per day); $r_2(t,S)$ is a tourist's expenses; and a is the effectiveness of the use of the time and means, characteristic for the state.

The use of insecticides to protect the forests in the vicinity of Lake Baikal against needle- and leaf-gnawing pests may serve as an interesting example in the determination of the effectiveness and damage of the anthropogenic impact.

The model described in [25] was used in the calculations. To assess undesirable side effects, such factors were taken into consideration as the probability of unsuccessful pest control, disturbance of the stability of forest biogeocenosis, and losses incurred by animal husbandry and bee-keeping, as well as recreation losses. The results of calculations performed for typical conditions have shown that the chemical control of forest pests in the Lake Baikal region is inappropriate.

The described model as a whole has already been used successfully for different assessments and decision-making processes.

The use of modules associated with the distribution and migration of pollutants in natural environments makes it possible to determine the relative significance of different sources in the pollution of Lake Baikal and their impact on the lake ecosystem (sources located at different distances, with discharges into different media).

Tentative computations with the model indicate that the Selenga River (its catchment area accounts for 60 percent of the Baikal basin) plays a dominant role in the undesirable input and accumulation in the lake of organic matter of allochthonous origin (these results are in good agreement with observational data).

At the same time, the effects of selfpurification processes in the Selenga on the Ulan-Ude – to – Baikal reach amount to (for the different ingredients) values of the order of 10^2 to 10^4 (i.e. an individual pollution source on the Selenga in the Ulan-Ude region contaminates the lake 10^2 to 10^4 times less than a similar source located on the lake shore). Computations indicate that long-range atmospheric transport contributes significantly to the pollution of the lake.

A serious requirement in the given model is the need to balance the modules (the accuracy and details of their output data should be approximately the same; this is an important requirement for all systems). The considered model has been further developed in [17, 29, *et al.*].

Studies were carried out to improve our understanding of the influence of economic activities in the region. A detailed joint inventory of the sources polluting the atmosphere and the natural waters was carried out, and the sources were analyzed. Models of the Lake Baikal ecosystem [5, 20] and models of the pollutant transport by air and water flows in the lake region [3, 32] were developed. The calculated air-mass trajectories made it possible to average the concentrations and to obtain the pollutant fluxes to the underlying surface.

The pollutant transfer within the lake water column was calculated in two stages. First, averaged stream functions were calculated taking into account data [32] on the dynamics of the wind-driven currents of Lake Baikal. Then, the pollutant fluxes between the 'chambers' into which the water column was divided were determined using a model of the chamber type.

Calculations were made of the Lake Baikal current pattern formed by atmospheric effects and also of the pattern of pollutant input to the lake from the Selenga, Barguzin, and Upper Angara Rivers, as well as from the Baikal pulp-and-paper works.

To forecast the change in the water quality and in the selfpurification ability of the lake ecosystem, some models of its pelagial were constructed. The intra-annual and interannual dynamics of the basic elements of the ecosystem, their energy budget, and species composition were investigated with the aid of these models. Lake ecosystem stability is maintained by the presence of detritus. The maximum deviation from equilibrium is brought about by factors affecting the phytoplankton. It has been shown that the spring maximum of the algal biomass depends to a great extend on the availability of mineral food, while the autumn maximum depends on the thermal conditions preceding the beginning of the period of maximum and during it. River runoff affects the plankton dynamics only in the vicinity of river mouths [5, 21].

The modeling indicated a lack of data and gaps in our knowledge of the characteristics of the lake ecosystem, and it suggested several leads for further research in the region.

It should be noted in conclusion that the above simulation-balance model of the Lake Baikal region is not only an example of the application of the analysis, but is also a practical tool of the comprehensive environmental analysis, for assessment of the current state, long-term forecasts, determination of the permissible anthropogenic loading on the entire region and justly on the Lake Baikal biome, on its unique ecological system under the conditions of a developing economy. The use of this model facilitates the planning of observations of the state of the lake and the factors of the anthropogenic impact, as well as the establishment of standards for this impact and the utilization of these data in managing the regional economy.

The principles of ecological modeling were developed and used in a number of Soviet works, so as to arrive at balance ecological-economic models taking into account interbranch and territorial aspects within the framework of the entire region [1, 4], individual large water bodies [6, 11], and various ecosystems [19, 26]. Monograph [27], article [35], and many other works deal with the mathematical modeling of ecological processes. The fourth Soviet-American Symposium on the Comprehensive Analysis of the Environment discussed joint studies of the environment of a large region [10].

In a number of non-Soviet countries, including the USA, ecological modeling aimed at estimating the anthropogenic effects and determining the permissible anthropogenic loading on a natural object has also been developed [34, 36].

For example, to interpret and describe the responses of the biota to anthropogenic impacts and to assess the intensity of such impacts on the natural environment, a multifactor analytical approach was applied in [23], and the use of simulation modeling to study the anthropogenic impact on the Lake George ecosystem (New York State) was described.

Lake George is a narrow, moderately deep lake (50 km long, maximum width 5 km, average depth 18 m, area 114 km^2, catchment area 492 km^2). The lake is oligotrophic.

To determine the impact of the local population and tourists on the lake, a multifactor analysis of dead diatom accumulations was carried out over the entire lake. An examination of the samples revealed the gradients of water quality and the state of the environment, leading to an interpretation of the relations between groups of algae, particularly those serving as indicators of oligotrophic and eutrophic conditions.

Two models were developed using the simulation method: an aquatic ecosystem model and a model simulating the land ecosystem of the territory adjacent to the lake.

The model of the aquatic ecosystem included all the significant ecological and physiological processes (20 variables were taken into account).

The effects of various loads on all the ecosystem components were estimated using the model. In the first case, the phosphate content of the lake water was decreased by 20 percent (compared with the norm). The system responds to change slowly; the biomass of the blue-green algae and fish decreases during the course of several years. Fig. 1.6 shows the response of this system to a consistent rise in the mean water temperature by 5°C; such a change leads to an increase in the biomass of the blue-green algae and to a decrease in the biomass of trout.

The described model was adapted to fit other lakes, and in particular European lakes (including the manmade Lake Slapin in Czechoslovakia).

In the land ecosystem model, changes in land use processes and in plant succession were simulated.

Ecological modeling techniques were also applied to rivers, and in particular to a comprehensive analysis of the upper estuary of the Potomac River, where eutrophication is the main problem (a similar situation is observed,

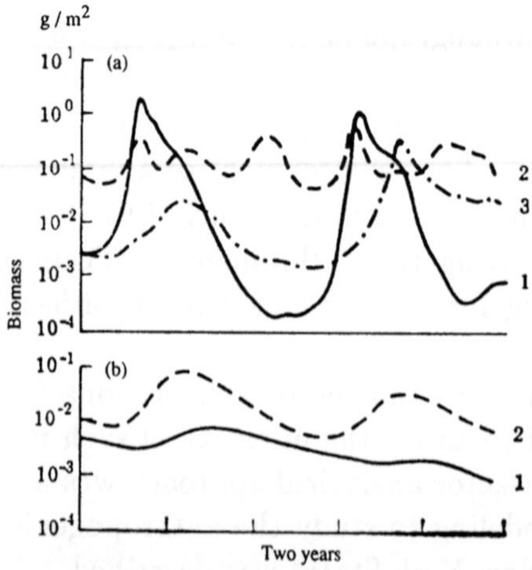

Figure 1.6. Experiment with rise in temperature:
(a) net phytoplankton (1), nannophytoplankton (2), and blue-green algae (3);
(b) predatory (1) and nonpredatory (2) fish.

for example, in Lake Erie and in more than 800 other lakes and reservoirs of the USA [33]).

About 3 million people live in the vicinity of the Potomac upper estuary, a region including Washington, D.C. The esthetic and recreational potential of the upper estuary was reduced substantially, because of vast fields of algae, which produce unpleasant smells, cover the shores and beaches, and impede boating. The possible use of the estuary as a source of water supply may well be hampered, due to the presence of toxic substances released by the blue-green algae.

Data obtained with the aid of the mathematical model, as well as from the results of field work, indicated that the yield of blue-green algae can be predicted on the basis of the nitrogen cycle, since it is the presence of nitrogen that mainly controls the yield. Similar methods have shown that, in order to reach the critical chlorophyll level of 25 mg/l (required for eutrophication control), the total phosphorus content of the water must not exceed 0.03–0.1 mg/l, while the corresponding limit for inorganic nitrogen must be 0.3–0.5 mg/l. Acccording to [33], the phosphorus content of the water in the upper estuary reached 1.2 mg/l.

The water-quality modeling was carried out using dynamic and universal models of the estuary. The dynamic model was used to simulate the water quality with a high time resolution. The model, which consists of a hydraulic part describing the tidal movements and a water-quality part, takes into account the basic transport mechanisms, i.e. advection and dispersal, as well as the sources and sinks of the basic ingredients.

The universal model was used to study the seasonal variations of the

Table 1.2. Maximum phosphorus and nitrogen loads in different areas of upper estuary of Potomac River (kg/day).

	Area	Phosphorus	Nitrogen
Ia.	Estuary 12.1 km below Chain Bridge	90	450
Ib.	Anacostia River before confluence with Potomac	40	140
Ic.	Estuary from 12.1 to 24 km (to Board Creek)	400	1580
II.	Estuary before Indian Head (to 48.3 km from Chain Bridge)	680	2600
III.	Estuary 72.4 km below Chain Bridge	900	4100

nitrogen and phosphorus distributions in the upper estuary; the model is based on an averaged solution of the fundamental equations of the balance of masses depending on tidal effects.

The integral permissible loads of nitrogen and phosphorus in different areas of the upper estuary were determined using the model (Table 1.2).

Note that in the 1970s the discharge of nutrients into the estuaries amounted to 10 900 kg/day of phosphorus and 27 200 kg/day of nitrogen, whereas under high runoff conditions the amount of nitrogen could reach 212 000 kg/day [33] (the model makes it possible to calculate the maximum loads under different hydrological conditions).

Seasonal requirements were developed for the permissible load and the discharges of nutrients. Special attention was given to maintaining a balanced structure of the ecological community in the upper, or fresh-water, part of the estuary.

It was estimated that the cost of ensuring an adequate water quality in the upper estuary of the Potomac would amount to 1.34 billion dollars over 50 years (64.8 million dollars per year or 13 to 24 dollars per capita annually).

The approach developed to determine the permissible load on the Potomac estuary can be applied in a general form to other lakes and rivers as well. However, the peculiarities of the morphological and limnological characteristics of other water bodies may well require a special analysis and interpretation of the data (as was shown for Lake Baikal).

CHAPTER 2

Permissible Loading of the Biosphere and the Establishment of Ecological Standards

2.1. Permissible Loading of the Biosphere

As already noted, a comprehensive environmental analysis makes it possible to take into account, when assessing the environmental load, all kinds of effects (natural and anthropogenic) on the different elements of the biosphere, i.e. individual organisms, populations, and entire biogeocenoses. In conjunction with a correctly organized functional monitoring system, this enables us to achieve an optimal man/nature interaction and forms a basis for a program of regulation of the environmental quality.

To develop such a program, it is first of all necessary to know the acceptable levels of environmental pollution, the impact of which could be considered permissible, both for man and for animal and plant populations, i.e. for entire biogeocenoses. It is obvious that an understanding of this problem gives us an idea of what kind of environmental quality should be attained [14], taking into consideration the direct effects of the impact and the occurrence of remote consequences in live organisms, and it also indicates what kind of quality can be permitted.

Whereas in the first stage of the comprehensive environmental analysis the effects of different factors and the consequences of these effects were analyzed (see Section 1.1), in the second stage the permissible impacts on biospheric elements at different levels are determined, and ecological standards (criteria) of permissible loadings are established.

To accomplish this, it is necessary to discuss some general approaches, to define the acceptability of the impact level, and to understand what is meant by 'ecological standards'. This entails a discussion of such concepts as high environmental quality and stability of biological (ecological) systems, as well as of the connection between the 'threshold nature of the impact' and the 'permissible loads'.

The concept of the 'permissible impacts and loads' on the natural environment is rather complicated. Any load originating in an ecological system due to some impact and able to alter the average (normal) state of the ecosystem is defined by us as an ecological load. Clearly, impacts not leading to a load in excess of the permissible load on an ecological or other natural system may be considered to be permissible impacts.

In some cases, the load can be identified with the impact; for instance, the concentration of some toxic substance in the organism of some animal, on the one hand, is a factor of the internal impact on the given organism, while, on the other hand, this concentration can be taken as a value characterizing the load on the organism.

The external concentration field is only a factor of the impact, and not of the load on the given organism (the load will depend on the residence time of the organism in the polluted environment, on the paths of pollutant arrival at the organism, etc.).

Probably, a load should be considered permissible if it does not produce undesirable effects or changes in the organisms (and biogeocenoses) inhabiting the Earth, primarily in man, of course, and if it does not lead to any (or any significant) deterioration of the environmental quality.

When formulating criteria for the permissible loads, we should first of all pay attention to anthropogenic impacts, under the influence of which the environment can change particularly rapidly. Of course, some natural events can also result in significant changes of the state of the environment for relatively short time intervals. These include various natural disasters on a large, or even catastrophic, scale, such as hurricanes, earthquakes, tsunamis, floods, and landslides. But in this work the problem of the permissible loads is considered from the point of view of anthropogenic impacts and their specific character and consequences.

The definition of the permissible loads pertains to relatively small changes in the environmental quality, not affecting the normal functions of ecosystems and individual populations, and it involves an assessment of the intensity of impact factors not leading to significant changes in the state of the ecosystems, but rather only to changes within a range called permissible; we will also be concerned with maintaining high environmental quality.

In this connection, the following issues should be discussed:

(1) high and acceptable environmental quality;
(2) permissible and critical responses of ecosystems and other biospheric elements to the impacts, as well as the state of ecosystems within the intervals bounded by such responses;
(3) permissible and critical intensities of impact factors at any point, and also averaged over space for an individual organism, population, community, ecosystem, or the biosphere as a whole, corresponding to the permissible or critical states of systems.
(4) permissible fluctuations of the environmental quality;

(5) permissible and critical loads on an organism, population, community, or ecosystem (and even on the biosphere as a whole).

It should be noted that, during the formulation of these definitions, two levels of the impact of hazardous factors on natural systems should be considered, namely the critical level of the impact (beginning with which the destruction or irreversible degradation of the given system, e.g., the extinction of some species, can occur) and the permissible level of the impact (limited by an arbitrary criterion of the maximum permissible impact), which is substantially lower than the critical level.

Let us try to give a definition of high environmental quality (there is as yet no generally accepted definition of environmental quality).

Without pretending to give an exhaustive definition of environmental quality, we suggest that high (or acceptable) quality of the natural environment for an ecosystem signifies [14]:

(a) the possibility of a stable existence and development of a historically established ecosystem, created or transformed by man, at the given place;
(b) the absence of adverse effects at present and in the future for any (or for the most important) population (especially for man) which occupies this place historically or temporarily.

There are ecological criteria characterizing a high quality of the natural environment: biological productivity (under the given climatic conditions), an optimum ratio of species, the biomasses of populations at different trophic levels, etc.

According to Shvarts's suggestions [46], a 'good biogeocenosis' should meet the following basic requirements:

(1) the production (biomass) should be high at all the basic trophic levels, without a pronounced excess of phytomass over zoomass;
(2) high productivity should correspond to high production; this is necessary for a rapid compensation of possible biomass losses at some trophic levels due to external impacts, since high production does not in itself guarantee an adequate compensatory activity of the biological systems;
(3) heterogeneity of the trophic levels and the ecosystem structure as a whole should contribute to a high stability of the biogeocenosis over a wide range of external conditions. Dynamic equilibrium of the biogeocenosis should ensure a state of homeostasis of its abiotic component, including both the hydrological regime of the territory and the gaseous composition of the atmosphere;
(4) the rate of biological selfpurification of the ecosystem should be a maximum, due to a high rate of exchange of matter and energy (ensuring biological turnover during the course of a few annual cycles);

(5) a very high 'reserve activity' (the possibility of a fast structural rearrangement of the community) should supplement a very high degree of ecosystem productivity and stability, which keeps the biogeocenosis in the optimum state when the environmental conditions change.

Satisfaction of these requirements for a good state of the biogeocenosis is necessary both for development in the natural environment and for development in an environment subject to anthropogenic impacts. This approach enables an evaluation of the permissibility of some load on the ecosystem. In this case, a high quality of the environment and the biological systems, i.e., the abiotic and biotic components of the biosphere, toegether with unity of these components forming the ecological systems, serves as the criterion.

The state of any environment of the abiotic component of the biosphere is determined from:

(1) a set of physical quantities characterizing the instantaneous state of the environment and the state averaged over time $[\theta(t)]$. For instance, the state of the atmosphere is characterized by a set of meteorological parameters: temperature, atmospheric pressure, air humidity, amount of precipitation, wind speed, etc.;
(2) data on the chemical composition, including small components $[\tau(t)]$. The state of the biotic component is also determined by biological properties, being characterized by a set of functional and structural indices giving an idea of the state of the organisms, populations, and ecosystems $[\xi(t)]$.

The state of the biotic component is, as a rule, the determining factor. In this case the impact of some factors may result in a change of $[\xi(t)]$, either directly or via a change in the physical or chemical properties of the environment (a change in $\theta(t)$ and $\tau(t)$).

Noted by Shvarts [46], "if a biogeocenosis is able to sustain itself (as a system) in the optimum state in an environment changed by man, this means that the degree of anthropogenic impact does not exceed the capabilities of the biological systems or undermine their capacity for homeostasis".

It is obvious that the state of the biogeocenosis is the most sensitive indicator of the state of the environment.

When formulating the concept of the permissible ecological load on a biogeocenosis, it is necessary to take into account different paths of biological progress. The possibility of achieving biological progress by different paths (at the levels of the organism and the population) to a great extent determines the structure of the biogeocenosis [46]. Protective mechanisms of populations have developed under conditions of an unfavorable combination of external factors along two lines: increasing the resistance of individual organisms and increasing that of populations. Unlike 'higher' animals, 'lower' animals display a great ecological variety of age stages (occupation of different ecological niches at different age stages of the development, as in the case of larvae and adult specimens of many insects). In this case, the possibility of extinction of species resulting from a sharp change in the environment is minimized. Such an effect determining the critical stage (from the point of view of sensitivity to external impacts) should be taken into consideration when determining the permissible (or critical) load on a population or ecosystem.

In the broadest sense, the permissible anthropogenic impact on the natural environment will consist of separate homogeneous and heterogeneous impacts, and it will not affect the environmental quality or alter the natural environment within the permissible range, i.e., it does not destroy the existing ecosystem, and it does not have an unfavorable effect on the most important populations, especially, of course, human beings.

As already noted, the permissible impact should be in line with the permissible load, which lies in the permissible range.

It is obvious, however, that the concept of 'permissible impacts and loads' is not only complicated but, probably, ambiguous as well. It depends on the range that can be considered permissible and on the objectives pursued by man while producing a conscious or unintentional effect on the environment.

The author considers that, from this point of view, all ecological systems can be tentatively divided into the following categories [14]:

(1) unique, or protected, ecosystems;
(2) widespread natural ecosystems;
(3) greatly transformed, or artificial, ecosystems.

In ecosystems of the first category, the extinction (death) of a biological species is undoubtedly out of the question. It is quite obvious, for instance, that the death of even individual endemics of Lake Baikal is altogether unacceptable.

In ecosystems of the second category, the basic requirements for ecosystem preservation and high environmental quality should be satisfied, although some (particular) changes are possible.

In specially transformed, changed, or even manmade, ecosystems of the third category, any justified anthropogenic changes complying with the objectives pursued by man will be possible in the future.

For a quantitative determination of the permissible load, it is necessary to know whether the ecosystem was in a normal state at the beginning of the impact (without any deviation from some background value, determined by averaging over a long time interval under normal environmental conditions) or in a state that had been changed under the influence of some impact factor.

Therefore, it is important to determine what state should be taken as the initial state: the average 'background' state (in the absence of the impact under study) or the state prior to the onset of a substantial anthropogenic impact. It is probably necessary to take as a basis the background value of the state of the biospheric element, not altered by any local impacts under mean (normal) hydrometeorological conditions. Moreover, regional and global background states of biospheric elements should be distinguished at all levels, i.e. at the levels of an individual organism, a population, a community, and an ecosystem.

The factors of an impact on the natural environment can also be divided into several categories: mechanical actions, electromagnetic radiations, magnetic and electric fields, and pollution. Mechanical actions, various kinds of radiation, and physical fields mainly affect the physical state of the environment, while pollution affects its chemical state. The biological characteristics change under the influence of any factors.

To explain the foregoing, let us introduce some function $\eta(R,t)$ of the state of an ecosystem or any biospheric element, characterizing some individual feature of the state of the ecosystem (according to a particular index) or its state as a whole (according to a set of indices or an integral index). The amount of biomass, the productivity, the metabolic rate, etc., can be used as the indices of the state of the ecosystem. These indices characterize the physical, chemical, or biological state of the ecosystem or the state determined by a complex of physical, chemical, and biological features.

It is obvious that function $\eta(R,t)$ varies in space and time.

The generalized function of the ecosystem state for a region can be written as

$$\eta^*(t) = \int_R \eta(R,t)\,\mathrm{d}R \tag{2.1}$$

Figure 2.1 shows the time variation of function $\eta^*(t)$, corresponding to a natural normal (or optimum) state (5) and to a state excited due to some earlier impact (4). The Figure also illustrates the change in the function caused by an anthropogenic impact on the ecosystem during the time interval under consideration (3).

As already noted, it is possible to determine for each population the permissible limits of the variation of state ($\eta^*_{p\,\min}$ and $\eta^*_{p\,\max}$, corresponding to the 'normal' physical, chemical, and biological state within the intervals from $\theta_{p\,\min}$ to $\theta_{p\,\max}$; $\tau_{p\,\min}$ and $\tau_{p\,\max}$; $\xi_{p\,\min}$ and $\xi_{p\,\max}$) and the critical limits ($\eta^*_{c\,\min}$ and $\eta^*_{c\,\max}$ are the maximum and minimum limits, respectively). In [40] deviations which can be liquidated during the course of time by the system itself are considered to be an acceptable measure of the deviation from the normal state of the ecosystem. Attainment of the critical values of the state leads to destruction or suppression of the given system.

The above functions mostly have two sets of values; however, in a number

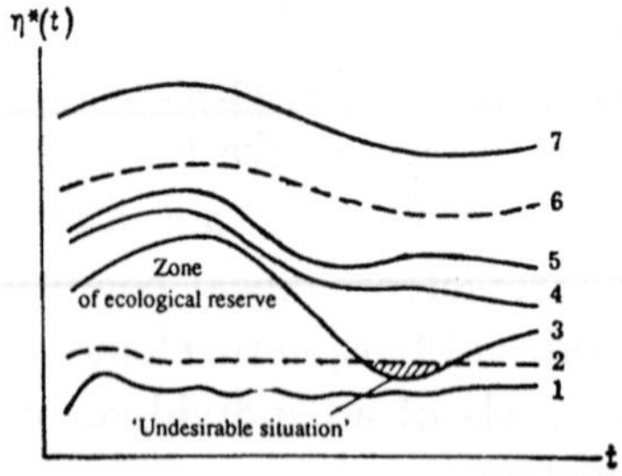

Figure 2.1. State function of biosphere elements $\eta^*(t)$ for different loads: (1, 7) critical minimum ($\eta^*_{c\,\min}$) and maximum ($\eta^*_{c\,\max}$) limits of change; (2, 6) permissible minimum ($\eta^*_{p\,\min}$) and maximum ($\eta^*_{p\,\max}$) limits of change; (3) state function of ecosystem with anthropogenic impact (η^*_{anth}); (4, 5) functions of excited ($\eta^*_{e\,2}$) and normal ($\eta^*_{e\,1}$) state of ecosystem.

of cases they have only one limiting state. For example, if the boron content of an organism is considered, then the permissible limits within which the state of the latter can vary are related to the maximum and minimum boron amounts, while for DDT the permissible and critical limits of state variations are determined only by its maximum content.

It is seen from Figure 2.1 that η^*_c has two sets of values (maximum and minimum), and that the permissible values lie between them; the time variation of $\eta^*_c(t)$ is caused by a variation of the external conditions (temperature, humidity, etc.). A set of states lying between the maximum and minimum values of η_c corresponds approximately to the homeostatic plateau.

The difference between the maximum permissible state and the actual state (and sometimes between the critical and actual states) characterizes the *ecological reserve* of a system. All other conditions being equal, the less the actual state differs from the average (normal) state, the greater will be the reserve.

Thus, the least value of $\overline{|\eta^*_p - \eta^*_{e2}|}$, averaged over the considered time interval, can be taken as a measure of the ecological reserve (at some moments of time this difference can also exceed the given value, but it should not exceed the value of $\overline{|\eta^*_c - \eta^*_{e2}|}$), the estimation being made both for the maximum values and for the minimum values (see Section 2.3 for more details).

Consider the case when only the maximum levels of the permissible η_p and critical η_c states exist. Let us assume that $\eta_p = \eta(q_p)$ and $\eta_c = \eta(q_c)$, where q_p and q_c are the permissible and critical impacts or loads on individual organisms of some population at the given point or on the population as a whole, but averaged over the entire area occupied by the population, for the average sensitivity of all individuals.

Then a criterion of permissibility can be established if the following condition is satisfied (the possible deviation of function $\eta(t, q_p)$ is defined):

$$\overline{|\eta(t, q_p) - \eta_{e1}(t)|} \leq \overline{|\eta_{e1}(t) - \overline{\eta_{e1}(t)}|} \tag{2.2}$$

where $\eta_{el}(t)$ is a quantity characterizing the natural normal state of an average organism at each time, and $\overline{\eta_{el}(t)}$ is a quantity averaged over a certain time interval.

Expression (2.2) means that the quantity q_p, at which the deviation from the average (normal) state of the given organism (population) does not exceed, on the average, the natural fluctuations (not disturbing the ecosystems), is the permissible load on an individual organism of some population at the given point (or on the entire population).* Evaluations should be carried out both for the maximum values and for the minimum values, characterizing the spread of the ecosystem state.

Let us introduce the function of the anthropogenic impact $\kappa_{\mathrm{anth}}(R,t)$ at a point (R,t) or for the region as a whole:

$$\kappa^*_{\mathrm{anth}}(t) = \int_R \kappa_{\mathrm{anth}}(R,t)\,\mathrm{d}R. \tag{2.3}$$

It is obvious that the state of the environment $\kappa^*_{\mathrm{anth}}(t)$ altered under the impact $\eta^*(t)$ will be defined by the ratio

$$\eta^*_{\mathrm{anth}}(t) = \epsilon\eta^*(t)\kappa^*_{\mathrm{anth}}(t) \tag{2.4}$$

where ϵ is the effect of unit impact on the state of the ecosystem; ϵ can be either constant (linear relationship between η_{anth} and κ_{anth}) or a function of κ (nonlinear relationship); if there is a threshold of the impact (κ_{th}), then the quantity $\epsilon = 0$ for $\kappa_{\mathrm{anth}} < \kappa_{\mathrm{th}}$.

Ecological problems related to the boundary conditions for the existence of biological organisms are also considered in [21].

2.2. The Threshold Nature of the Effect and the 'Dose–Response' Relation

The thresholds of harmful effects of an impact and the character of the 'dose–response' relation are of great importance when estimating the permissible impact of different factors on biospheric elements.

Here the threshold of the effect of an impact on the biological systems does not mean a general threshold for any alterations in the ecosystems subject to the impact, but rather a biotic response lying outside the limits of the usual physiological fluctuations observed during the process of homeostasis.

The author of [34] considers that even the simplified expression

$$D_r = D_o - (D_e + D_m) \tag{2.5}$$

(where D_r is the dose of harmful substance which has reached the receptor; D_o is the dose of harmful substance introduced into the organism; D_e and

* It should be taken into account that these deviations may be very small, if the external conditions vary over a wide range.

D_m are the doses of harmful substance, respectively, released from the organism and detoxified during the movement of poison toward the receptor) indicates the threshold nature of the effect of an impact, since for a relatively constant resistance of the organism, i.e., provided that $(D_e + D_m)$ does not decrease proportionally with decreasing D_o, the right-hand side of relation (2.5) becomes > 0 beginning with some value $D_o > D_{th}$ (where D_{th} is the threshold dose).

Experimental data on different substances show that the above condition is satisfied.

In addition, when an organism is subjected to adverse impacts, the processes of adaptation and continual restoration after the injury manifest themselves (e.g., activation of detoxication and of adaptive reactions). An injury develops only when the rate of the damage processes exceeds the rate of the adaptation and restoration processes [34].

It has been noted in a number of works [25, 34] that an increase in the amount of harmful substances entering an organism can lead to physiological adaptation, while the existence of the threshold itself (its value) may depend on the adaptability of the organism to some pollutant. The adaptability is the ability of the organism to endure the effect of the pollutant without pronounced unhealthy effects. Adaptation means growth of the adaptability for prolonged impact of some factor, if the intensity of this impact is not high.

Adaptation to the continuing impact of a certain factor at different biological levels can be demonstrated using several examples. For instance, under conditions of a radiation background elevated to a certain level (in regions with high mountains, outcropping of radioactive rocks, etc.), normal life proceeds for many human generations; bacteria adapt very rapidly to antibiotics, populations of insects to insecticides, etc. [34].

The concept of a threshold suggests a high environmental quality and complete safety for human beings and any populations, on condition that the pollution in this environment is below a definite level, whose effect on any organisms is below a certain threshold value.

However, the problem of the threshold nature of an impact is not so clear when we consider not the toxic effect of a substance but rather a chemical or radiation mutagenesis.

It has been noted [5, 25, 43] that the threshold of mutagenic action is probably absent for some factors of an impact and for some kinds of genetic effects. Hard electromagnetic radiation and various chemical substances induce alterations in the hereditary material, which can originate even as a result of a single simple chemical event or the effect of a single pair of ions. Thus, the formation of a single pair of ions or the appearance in the cell of a single molecule of a chemical mutagene may be enough to induce a point (gene) mutation resulting from a change of only one base in DNA

[43]. Such mutation changes may lead to severe consequences, right up to a lethal end. In this case the 'dose–response' curve is extrapolated for zero dose (of the impact) to the zero point, while the appearance of mutations in the organisms is probabilistic.

The degree of mutation risk during an exposure to chemical substances is often compared with the action of radioactive radiation; the comparison is carried out by introducing an equivalent (with respect to the effect) dose of hard electromagnetic radiation (for this purpose, special units of dose equivalence are introduced) [32, 44, 58].

The absence of a threshold of the effect during induction of one-impact mutation events is confirmed by direct experiments.

The following point of view with regard to this problem in the case of mutagenesis was formulated in [45]: the mutagenic action has no threshold; the countereffect of the protective cell systems to weak impacts leads to inactivation, elimination of the given substance, and healing of the injuries; the response of the organism to the impact has no threshold in the widest range of the impact either; the appearance of a pronounced pathological symptom has a threshold: it originates only when the protective capabilities of the cells and the organism have been exhausted.

If an injury of the chromosomal material results from two or more independent breaks of chromosomes (as a result of the action of ions from independent tracks or the multi-impact attack of chemical agents), then the frequency of these injuries grows nonlinearly with an increase in the dose, and the effect of the appearance of such structural changes has a threshold (threshold dose) [5].

In general, a definite biological effect (designated as harmful to an organism or a system) of an impact may not be observed until the dose of this impact reaches some level which is commonly also called a threshold dose. It could be defined as the minimum dose required for the onset of the given harmful effect.

Such a representation of the threshold nature of an impact suggests that below this dose there will not occur any unfavorable change associated with a pathological symptom (probably lower impact thresholds exist for some reactions preceding the appearance of the symptom).

It is known, for instance, that some dose levels have practically no adverse effect on animals or people, although the effect may be statistically significant.

A hypothesis of a threshold number of damaged systems [1], i.e., a hypothesis of a threshold of an adverse effect, is suggested even for ionizing radiation, together with the statement that there is no threshold for primary responses to radiation.

In this sense, the theory of the threshold nature of the effect of an impact assumes not the presence of any biosubstrate responses to the impact of external factors, but rather responses having a biological (including medical) significance, constituting an injury index.

The harmful effect of a substance exceeds the threshold when minimum concentrations of this substance in the external environment lead to changes in an organism (under the given conditions of arrival of the substance), which go beyond the limits of the physiological adaptive responses, or else to a latent temporarily compensated pathology [34].

It follows from the above that, if it is impossible to avoid the arrival of harmful substances to the environment, then their content in the environment must have a toxicological limit placed on it, on the basis of a determination of explicit thresholds of the harmful effect. The establishment of public-health standards for environmental pollution is based on the above conclusion. If the rates of decay and elimination of the substance from the environment do not exceed the rate of its arrival, then the permissible amount of the substance in the environment is established taking into account this dynamics, and also taking into account the thresholds of the adverse effects of toxic substances determined experimentally.

It is much more complicated to establish standards for environmental pollution in the absence of a threshold of the harmful effect; in this case, a scientifically justified approach would consist in a blanket prohibition of the discharge of such substances into the environment. Damage from the impact of any substances can be taken into account using the approaches described in Chapter 4.

At the same time, it should be noted that often the level below which, according to available data, the 'dose–response' relation does not yet manifest itself is taken in practice as the threshold level of the influence of some impact on any organisms. Experience has shown that, the more information and accumulated knowledge we have (and the better the observational instruments), the lower will be the threshold level. Therefore, the threshold value is often a concept referring to the level where uncertainty begins, rather than to the reality lying behind that level [22].

A comprehensive environmental analysis requires not only complete data on various factors and the averaged values of their influence on the state of individual organisms, populations, and ecosystems, but also a correct interpretation of these data. The 'dose–response' matrix may be inadequate for this (even taking into account multiple paths of arrival of stressors and their effects on an organism). In [22] the concept of a 'susceptibility range' was proposed, so as to construct a three-dimensional surface of the 'dose–response' interaction. Such a graph makes it possible to find the response to a specific dose of a particularly susceptible subgroup of people and various organisms. The concept of a 'susceptibility range' is closely associated with the concept of a 'response spectrum'. The author of [22] also presents a classification of the categories in the spectrum, as a function of their intensity (total pollutant load on an organism, physiological changes whose values cannot be clearly determined, sickness, and mortality). The explicit threshold of a harmful effect is associated with the appearance of pathological changes.

A still more complicated situation occurs when we consider the impact not on an individual population but rather on the biogeocenosis as a whole. Any impact is ambiguous, even for an individual organism, if we wish to estimate the total, integral effect on the organism, since some factor or other exhibiting a pronounced symptom of the impact can ultimately lead to a positive desirable effect for the organism as a whole.

A similar approach can also be used to assess the impact of one or several factors on natural ecosystems. If such an impact does not cause a worsening of the indices of a 'good biogeocenosis', then it can be considered to be permissible [46].

The threshold nature of the effect of an impact on a community (biogeocenosis) should be considered and evaluated, proceeding from the above considerations concerning the final integral effect on the entire system (here we are referring to the threshold of a harmful impact, causing the ecosystem responses to go beyond the limits of the normal physiological 'system' fluctuations or even causing the death of the system or population [3, 4], while attaining the critical number (density); we do not refer to the threshold for the onset of individual biological responses).

2.3. The Stability and Reserves of an Ecological System

An ecological system may be considered (see, for instance, [23]) as an aggregate of several sets, the first of which is an aggregate of objects of the system or an aggregate of individuals; the second set can be the cardinal structure of a set of objects, constituting a quantitative distribution of individuals among groups; the third set (organization of the system) is a set of bonds of the system (trophic, informational, and spatial links between objects); the fourth set includes the intensities of the bonds entering into the organization of the system (measured, for example, by the amount of energy or biomass in the case of trophic bonds and by the mass of some element in the case of chemical bonds).

Thus, an ecological system is a set of objects with their cardinal structure and with specified bonds of different types and intensity between the objects [23]. Note that such an approach is also possible with respect to the definition of any other system.

An ecological system is considered in [23] as a representation of the possible states of an environment in the above system (an aggregate of sets).

The function $\eta(R,t)$ varies in space and time, and when the time relation is considered, a distinction should be made between the usual physical time

t, the time of development T corresponding to age-related, seasonal, or successional changes, and the evolutionary time τ corresponding to genetic or large-scale geophysical (geological) changes in the environment (climatic changes). For instance, the biomass C or production of the system dC/dt may be considered as an index of the efficiency of the functioning of the ecosystem.

The ecological resistance of a system is usually defined [36] as the ability of the ecosystem to resist disturbing factors for a sufficiently long time without a dying out or degradation of individual components of the ecosystem (resistance according to Lyapunov). The resistance according to Lagrange, on the other hand, is the limitation at the top and at the bottom of all trajectories emanating from some definite domain of the model phase space (see our approach, described in Section 2.1). This concept is developed in [24].

The authors [24] dwell on two alternatives of the concept of resistance according to Lyapunov, as applied to the analysis of dynamic models of an ecosystem: Lagrange resistance and 'hierarchical' resistance. In the latter case, the instability of some subsystem is stabilized by a block located hierarchically above it.

The measure of resistance S of a biological system in [23] is defined as the ratio between the measure of a possible change in the required properties of the system ΔR and the measure of the corresponding impact ΔF, i.e., $S = \Delta F/\Delta R$; thus, the resistance of the system is determined by the quantity $1/\Delta R$.

A change here refers to deviations from some accepted norm (which can itself be a function of time) which exceed the given range; an impact refers to a change in the state of the environment, that is, in the inputs to and outputs of the ecosystem. Properties undergoing structural, organizational, evolutionary, and functional changes are placed in the category of 'specified properties'.

It is suggested in [23] to use the concept of relative resistances, whereby we compare the state of the system according to its remoteness from a 'red line' dividing the permissible and unacceptable values characterizing the states. This approach is similar to that described by us earlier [14].

In some determinations of the resistance of an ecosystem, only the measure of change of the specified properties during the impact on the ecosystem is considered (here the concept of 'resistance' is practically equivalent to the concept of 'stability'). Stability is the tendency of the system to remain approximately in conditions of equilibrium or to return to these conditions after a disturbance [53]. This concept implies constancy (absence of changes), persistence (associated with survival), elasticity, flexibility (associated with the rate of system return to the state preceding the disturbance), and amplitude (characterizing the disturbance from which restoration is possible).

The diversity of an ecosystem is often taken as a measure of its resistance. This is due to the good correlation between the diversity and the stability (resistance) of an ecosystem, which is also confirmed experimentally (see, e.g., [85]). The diversity is often determined by the number of species per unit area or volume. This term actually includes two concepts [53]: (1) the abundance of species (or other groups) and (2) the homogeneity of the distribution or relative number of individuals within each species (or group). Numerous attempts have been made to select an index of diversity, i.e., a single number characterizing the multicomponent quantity (both the number of elements in the system and their distribution among the different groups).

In [80] the index of diversity is taken simply to be the number of species l in the ecosystem and in [55] it is a quantity characterizing the inhomogeneity of the distribution of individuals among the species:

$$-\sum_{i=1}^{l} \frac{N_i}{N} \log \frac{N_i}{N} \tag{2.6}$$

where N is the number of individuals in the system and N_i is the number for the ith species. The above quantity will be greater, the higher the number of species in the system; it decreases with a decreased number of individuals in the groups (species).

Odum [79] used Simpson's index, equal to $1 - \sum(P_i)^2$, where P_i is the probability of occurrence of each species (based on a count of the number of individuals, and sometimes of the biomass as well).

In [63] the diversity is interpreted as the probability of interspecific collisions. The enumerated indices were based on the diversity of the cardinal structure of an ecosystem. In [74] the indices suggested are based on the diversity of the organization of the system, that is, of its bonds.

Returning to the problem of the resistance, one can state that, the greater the diversity (and its index), the higher will be the resistance. Thus, the more species l there are in an ecosystem, the higher will be its resistance, since there is a high probability that species adapting to the changing conditions will be present in the system. Diversity indices (see, e.g. [23]) present a more subtle characterization of the resistance; here the resistance grows with an increase in the number of species and it drops with an increase in the number of low-population species.

A great diversity of bonds increases the resistance of the ecosystem; however, a high resistance often correlates with a so-called 'weakness of bonds' (thus it is important to take into account the intensity of the bonds).

Of interest, too, is an approach using the methods and apparatus of the theory of reliability when defining the resistance of ecological systems [23]. For example, the higher the number of simultaneously functioning channels, the greater will be the reliability of the system (this is identical to the definitions of resistance introduced when the diversity was considered).

If the given approach is used, ecological systems can be considered as systems whose arrangement is dictated by the requirement of maximum reliability, while the optimum conditions for the existence of the systems are those providing the highest reliability for their existence in the future [39]. The basic properties of the systems can be defined as functions of the reliability. Also important is the inverse process: determination of the degree of reliability and resistance of a given ecosystem on the basis of its measured and studied properties.

On the basis of the theory of reliability, we can denote the time interval T within which the system functions normally and preserves its structure; for this, the ratio T/τ should be close to one (where τ is the time scale characterizing the time of development or evolution).

It is interesting to consider the problem of the critical level of the state of the system and its ecological reserve, in the light of the described aspects of resistance and reliability and the definitions introduced earlier.

The critical state of the system $\eta_c(t)$ (the 'red line') is not constant. Rather it is a function of the external natural conditions, anthropogenic impacts, and the state of the ecosystem itself as a whole during the time intervals $(t-1)$ preceding the considered (t):

$$\eta_{ci} = F[\eta_n(t-1), C, Q] \tag{2.7}$$

where $\eta_n(t-1)$ is the normal state of the ecosystem at time $t-1$; C is the vector of the anthropogenic impacts on the system; Q is the vector of external natural impacts on the system; and index $i = 1, 2$ refers to the sets of maximum and minimum values of η_c, respectively.

The dependence of η_{ci} on the external natural conditions Q is explained by the obvious fact that the maximum capabilities of the ecosystem can alter substantially when these conditions change (e.g., with variations in the temperature or water content), and the range $|\eta_{ci} - \eta_n|$ will decrease (or increase).

The quantity η_{ci} can also change under the anthropogenic impact; this effect is closely related to the dependence of η_c on $\eta_n(t-1)$.

A change of η_n during the time intervals preceding the interval under consideration can significantly affect the quantity $\eta_{ci}(t)$, since a weak, 'sick' system is more 'critical'.

In this case, while considering the variations of $\eta_{ci}(t)$ and $\eta_n(t)$ with time, we see that the curves in the plots of these functions will gradually draw closer to each other; for such a system, not only the difference $|\eta_{ci} - \eta_n|$ will decrease, but also the difference $(\eta_{ci} - \eta_{c2})$.

Taking into account the foregoing, let us determine the ecological reserve of a system as the minimum value (for either the maximum or the minimum value of i):

$$\frac{1}{T}\int_0^T [\|\eta_{ci}(t) - \eta_n(t)\|]\,dt. \tag{2.8}$$

As already noted, it is necessary to take as the quantity T a lifetime of the system comparable with the time scale characteristic for the time of development or evolution of the system.

If T/τ is much less than unity (i.e. $T \to T_0$), then T_0 should be taken as the integration limit; in this case the ecological reserve of the system is equal to the resource of this ecosystem (by analogy with any system having a limited resource and lifetime T_0):

$$\frac{1}{T_0}\int_0^{T_0} [\|\eta_{ci}(t) - \eta_n(t)\|]\,dt. \tag{2.9}$$

Here also the minimum (absolute) value of the integral for $i = \overline{1,2}$ is taken, and the quantity T_0 is the lifetime of the system.

It is quite obvious that in the latter case an additional natural or anthropogenic impact leading to a ratio $T/\tau < 1$ belongs to the category of impermissible impacts.

The above-mentioned approach is applicable both to local communities and to ecosystems on large scales.

For instance, it is obvious that for the Lake Baikal ecosystem (notwithstanding the presence of some anthropogenic loading) $T/\tau \sim 1$, whereas for the Aral Sea ecosystem $T/\tau < 1$.

The potential resistance of a system can also be evaluated from the degree of homeostasis, which determines the ability of a biological system to resist changes and to preserve a state of equilibrium. A system is resistant as long as its state is within the limits of the homeostatic plateau (destruction of the system occurs outside of homeostasis [26]).

The degree of homeostasis is defined as [23]

$$G = \frac{\sigma_x}{\sigma_y}$$

where

$$\sigma_x = \frac{1}{n}\sum_{i=1}^{n}\frac{\Delta x_i}{x_i}$$

(a similar expression is also written for σ_y). Here x is the impact on the system, and $y = y(x)$ is a function of this impact. In [38] the ratio $\delta = \frac{\sigma_s}{\sigma_f}$ served as a measure of homeostasis, characterizing the ecosystem resistance. Here σ_s and σ_f are, respectively, the structural and functional indices of the ecosystem.

The resistance of a dynamic system, and adaptive changes of populations as applied to biological monitoring, are considered in [30, 35].

2.4. Ecological Approaches to Establishing Standards

When establishing ecological standards for anthropogenic loads, it is necessary to take into account the capabilities of the biospheric elements and ecological systems; this work should include a determination of the required environmental quality, the permissible loads on the biosphere and its elements, as well as the establishment of standards for limiting the anthropogenic impacts (limitation of the sources of the impact, taking a rational account of local geophysical features, etc.).

The approaches used to determine the permissible impacts, and, consequently, the permissible loads on a community or ecosystem, differ from the public-health approach, adopted to protect people and envisaging the obligatory protection and safety of each person.

When establishing standards for the maximum permissible concentrations for man (the public-health approach), we use a very strict criterion, namely the absence at present and in the future of any deviations of the state of man from the normal state, or any pronounced responses of man to the impact.

While studying the possible and existing loads on different natural complex systems and while developing approaches to determining the permissible loads on such systems, the problem arises of establishing ecological standards for different impacts and loads. The goal of the establishment of ecological standards, which represents a stage in the strategy of environmental quality control, is to protect ecological systems and biological communities as a whole.

If the problem is stated in this way, a loss of individuals from the populations does not present a danger to an ecosystem or a community (provided it does not reduce the potential productivity, the stability, or the diversity of the system).

Here we have to use as a basis the concepts of the resistance and the reserve of an ecosystem discussed earlier (Section 2.3). Suggestions for assessing the normal and pathological states of an ecosystem, necessary for establishing ecological standards, are also given in [42].

In order to establish ecological standards, it is necessary first of all to understand the effect of pollutants on individual species which are of primary importance for man or for an ecosystem, as well as to consider the resistance and reserves of the system as a whole. Since, even if some of the most sensitive species suffer, this can lead to a change in the structure of the biological community, to a reduction of stability. In the case of unique ecosystems, it is important to define special high requirements both for the state of the biota and for the features of the abiotic component of the biosphere in specific places and conditions, for example, in the areas and periods of fish spawning, etc.

Let us stress once again one feature of the determination of the permissible load on an ecosystem, namely that some individual species that is very sensitive to a given impact may prove to be a 'critical link' of the entire ecological system, so that the permissible load on the ecosystem as a whole will be determined by the load on precisely this species.

It is probably more appropriate to refer here to the 'nonresistance' of the ecosystem, since, even though its overall resistance is adequate, the disturbance of some (critical) link or the extinction of some (critical) species can lead to the death of the entire system.

In the Soviet Union, a system of public-health standards for the maximum permissible concentrations (MPCs) of pollutants in different parts of the environment has been developed. However, the establishment of limits for permissible loads under the impact of pollutants on the populations of animals and plants forming ecosystems has been started only comparatively recently. For instance, MPCs were established for fishery water bodies, that is, for fish populations of commercial value. Quite recently the establishment of standards for the impact of pollutants on forest ecosystems was started. At present, however, the need has become obvious to take appropriate steps to assess the permissible loads on natural ecosystems, i.e., to establish ecological standards. The establishment of ecological standards implies a determination of ecologically justified standards for the permissible loads of pollutants on a community, an ecosystem, or a region, as well as for the permissible rate of input of such substances to the environment [15].

The need to establish ecological standards is sometimes caused by the very conspicuous differences between the responses of man and some types of fauna and flora to the same impact. Public-health standards of MPCs, protecting people from the impact of some harmful substance, are often not strict enough, and they do not protect some species of plants and animals. Consequently, the ecosystem as a whole is not protected from the impact of these substances either.

In this case the approach to establishing standards of the permissible impacts on ecosystems and to determining the maximum permissible loads on natural objects should differ from the public-health approach, which does not allow any harmful impact or damage even with respect to an individual organism. As already noted, the loss of an individual is not dangerous from the point of view of ecosystem protection, provided that the stability, diversity, and productivity of the system are preserved.

In the light of these considerations, the following question arises: What levels (intensities) of the individual responses of the biota should be used as a basis for the ecological standards for the permissible impacts?

If we know the value of the critical impact q_c, as a quantity leading to the destruction of the given population (in the case of an adult population, q_c could be the concentration of $LC_{50} \approx LC_c$ which is lethal for 50 percent of the individuals of the population; this value is probably much lower for spawning areas), we can determine the safety factor $K_{mi} = q_p/q_c$. This factor can to a great extent serve as a measure of the resistance of the given population to external factors. Evidently, the public-health approach will also be unacceptable here, because with this approach the standard for the permissible concentration of a pollutant in the environment differs substantially (by many orders of magnitude) from the value of LC_{50} (for instance, for ionizing radiation and for some sources the ratio is 1 : 2000 or more). In [9, 77] a safety factor $K_{mi} = 0.05$ to 0.1 is introduced during the assessment of ecological standards for some populations, i.e., the permissible concentrations of pollutants affecting these populations which were suggested by the authors should be only 1/10 to 1/20 of LC_{50}. It is obvious that this approach should on the whole be considered reasonable, starting with some ecological level (the critical and permissible loads at lower levels of the trophic chains could probably be identified, provided that the population occupying this level is not unique).

Therefore, let us formulate the following requirements as a basis for the permissible ecological standard [47]:

(a) For every population m, starting with the level m_0,

$$\frac{\int_B C_i(R) n_{m \geq m_0}(R) \mathrm{d}R}{K_{mi}(LC_c)_m} \leq 1; \tag{2.10}$$

(b) For a small part of the organisms of each population it is possible that

$$C_i(R)/(LC_c)_m \geq 1, \tag{2.11}$$

where C_i is the concentration of the ith ingredient; $n_m(R)$ is the normalized distribution of the organisms of the m-population in space; $(LC_c)_m$ is the critical load (concentration) for the m-population. Here we should use as a guideline the critical number (density) of the population below which the population is doomed to extinction (see, for instance, [3, 4]).

Some interesting approaches to assessing the effect of pollution on various fish and bird populations, as well as some considerations about determining the permissible impacts on ecosystems, are presented in [9], which includes a determination of the maximum permissible pollution levels in water bodies for different fish species. These were found by determining the concentration lethal for 50% of the fish (LC_{50}) exposed for four days, and by using a safety factor of 0.05 to 0.1. In [77] the safety factor was determined by dividing the value of the concentration producing no effect by the 96-hour LC_{50}. The authors of [77] used data obtained in a study of the chronic toxicity during the entire life cycle to determine the maximum permissible toxicity. It turned

Table 2.1. High toxicities and computed safety factors for some pesticides and heavy metals.

Toxic chemicals	Fish species	96-hour LC_{50}, mg/l	Safety factor K
	Pesticides		
Diazinon	Black-head pymephales	6	0.0005
Captan	" "	0.065	0.10
Butoxyethyl ether 2.4-D	" "	5.6	0.05
Carbaryl	" "	9	0.023
Methoxychlor	" "	0.0075	0.107
Malathion	" "	10.5	0.02
	Sunfish	0.08	0.04
	Loach	0.20	0.02
Lindane	Black-head pymephales	50	0.5
	Loach	26	0.38
	Heavy metals		
Chromium^{+6}	Black-head pymephales	33	0.03
	Loach	50	0.01
	Rainbow trout	69	0.003
Copper	Black-head pymephales	0.47	0.03
	Sunfish	1.1	0.02
	Loach	0.1	0.009
Cadmium	Black-head pymephales	31	0.001
	Sunfish	20	0.0015
	Oceansunfish	20	0.0025
Methylmercury	Black-head pymephales	0.04	0.006
	Loach	0.096	0.003

out that the safety factor for a given compound varied only slightly from species to species, and, consequently, it could also be extended with some caution to fish species that were not included in the special experiments.

Some values of LC_{50} and the safety factor K are presented in Table 2.1, which was borrowed from [9].

Violent responses to high concentrations, as well as consequences associated with a chronic impact of low concentrations, are used during the organization of biological monitoring of pollutions and other impacts. Short-term indicators pointing to the possible appearance of long-term and serious effects, which may show up only much later (sluggish effects), may prove to be of primary importance for establishing ecological standards, as well as for biological monitoring. For instance, when some fish species (loach, in particular) were exposed to copper and zinc, an increased frequency of sounds uttered by the fish, a lower feeding activity, and an elevated motor activity were noted after only a few hours (copper concentration of 6 to 15 mg/l); blood changes were registered six days later; only after exposure to such copper concentrations for eight months did an effect on the growth, reproduction, and survival of fry show up. The effects of different copper

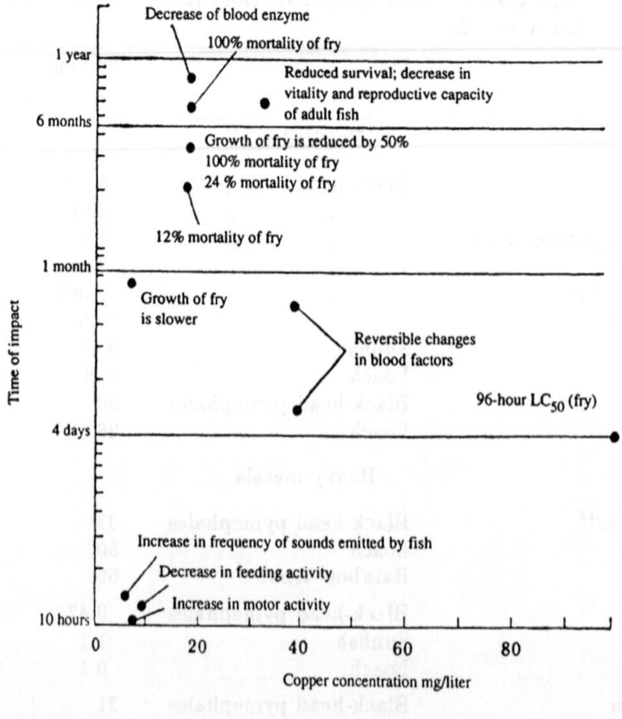

Figure 2.2. Some effects of different copper concentrations on loach and times of their appearance.

concentrations on the population of loach, and the times of appearance of these effects, are shown in Fig. 2.2 [9, 75]. Experiments have shown [65] that the effect of harmful substances on reproduction and survival at early life stages can show up long before regular observations reveal changes in the density of the population under conditions of quite considerable natural fluctuations.

An increase in the density of a population under natural conditions is ususally accompanied by an increase in death rate, while a decrease in density corresponds to a lower death rate. However, no decrease in the mortality of bird populations is observed when their density decreases under the impact of pollution [9].

It seems that the ecological standards cannot be universal for any ecosystem type, or for any physical-geographical conditions.

In view of the foregoing, it is of great importance for the establishment of ecological standards to carry out a series of studies of the nature and laws governing the distributions, accumulation, destruction, bioaccumulation, and trophic transformations of pollutants in ecosystems, as well as their passage from one medium to another on local, regional, and global scales. It is shown in [18], for instance, that the accumulation capacity of a marine medium with respect to different pollutants increases appreciably, in connection with biosedimentation and microbiological oxidation; this reduces the anthropogenic load on the ecosystems and should be taken into account when establishing ecological standards (see Section 11.4).

Let us now consider an extremely important circumstance. Existing public-health standards define the maximum permissible (for man) concentrations of harmful substances in a specific environment, namely atmospheric air and drinking water. These standards do not take into account at all (or only partly) the effects of the chemical and biological accumulation of harmful substances having unacceptably high concentrations, resulting from their passage into some other medium of the environment, e.g., from air to water, from water to the biota, etc.; their accumulation in the food chains and their transformation into more toxic forms during migration are not taken into consideration either. Thus, if the concentration of a harmful substance in the environment is below the permissible maximum, this does not mean that for the same amount of this substance in some other medium its concentration will not exceed the MPC. The public health standards, in their present form, are appropriate mainly for cities, settlements, and sites of water use, i.e., for places where secondary natural processes are unimportant. These 'secondary' effects, which are insignificant when establishing public health standards, turn out to play an important, and even decisive, role when assessing the impact of harmful substances on ecosystems. Often it is not the initial concentration of harmful substances in an environment which is the determining factor, but rather their accumulation and transformation in the critical links of the ecosystem, whereby other concentrations and secondary products having different toxic properties are observed.

Special emphasis in the establishment of ecological standards should be placed on stable pollutants, and to the possible passage of pollutants from medium to medium, as well as the diverse ways they can impact populations and ecosystems.

Some stable pollutants are especially dangerous, because they tend to accumulate in the food chains; this can lead to hazardous effects at high trophic levels. Some such substances are chlororganic pesticides (DDT, deldrin, etc.), polychlorinated biphenyls (PCBs), which are also organochlorines, some heavy metals (mercury), and radioactive products. For instance, PCBs enter into the composition of some lubricants, hydraulic fluids, and synthetic resins, and they enter the environment when they are discarded and not completely burned.

The coefficient of PCB accumulation by aquatic invertebrates and fish can be as high as 10^3 to 10^5, while for birds and mammals associated with an aquatic environment it reaches 10^7 to 10^8 (compared to the PCB content of the aqueous medium [78].

In industrial regions the PCB concentrations in fish bodies can reach 10 to 800 ppm, while in unpolluted areas they do not exceed 1 ppm. These values are higher than the levels at which adverse effects on organisms are already observed. The toxicity of the PCBs increases with long-term exposure. The toxic concentration for a 15-day impact on different organisms ranges from 30 to 50 mg/l [83].

The PCB content (12–17 ppm) of salmon eggs in Lake Michigan, USA, causes a low reproductive capacity [86].

In order to work out the permissible ecological loads and to set up monitoring systems we must use models taking into account the rates of transport, degradation, bioaccumulation, and trophic transformations; the models should also include the pollutant sources, the routes of pollutant transfer, and the effects on regional and global scales.

Only research into the fate of a pollutant, from its source of discharge (via the processes of its physical, chemical, and biological transformations and its interactions with other diverse factors of the environment) up to its entry into a living organism affected by it, could ensure the establishment of scientifically justified ecological standards of permissible effects on the living component of the biosphere, on entire ecosystems.

Studies [47] and an analysis of numerour literature data have shown that, during the establishment of standards of the impact on natural ecosystems, the determining feature of the pollutant behavior in the environment consists in a possible considerable accumulation in some zones of the environment, together with a transformation to new more toxic forms, due to geophysical, geochemical, and biological processes. This may cause unexpected, and serious at first glance, effects to show up with great sluggishness (compared to the moment of discharge of the pollutant into the environment). This aspect, which will be considered in Section 2.5, can be illustrated using the example of such pollutants as sulfurdioxide, mercury compounds, and pesticides.

It is known that hazardous substances are, as a rule, present in the environment in the form of complex mixtures and compounds of a great many different ingredients. A comprehensive analysis of the behavior of hazardous substances in the environment should, therefore, be carried out, together with a study of their combined effect on all the elements of nature (plants, land animals, and water animals, as well as the abiotic component of the biosphere), in order to determine the maximum permissible loads on the given natural objects (systems) as a whole [19].

We should stress here the complex diverse effects observed in the presence of many pollutants with different paths of penetration and impact, effects strengthening the impact of some factors in the presence of others and effects of synergism against a background of changing external conditions.

A comprehensive analysis should take into account the entire complex of interactions in the environment, including the above effects. It is obvious that, in order to take into account these effects together with some other parameters, allowance should be made for the temperature regime and its changes due both to natural factors and to anthropogenic factors (an example is the hot-water discharge from a power station). At an elevated water temperature, fish require more food, their reproductive ability is lost, their activity diminishes; under such conditions, the toxicity of some pesticides for fish increases; in addition, the rate of methylated mercury accumulation increases [9, 27]. Heating of the water accelerates the eutrophication of a water body.

The interaction of wastewater containing mercury with large populations of aquatic bacteria, resulting in the formation of highly toxic methylated forms of mercury, is another example of a complex interaction of different factors which ultimately intensifies the impact on living organisms.

The strengthening or weakening of several factors of an impact may occur either at the stage of physical or chemical interaction of those factors (neutralization of acids, reactions leading to the appearance of smog) or at the stage of response of the biological systems to the impact.

If the (vector) function of the response of a biological system $\mathbf{Y} = (Y_1, Y_2, \ldots, Y_n)$ is written [41] as

$$\mathbf{Y} = f(x_1, x_2, \ldots, x_n) \tag{2.12}$$

where $\mathbf{X} = (x_1, x_2, \ldots, x_n)$ is the vector of the external effects, then by expanding the function $f(x_1, x_2, \ldots, x_n)$ as a power series we get the following approximation:

$$f(x_1, x_2, \ldots, x_n) = a_0 + \sum_i a_i x_i + \sum_{ij} a_{ij} x_i x_j \tag{2.13}$$

where a_i are empirically found quantities whose values determine the response of the system to a change in the external impacts.

If the result of the impact of each factor on the biological system, in the form of a response of the system, did not depend on the effect of any other factor, and if the total effect were additive (i.e., if it was the sum of all the individual impacts), then the function could be approximated by an expression of the type

$$Y = a_0 + \sum_i a_i x_i \tag{2.14}$$

provided, of course, that the effect is linear, if it depends only on the magnitude of the impacting factor, and does not increase per unit impact (this is an important assumption concerning linearity of the impact effect).

A more complicated situation is represented by the possible appearance of synergism, whereby the effect of one kind of impact increases in the presence of some other kind, i.e., the effect of the combined impact is greater (examples of such effects were presented earlier), and by the phenomenon of antagonism, whereby the adverse effect of one of the factors is reduced; for two factors x_1 and x_2, assuming linearity of the effect,

$$Y = a_0 + a_1 x_1 + a_2 x_2, \tag{2.15}$$

in the case of synergism

$$Y = a_0 + a_1 x_1 + a_2 x_2 + a_{1,2} x_1 x_2, \tag{2.16}$$

and in the case of antagonism

$$Y = a_0 + a_1 x_1 + a_2 x_2 - a_{1,2} x_1 x_2, \tag{2.17}$$

For instance, in [41] the example is given of the weakening of the impact of zinc in an aquatic ecosystem (on the primary production of phytoplankton) in the presence of chromium compounds (phenomenon of antagonism); the regression equation describing the results of this experiment is similar to Equation (2.17).

It should be noted in conclusion that, when determining the permissible ecological load on a biological system or biogeocenosis, we must take into account already existing effects of anthropogenic impacts on the environment.

Many impacts, including the discharge of various substances into the environment, have both negative and positive aspects. For example, an increase in the CO_2 content of the atmosphere, on the one hand, sets up conditions for an increase in biological productivity and leads to the formation of highly productive communities, while, on the other hand, it may lead to undesirable climatic changes resulting from the so-called 'greenhouse' effect in the troposphere. In general, human activity can contribute to an overall rejuvenation of the biosphere, can accelerate the process of evolution, and can create objective prerequisites for the development of flourishing biogeocenoses [46]. As already mentioned, the deterioration ('degradation') of the state of the environment caused by the anthropogenic impact is not a consequence of man's general industrial development and technological progress, but rather is a result of ecological errors during the planning and implementation of major projects and of an inadequate level of technological development.

According to Shvarts [46], the simplification and rejuvenation of biogeocenoses, the changes in the structure of individual links of the food chains, and the increased role of animals as destroyers of primary organic matter, do not constitute a simple degradation of the biosphere, but rather its evolution under new conditions, and by no means should all consequences of this kind be considered undesirable.

2.5. Establishment of Standards, Taking into Account Diverse Pollutant Impacts

Public health standards associated with environmental pollution are, as a rule, established assuming that only one component of the environment is being polluted. This is how standards for the maximum permissible concentrations of pollutants (MPCs) were set up for the atmospheric air and the surface waters.

Under present conditions, however, often not only localized pollution sources are encountered, but also extended sources, whereby the components of the environment are contaminated nonuniformly on the scale of an area or region. In such cases the impact of the pollution on populations and ecosystems is assessed assuming a spatially distributed source, subject to complicated transformations from the viewpoint of a continuous redistribution of pollutants due to their transformation, migration, etc.

Under these conditions the establishment of standards for permissible pollution levels, even for individual organisms and populations, acquires an ecological character. This is because the establishment of such standards requires a good understanding and a careful consideration of the paths of distribution, transportation, and possible accumulation of the ingredient, as well as of the character of its interaction with the biotic and abiotic components of the biosphere, and of the correlations between the concentrations of pollutants in different media and the possible rates of arrival of this ingredient from different media to the organisms of animals and humans (allowing for all possible paths of arrival) for established equilibrium or in its absence.

When establishing ecological standards, it is also necessary to have an idea of the structure of the ecosystem, in order to understand what actually determines the permissible load: the load on the most sensitive population or the load on a complex of populations.

In this section we will present some approaches to the establishment of ecological standards for an organism or a population, taking into account the wide distribution of some pollutants and the diverse paths of their arrival at organisms (this approach can be used to set up ecological standards both for a community and for an entire biogeocenosis).

For instance, the concentrations of sulfur dioxide and mercury compounds emitted from high sources (high chimneys) rapidly reach the existing permissible levels for the surface air, due to diffusion. However, later, as a result of chemical conversions and accumulation (after deposition on vegetation and soil or washing out into rivers and lakes), their relative toxicity and potential hazard for many organisms and for man are seen to increase.

A similar situation is observed when pesticides migrate. In the following, we will pose a problem associated with ecological standards for pollutants of this kind, and we will analyze the above examples in detail.

This problem, which was developed at the Institute of Applied Geophysics of the USSR State Committee for Hydrometeorology, is treated in detail in [16, 47].

Suppose that an ecosystem consists of N components (populations or elements of the environment). Let us use the following notation: $\mathbf{X} = (x_i, x_2, \ldots, x_N)$ is the vector of the state of the entire ecosystem; c_{ij} is the concentration of the jth ingredient in the ith component ($i = 1, \ldots, N$; $j = 1, \ldots, M$) (the concentration is here deliberately distinguished as an independent parameter, although it refers in essence to the abiotic component of the system and may be included in the vector of state of this system); $\mathbf{C}$ is the vector of pollutant concentrations in the ecosystem. The dynamics of the state of the ecosystem is described by the following equations:

$$\dot{x}_i = F_i(\mathbf{X}, \mathbf{C}, \mathbf{Q}), \tag{2.18}$$

$$\dot{c}_{ij} = \Phi_{ij}(\mathbf{X}, \mathbf{C}) + E_{ij}\beta_i(x_i) \tag{2.19}$$

where F_i is a function describing the change of state of the ith component; Φ_{ij} is a function describing the variation in the concentration of the jth pollutant in the ith component (resulting from any conversions); E_{ij} is the anthropogenic arrival of the jth pollutant at the ith component of the environment; $\beta_i(x_i)$ is a function taking into account the change in concentrations in the ith component per unit mass of arriving pollutant; $\mathbf{Q} = (Q_1, \ldots, Q_N)$ is the vector of external impacts (aside from pollution) affecting significantly the state of the system (change in water content, temperature, areas of vegetation cover, etc.).

Let us designate the damage caused to the ith component by pollution as $D_i(x_i, c_i)$ (it depends on the state x_i).

Our problem now consists in establishing standards for the values c_i and E_i, $\mathbf{C}$ and $\mathbf{E}$, of the concentrations and intensity of input of the given ingredient to the ith component and to the entire system, respectively, taking into account the actual values of x_i, $\mathbf{Q}$, and β_i (and, of course, taking into account the maximum yield).

For this, we have to observe the following conditions for each component of the ecosystem:

$$D_i(x_i, c_i) \leq D_i^0 \tag{2.20}$$

where D_i^0 is the given maximum permissible or critical (depending on the statement of the problem) impact (e.g., the arrival of pollutants at the given system); the quantity D_i^0 will correspond for the given state to the maximum permissible value c_i^0, while the permissible concentration c_i in the ith medium of the ecosystem satisfies the condition

$$O \leq c_i \leq c_i^0(x_i). \tag{2.21}$$

It is known that the amount of waste (discharge into the environment) for a fixed technological process is practically proportional to the function of production 'effectiveness' (i.e., the production output) V. Consequently, it can be formally considered that the effectiveness is a function of the discharge into the environment and that it also depends on $\mathbf{Q}$.

In this case the requirement for maximization of the useful effect is written as

$$V(\mathbf{Q},\ \mathbf{X},\ \mathbf{E})\,\mathrm{d}t \to \max, \tag{2.22}$$

where $\mathbf{E}$ is the vector determining the input of pollutants to the ecosystem.

In a definite interval of values of E_{ij} $(0 \leq E_{ij} \leq E_{ij}^0)$ function V will increase monotonically with respect to each argument E_{ij}, up to certain values of E_{ij}, when the need for decontamination (or environmental damage) reduces the magnitude of V.

It should also be kept in mind that the effectiveness function may be significantly affected by other external factors (including other pollution); this is why, in Equation (2.22), V also depends on $\mathbf{Q}$ and $\mathbf{X}$.

Equation (2.22) provides an answer to one more general question, the question of why it is necessary to determine the maximum standard for permissible loading of the environment and, consequently, for the impact. It is because of the requirement of maximum effectiveness under conditions of inadequate technology, which leads to the introduction and use of the maximum permissible loads and impacts.

Let us consider, for the sake of simplicity, the input of a single pollutant to a single environmental medium, and let us suppose that the change in concentration is linearly related to the concentration of the substance in the environment (this case is of importance for consideration of the turnover of pollutants in the abiotic components of the ecosystem):

$$\dot{\mathbf{C}} = A\mathbf{C} + \mathbf{E} \tag{2.23}$$

where $A = \|a_{ij}\|$ is the matrix of the coefficients of transfer between environmental media.

The components of vector $\mathbf{E}$ are determined as $E_i = k_i \beta_i E$, where E is the total discharge of the pollutant into the environment; k_i is the portion entering the ith medium; and β_i is the normalizing multiplier.

Consider the case of a steady state of the ecosystem.

Let us solve the problem on condition that $E \to \max$ (see (2.22)) and also that

$$-A^{-1}kE \leq c_i^0, \tag{2.24}$$

where c_i^0 is the maximum permissible concentration in the ith medium; vector $\mathbf{k}$ is determined from the condition $\mathbf{E} = \mathbf{k}E$. The solution of the problem is written (for the optimum value E^{opt}) as

$$E^{opt} = \min \left[\frac{c_i^0}{(-A^{-1}\mathbf{k})_i} \right] \tag{2.25}$$

for medium i, where the maximum permissible concentration is attained most rapidly and which is, therefore, the limiting environment l for the given impact.

Similar arguments might be adduced for each component of any pollutant (or for a combination of pollutants, taking into account their joint impact).

Thus, the limiting concentration c_l^0 for environment l

$$c_l^0 = -A^{-1}\mathbf{k}E^{opt}; \tag{2.26}$$

the environment for which

$$(c_l)_i = (c_l^0)_i$$

is the limiting environment.

The use of this approach to establish ecological standards is considered further, using as examples environmental pollution by sulfur dioxide, pesticides, and mercury compounds. These ingredients were selected not only as an illustration of the applicability of the described approach, but also because these substances are hazardous due to their wide distribution and the many ways they can impact man and the biota. The substances mentioned above also cover a broad spectrum of 'residence times' in the natural environment, characteristic for pollutants of anthropogenic origin.

Sulfur dioxide has the shortest residence time in the atmosphere (of the order of several hours), due to chemical conversions. Therefore the possibility of pollution by sulfur dioxide and its detrimental effect are, as a rule, local, or sometimes regional, in character.

The transfer of sulfur dioxide from the atmosphere to the earth's surface occurs due to washing out by precipitation and absorption by the underlying surface. Moreover, the amount of sulfur dioxide in the atmosphere decreases rapidly, due to chemical conversions; sulfur dioxide easily oxidizes to sulfur trioxide (SO_3). In the liquid phase this leads to the formation of sulfuric acid (H_2SO_4). The SO_2 oxidation reaction is extremely fast, as is confirmed by the fact that only traces of sulfurous acid (H_2SO_3) are discovered in precipitation which has just fallen. Sulfur compounds are mainly represented in precipitation either by sulfuric acids or by its salts (sulfates $MeSO_4$), which are formed during chemical reactions of various pollutants with sulfuric acid. Sulfates are stable compounds. Aerosol particles consisting of sulfates remain after the evaporation of drops in the atmosphere. These particles are also eliminated from the atmosphere by washing out and by wet and dry deposition.

All the above-mentioned compounds are toxic. Sulfur dioxide has a detrimental effect on vegetation, and it causes diseases of the respiratory organs of humans and animals. The effects on man's respiratory organs manifest themselves at a chronic impact of concentrations of 100 $\mu g/m^3$ [49]. Mosses and lichens are very sensitive to the effect of SO_2; pine needles are affected at a concentration as low as 20 $\mu g/m^3$, and the foliage of broad-leaved trees is damaged at a concentration of 100 $\mu g/m^3$ [49, 70].

The conversion products of sulfur dioxide produce a stronger effect. Besides the above direct impact on animals and plants, other serious negative consequences of the impact of sulfuric acid should be mentioned, such as acidification of the soil and of the water in fresh-water bodies, an accelerated corrosion of metal structures, and destruction of limestone constructions. The soil and water acidification also has an adverse effect on the flora and fauna. For instance, many rivers and lakes in Norway and Sweden have lost their value as fishery zones, due to the long-range transport of sulfur compounds. The soil fertility is sharply reduced by acidification, an effect which is particulary marked in regions with acid soils (e.g., in the northwestern and central parts of the European USSR). The effects of sulfates begin to manifest themselves at a concentration of 6 $\mu g/m^3$ [82], and at 10 or 12 $\mu g/m^3$, the incidence of human respiratory diseases increases to 30% [61].

Thus, the impact of sulfur compounds is very many-faceted and quite toxic. The situation is aggravated by the fact that atmospheric emissions of sulfur dioxide are massive in character and ever-increasing throughout the world (they amounted to 70 million tons in 1950 and 120 million tons in 1975), and measures to control these emissions are practically just beginning. According to some data, sulfur dioxide emissions will more than double by the year 2000, and they will reach an astronomical level: about 300 million tons [12, 68].

This means that the problem of sulfur dioxide emission into the atmosphere is very complicated, and the standards for these emissions, and for the content of sulfur dioxide in the atmosphere, should be established taking into consideration all the above effects (i.e., it is not enough to establish standards for SO_2 only on the basis of the public-health standards for the maximum permissible concentrations of SO_2 in the atmosphere).

To develop an ecological approach to the establishment of standards for the SO_2 content and emissions, it is necessary to carefully examine the modes of chemical conversion of sulfur compounds and the paths of their passage from medium to medium. In addition, we have to determine the ratios of the concentrations of sulfur compounds in the different media under conditions of equilibrium and take into account their relative toxicity.

Ultimately, the realization of the approaches described during the statement of the problem will be possible, enabling not only a determination of the ecological significance of the individual components of some compound (allowing for conversions), but also a determination of the limiting environment. At present, however, the permissible concentrations of H_2SO_4 and sulfates in the soil, as well as in river and lake water, are not yet known, and this should probably be the next stage in the development of the problem of the permissible pollution of the soil, air, and water by sulfur compounds.

An approach involving a determination of the limiting environment will be considered below, using as an example the pollution of agricultural fields by pesticides. Pesticides are hazardous by virtue of their direct toxic and genetic effects, as well as because of secondary effects associated with upsetting the biological equilibrium in ecosystems. This effect is most prominent for stable pesticides (mainly organochlorines), which accumulate in the organs of animals at the end of the trophic chains, even if the contents of these substances in the environment is very low; this has already led to the death and degradation of some populations [56]. A dramatization of the situation that occurs in nature in this case, was presented articulately as early as 1963 in Rachel Carson's book *Silent Spring*.

The problem consists in determining the final distribution of pesticides among the different elements of the environment and finding the critical environment from which their arrival at the organisms of people, animals, and plants is accompanied by the greatest detrimental effect. This distribution depends greatly on physical-geographical and various local conditions.

Some estimates are given in [47] of the distribution of pesticides in one of the valleys of Soviet Central Asia: a region where pesticides are widely used and physical-geographical conditions promote an intensive spreading and redistribution of pesticides among different parts of the environment. The system of differential equations describing the dynamics of pesticides in the different parts of the environment (atmosphere, soil, and water) can be written in the following generalized form:

$$\dot{C}_i = \frac{(1-k)P}{m_i} + \sum_j \frac{m_j c_j}{m_i \tau_{ji}} - \frac{c_i}{\tau_i}, \tag{2.27}$$

where j refers to all media except the ith medium; c_i, c_j are the concentrations of pesticides averaged over the time of their use in the ith and jth media, respectively; P is the amount of pesticides introduced into the environment during the time of averaging (for instance, during a season); m_i is the mass of the layer of the medium i in which the concentration of pesticides is averaged; m_j is the mass of the layer of the medium j, equal in area to the mass of the layer of medium j; k is the portion of the pesticides directly entering all the media during spraying of the fields, except for medium i (if i is the soil, then k is the portion of the pesticides entering the atmosphere, $(1-k)$ is the portion of the pesticides deposited on the fields

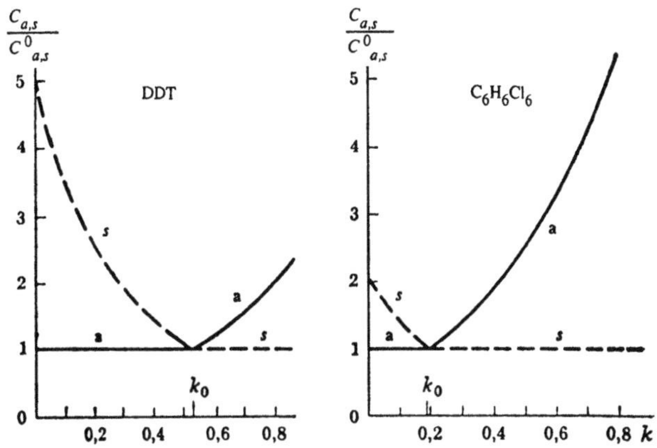

Figure 2.3. Calculated values of limiting concentrations of pesticides (DDT and $C_6H_6Cl_6$) in soil (s) and atmosphere (a) for case when pesticide concentration in atmosphere is C_a^0 (MPC) at $k \leq k_0$ and in soil C_s^0 (MPC) at $k \geq k_0$. (Here k is portion of pesticides entering atmosphere.)

during spraying; if i is the part of the soil not reached by pesticides during treatment, then $k = 1$ and $(1 - k) = 0$); τ_i are the total lifetimes of the pesticides in the different media (e.g., the lifetime of DDT in the atmosphere is 6 to 30 days, in water it is more than a month, and in soil it is more than a year); τ_{ji} is the time needed for the pesticides to pass from medium i to medium j (if essentially no passage occurs, as, for instance, from water to the atmosphere, then $\tau_{wa} = \infty$). Thus, the time needed for the passage of DDT from the atmosphere to the soil amounts to several days, from the soil to the atmosphere it is several years, and from the soil to the water it can be up to 30 years.

The limiting environment is determined on condition of maximization of the effectiveness, which in the present case is determined by the amount of pesticides applied (see beginning of this section):

$$(1 - k)P \rightarrow \max \tag{2.28}$$

with the limitation

$$0 \leq c_i \leq c_i^0 \tag{2.29}$$

where c_i is a steady-state solution of system (2.27); c_i^0 is the maximum permissible concentration in the ith medium.

Hexachlorocyclohexane and DDT were selected as the pesticides to be modeled (the use of DDT is prohibited in the USSR, but its behavior is well known). The initial values of the model parameters are presented in [47]. The results of the calculations showed for which situations the soil or the atmosphere is the limiting medium. Figures 2.3 and 2.4 show the

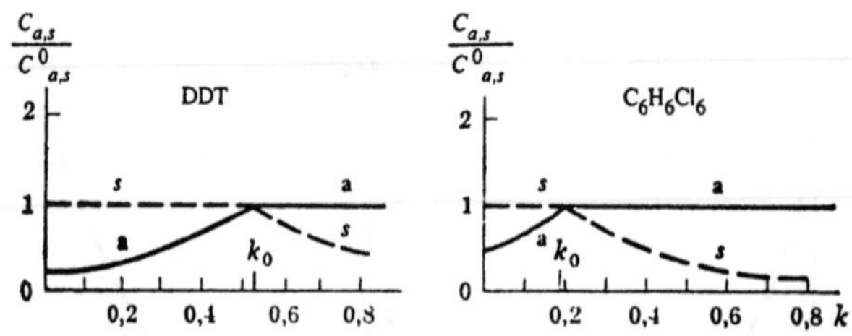

Figure 2.4. Calculated values of limiting concentrations of pesticides (DDT and $C_6H_6Cl_6$) in soil (s) and atmosphere (a) for case when pesticide concentration in atmosphere is C^0_a (MPC) at $k \geq k_0$ and in soil C^0_s (MPC) at $k \leq k_0$.

calculated concentrations of the above pesticides in the atmosphere and soil (in units of MPCs) as a function k (the portion of the pesticides entering the atmosphere during spraying of the fields). A change of the limiting medium occurs at a certain value k_0; soil is the limiting medium when $k < k_0$, and the atmosphere becomes the limiting medium when $k > k_0$ (if $k = k_0$, then the atmosphere and soil are simultaneously the limiting medium). k_0 equals 0.53 and 0.19 for DDT and $C_6H_6Cl_6$, respectively. It follows from the above that, if the MPC is observed in one of the media, then the concentration in the other (limiting) medium can exceed significantly the established standard of the maximum permissible concentration. In the above examples, the pesticide concentration in the water medium remains below the established MPCs; however, significant ecological damage can be caused, even for these concentrations of pesticides in biological systems.

Environmental pollution by heavy metals, and by mercury compounds in particular, is perhaps the most striking example of the need for development of an ecological approach to the establishment of standards and the use of a general approach (described at the beginning of this section) for determining the limiting medium.

This is because, on the one hand, mercury compounds are very hazardous to organisms, while, on the other hand, it is precisely this example which shows clearly how, during emissions into one medium (primarily the atmosphere), another medium (the surface water) becomes limiting (that is, critical with regard to man and ecosystems).

Anthropogenic mercury emissions into the surroundings have reached a very high level, and although the values in the literature vary somewhat, the total annual atmospheric emission of mercury probably exceeds 5000 tons due to the burning of solid fuel and oil, and about 10 000 tons from the processing of raw materials [7]; the emissions consist of metallic mercury (vapor) and its inorganic compounds. The mercury discharge directly into the hydrosphere is much less than that into the atmosphere.

The anthropogenic input of mercury to the environment via different chains is much greater than (approximately by an order of magnitude) the input due to natural processes (the weathering of rock and soil, the vital activity of organisms).

Mercury is present in the natural environment in the form of metallic mercury, inorganic compounds, and organic compounds. The first two forms of mercury predominate in the atmosphere.

The total mercury content of the surface biomass is 5×10^5 tons; 4×10^4 tons per year are involved in the biotic turnover [2]; about 10^4 tons per year are present in aquatic ecosystems [7]. When inorganic mercury enters aquatic objects, its methylation occurs. This process leads to high mercury concentrations in water bodies, and it promotes the accumulation of mercury in hydrobionts (fish and other marine organisms).

At one time the discovery of mercury (in the form of methyl mercury) in fish was quite unexpected [87]. Later it was shown that this form of mercury is produced by microbial methylation [64]. Aerobic and anaerobic microorganisms and fungi are able to methylate mercury [88]. Monomethyl mercury and dimethyl mercury are the basic products of the biological methylation of mercury. Dimethyl mercury, which has a low water solubility and a high volatility, easily enters the atmosphere, where UV-radiation transforms it into metallic form. Monomethyl mercury accumulates in living organisms and, therefore, it is the most dangerous for aquatic ecosystems. The ratio of monomethyl to dimethyl mercury depends considerably on the pH of the water; for ph > 8 to 9, all the mercury is dimethyl, while for pH < 6.0 all the mercury is monomethyl.

It will be recalled that, due to the washing out of sulfur compounds from the atmosphere, the pH of river and lake water decreases sharply; this can lead to a significant shift in the ratio between the organic mercury compounds in the aquatic environment toward monomethyl mercury, and to a fixing of this form in aquatic ecosystems.

Therefore, although anthropogenic mercury mainly enters the atmosphere, later, via migration and transformation, it accumulates in hydrobionts in the form of highly toxic organic compounds, the coefficient of accumulation being very high. In this case, it is the aquatic medium which may become the limiting environment.

The detrimental impact of mercury on the human health is caused by toxic, genetic, and teratogenous effects. Some cases have been reported (e.g., from Japan [72]) of the poisoning of people by fish and other marine products used for food. It was found, too, that the deaths of a great number of landbirds are associated with the consumption of seeds treated with mercury compounds [87]. Even very small amounts of mercury in the form of organic compounds lead to a decrease of photosynthesis in phytoplankton, and the reaction of diatoms is still more pronounced (at mercury contents of 0.1 to 1 μg/l) [69].

The toxicity threshold of inorganic mercury compounds for fish varies from 6 to 10 μg/l [71].

In Sweden a mercury concentration not exceeding 0.3 μg/day was recommended as the maximum permissible value for man [71]. In the USA and Canada, mercury concentrations of 0.5 ppm (dry weight) are taken as maximum permissible. The coefficients of mercury accumulation in fish (especially in predators) are 10^3 or more. For example, the Swedish authors of [66, 87] have shown that mercury concentrations in the meat of pike can reach 6000 ng/g (for mercury concentrations in the water of Swedish rivers of up to several tenths of a nanogram per gram [66]). This means that the concentration of organic and inorganic mercury compounds in water should not exceed at least 0.5 μg/l [6] (in the USSR the MPC is now taken to be 5 μg/l for water bodies used for sanitary and domestic purposes [29]). A value of 0.5 μg/l will thus be used in further calculations as the maximum permissible concentration.

It is noted in [87] that at present the study of many aspects associated with the problem of environmental mercury pollution and its penetration into individual organisms, as well as biochemical studies, have advanced significantly. However, the complex subject of emissions of mercury into the atmosphere, and its transport, precipitation, and further migration, has remained almost untouched by researchers.

In [47] an attempt was made to combine all sides of this problem, from a study of the atmospheric mercury emission and distribution to a determination of the limiting media and the mercury concentrations under different conditions.

The dynamics of mercury compounds in the natural environment (under conditions of equilibrium) can be described by a system of differential equations [29]; these equations in generalized form are:

$$\dot{O}_i = k_{\text{anth }i} E_{\text{anth}} + k_{ni} E_n - \sum_j \frac{Q_i}{\tau_{ij}} + \sum_j \frac{Q_i}{\tau_{ji}} \tag{2.30}$$

where E_{anth} and E_n are the mercury inputs of anthropogenic and natural origin, respectively, to the parts of the environment; k_{anth} and k_{ni} are the portions of anthropogenic and natural mercury entering the ith environment; j is any part of the environment except the ith part; Q_i and Q_j are the total mercury contents in environments i and j (able to migrate); τ_{ij} is the mean time for the removal of mercury from the ith to the jth environment (it is 3×10^{-2} years for removal from the atmosphere to the underlying surface, 360 years from the soil to the atmosphere, 850 years from the soil to the surface water, and 35 years from the surface water to the ocean [47]).

The atmosphere, the soil, the surface water, and the ocean water are considered to be natural environments (or parts of the environment).

The model parameters are assumed in [47] to be as follows: $E_n = 6.3 \times 10^3$ ton/yr, and in the absence of anthropogenic sources the mercury contents of the atmosphere Q_a^0, the soil Q_s^0, and the surface water Q_w^0 are 3×10^2, 3.6×10^6, and 1.5×10^5 ton, respectively.

Turning now to the concentrations C_i, we can write $C_i = (C_i)_0 Q_i / Q_i^0$, where $(C_i)_0$ is the natural mercury content of the ith environment (e.g., $(C_a) \approx 10^9$ g/m^3).

Assuming limit values $[C_a]_L = 3 \times 10^{-7}$ g/m^3 (which conforms with the MPC of mercury in atmospheric air) and $[C_w]_L = 0.5 \mu$g/l, we can estimate the maximum permissible emission (according to the approach described at the beginning of this section):

$$E_{\text{anth}} = \min \left[\frac{\dfrac{[C_a]_L Q_a^0}{(C_a)_0} - \alpha_{12} E_e}{\alpha_{11} k_a + \alpha_{12} k_{\text{th}}}; \quad \frac{\dfrac{[C_w]_L Q_w^0}{(C_w)_0} - \alpha_{32} E_e}{\alpha_{31} k_a + \alpha_{32} k_{\text{th}}} \right] \tag{2.31}$$

where α_{ij} are the elements of the matrix A^{-1} (here A is the matrix of system (2.30), and $k_{\text{th}} = 1 - k_a$).

From (2.31) we obtain $E_{\text{anth}} = 3 \times 10^4$ to 4×10^4 ton/yr.

Hence the limiting surface concentration of mercury in the atmosphere, at which its concentration in water is 0.5 μg/l, amounts to 10^{-8} g/m^3. This value is 30 times lower than the MCP for the atmosphere. Therefore, for steady-state equilibrium, the mercury emission, leading everywhere (over a large region) to atmospheric mercury concentrations conforming to the MPC, will cause the formation of mercury concentrations in fresh-water bodies which are 30 times higher than the MPC values guaranteeing the safe use of fish for food.

It should be noted that, in connection with the very long time (hundreds of years) required for the mercury to pass from the soil to the water bodies, considerable sluggishness is observed in the growth of the mercury concentrations in these water bodies (even after the cessation of atmospheric mercury emissions), followed by a very gradual decrease of the concentrations in water (especially in stagnant water).

Figure 2.5 illustrates this situation. Figure 2.5a shows the change in the anthropogenic mercury content of the atmosphere for a rate of emission a ton/yr (as a time function); the time (years) needed to remove half of the mercury from the soil T_1 is taken as time unity on the time scale. It is assumed for purposes of illustration that the emission into the atmosphere stopped at a time 0.2 aT_1. Figures 2.5b and 2.5c illustrate changes of the mercury concentrations in the soil and in the water bodies. Elimination of the mercury from the biota after it stops entering the water occurs relatively rapidly; for instance, the time for the half-elimination of mercury from fish is 2.5 years [88]. The limiting value of the mercury content of the soil

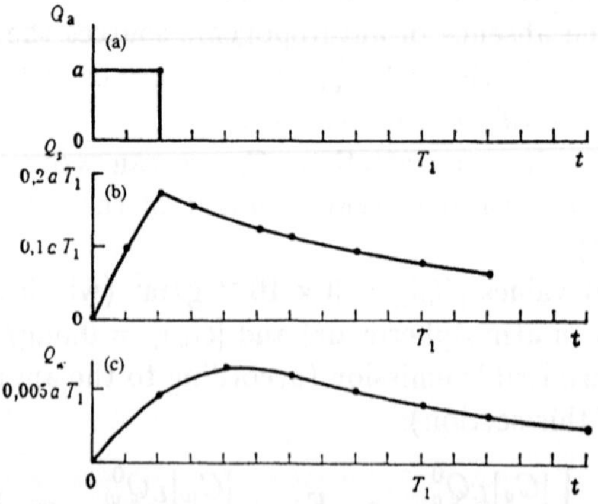

Figure 2.5. Dynamics of change in anthropogenic mercury content of atmosphere (a), soil (b), and water bodies (c): *a* is the rate (tons per year) of mercury input to atmosphere; T_1 is the time for half-elemination of mercury from soil (850 years).

is determined similarly; it is only 1.5 times greater than the background mercury concentration in the soil.

The authors of [47] also present some approaches to determining the limiting concentrations in the atmosphere on a regional scale (in the absence of equilibrium, when it is necessary to take into account the time of 'aeration' of the region, the water-exchange time, and the local conditions, including the number of water bodies, characterized by the quantity α, i.e., the ratio of the surface area of the water bodies to the territory of the region). In [47] a system of differential equations is given, describing the dynamics of the change in the mercury content of the environment within the framework of a regional model, and the parameters of this model are presented. Figure 2.6 shows plots of the limiting mercury concentrations in the air, as a function of α, for different depths of the water body at $k_a = 1$.

In the calculations on a regional scale, the limit mercury concentrations in the water were assumed to be equal to the MPC for water bodies intended for sanitary and other domestic purposes, i.e. 5 μg/l. Even with such a 'moderating' assumption (and, allowing for mercury accumulation in hydrobionts, it would be necessary to assume a limit mercury concentration in water of 0.5 μg/l, as in the previous case), the MPC in shallow water bodies already reaches its limit values (for small α) when only some hundredths of the MPC in the air are present.

The impact of substances in different environments, as they spread and become transformed, was studied using an approach similar to the one in this section, in the 'dose-commitment' ('expected' dose) model. This model was developed in order to describe the effects of ionizing radiation (for diverse paths of propagation and impact of radioactive substances) by the

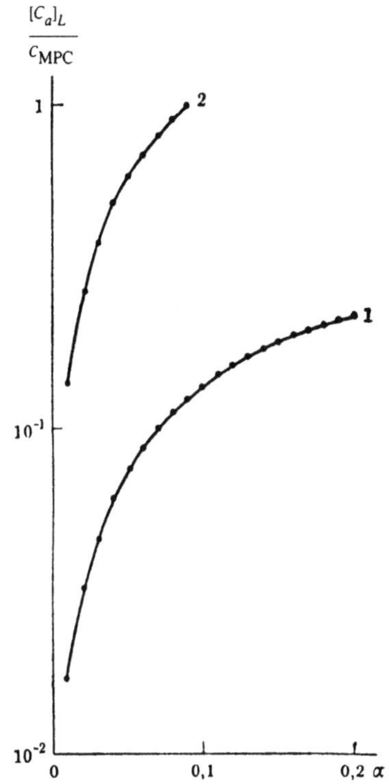

Figure 2.6. **Limiting mercury concentrations in air (in fractions of the MPC) as a function of α for different depths of the water body, $k_a = 1$: (1) 10 m; (2) 80 m. Values of $[C_a]_L$ given in units adapted for MPC for air.**

United Nations Scientific Committee on the Effects of Atomic Radiation (UNSCEAR). This procedure is used to study steady-state situations, when the pollutant enters from one large reservoir (some part of the environment or element of the biosphere) into another until it leaves the system (due to conversions or else because it passes to some place where it practically comes to rest, for instance, in sea-bottom sediments). To construct such a model, we have to know the coefficients of distribution of the pollutant among the parts of the environment and its residence time in each reservoir.

The concept of 'dose-commitment' used to determine the radiation dose for the human population, caused by radioactive emissions into the environment, involves integration during an infinite time period of the mean dose rate $R(t)$ (for a certain organ or for the entire body):

$$D = \int_{-\infty}^{+\infty} R(t)\,\mathrm{d}t, \tag{2.32}$$

referred to some definite population, as a result of both external and internal irradiation.

Before the formulation of the 'dose-commitment' concept, a similar ap-

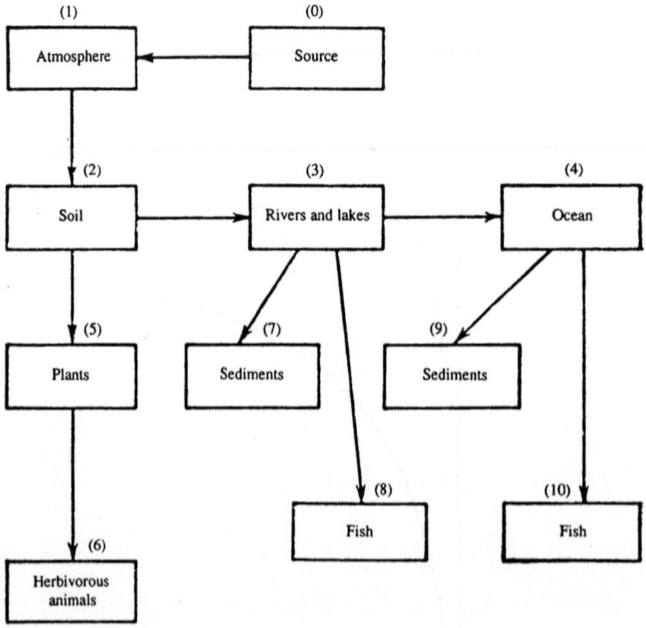

Figure 2.7. Spreading of pollutants in natural environment (and biota).

proach was used by the author to calculate the dose loads of radioactive products propagating over long distances [8, 13].

By tracking, along the various chains (conversion chains or food chains), radionuclides released by a large-scale source, it was possible to determine the dose loads on a human or other population at different distances from the source, due to the input or impact from the parts of the environment under consideration, as well as to find the limiting environment and the limiting concentrations of individual radioactive products.

It should be noted that the above-described 'dose-commitment' approach should be used with some caution. If the genetic effect on a population is determined, then integration over an infinite time is correct. If only somatic effects are taken into account, it is necessary to isolate the effects on each generation and to take into account the harm done (e.g., a decrease in the size of the population) to subsequent generations.

The 'dose-commitment' concept was also applied to the case of general pollutions [51]. Then it was supposed that a dose means the absorption of certain substances by a sensitive receptor.

In this case the dose was determined as the integral (over infinite time) of the average concentration $C(t)$ in the receptor or reservoir (element of the biosphere):

$$D = \int_{-\infty}^{+\infty} C(t)\,\mathrm{d}t. \tag{2.33}$$

For example, examining the chain of conversions (Figure 2.7), we find it necessary to establish the coefficient of transfer between the links (reservoirs) of this chain. The coefficient of transfer P_{ij} from reservoir i to reservoir j is defined as the ratio of the average contamination level (integrated over time) in the jth reservoir to the average level in the ith reservoir. Figure 2.7 shows the transfer of pollutants in one direction (movement is usually possible in both directions). In accordance with the above, for instance, the total pollution entering the ocean via two routes $[(0 \to 1 \to 2 \to 3 \to 4)$ and $(0 \to 1 \to 4)]$ is defined as the sum (for a source intensity I)

$$IP_{01}P_{12}P_{22}P_{34} + IP_{01}P_{14}.$$

This approach proves its worth particularly well when the residence time of the pollutant in the reservoir is very long. It is applicable, for instance, to the case of pollution of the environment by mercury, which can stay for a long time in soil, surface water, or sea water.

The 'dose-commitment' method was also used in [54] to calculate the impact on man of methyl mercury (entering the system via consumed fish).

In order to assess the hazard (and damage) of mercury for man, we have to integrate the amount of mercury entering the human organism via different routes.

The mercury input $q(t)$ to a critical organ of the human body is defined as

$$q(t) = \int_0^t IC(\tau)AR_S(t-\tau)\,\mathrm{d}\tau, \tag{2.34}$$

where I is the amount of water, air, and food used by the person in unit time; $C(\tau)$ is the pollutant concentration in the environment at time τ (taking into account the decomposition and conversions of the given substance); A is the portion of the pollutant entering into the critical organ (in the case of mercury, into the blood) during food consumption and respiration; $R_S(t)$ is the portion of the pollutant remaining in the human body during a time t after entering into the organism.

The dose or total impact during time t is estimated as the integral of $q(t)$ per unit mass (or per entire critical organ or entire organism):

$$D(T) = \int_0^T \frac{q(t)}{m}\,\mathrm{d}t \tag{2.35}$$

(m is the mass of the critical organ).

Calculations made for monomethyl mercury ($CH_3^-Hg^+$), when a person consumes fish and mercury accumulates in the blood, lead to such quantitative results [82]. Consequently, in order to calculate the doses (impacts) using this method, it is necessary to know the distribution of the substance among the parts of the environment.

In [76] a model of the behavior of mercury in an aquatic environment, and its penetration into fish and then into man, is described. This model complements well the above model of mercury diffusion within the various compartments of the natural environment.

Experiments have shown that monomethyl mercury can be synthesized under the impact of bacteria *in situ*, and probably in water as a result of photochemical reactions [50].

When examining the transport of mercury in water, it is first of all necessary to isolate such elements as water and sediments (suspended and deposited). The following issues should be considered: transport within the biota, mercury transformation by invertebrates and higher plants, routes of mercury penetration into fish (herbivorous and predatory) and man (fish are the basic source of the possible entry of mercury into man, and their disappearance from water bodies is a striking ecological effect of mercury pollution of these water bodies) [60]. Monomethyl mercury, as the most toxic form, should be considered separately during the construction of a model; the other forms can be united into inorganic mercury. A pollutant moves in a water body by different means: it can be transferred by the biomass or it can move by adsorption or desorption (the biomass staying motionless).

The modeling in [76] is carried out at several levels simultaneously; one set of variables is introduced for the biomass, and another is introduced for each of the two chemical forms of mercury. Therefore, the 'minimum' model incorporates 18 variables for six components: water (W), suspended particles (SS), sediments (S), invertebrates (I), plants (P), and fish (F), in each of which the paths of the biomass and two forms of mercury are traced. The number of interactions here reaches 80. However, some interactions can be neglected in the model (e.g., the mercury uptake by plants directly from sediments is negligible). After simplification, the number of variables in the model is reduced to 15 and the number of interactions to 40 (Figure 2.8). The mass of each component is denoted by X_i, and the coefficient of transport of the substance from X_j to X_i is denoted by P_{ij}.

Let us describe the mercury dynamics inside a water body in accordance with the presented model, using as an example the methyl mercury content X_{12} of suspended particles (the dynamics of the mass of the particles X_3 is assumed known).

There exist three basic mechanisms for the transport of X_{12}: uptake of methyl mercury from the water, decontamination (passage) into the water, and absorption by invertebrates.

The rate of methyl mercury uptake from the water is proportional to its concentration in the water ($C_{10} = X_{10}/X_1$) as well as to its total amount X_{12}. Hence

$$\frac{dX_{12}}{dt} = P_{12,10}X_{10}X_{12}(C_{12,\max} - X_{12}/X_3)X_1. \tag{2.36}$$

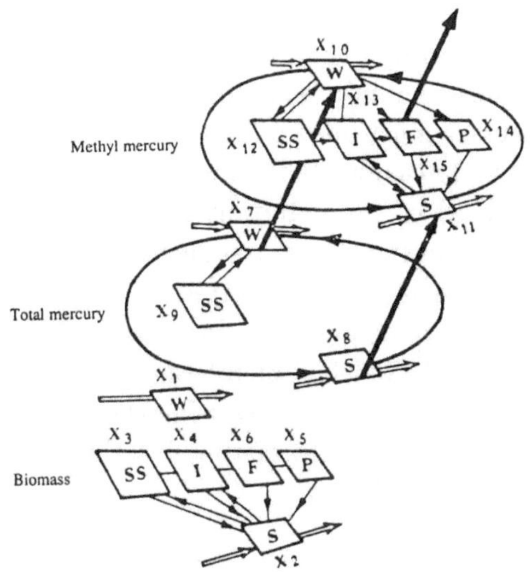

Figure 2.8. Transport of mercury inside a water body. Upper arrow shows route of mercury migration to man.

A similar term for the case of passage into the water takes the form

$$-P_{10,12}(X_{12}/X_3)(C_{10,\mathrm{max}} - X_{10}/X_1).$$

Elimination with the aid of invertebrates will be described by the term

$$-P_{13,12}(P_{4,3}X_4X_3/X_{3,\mathrm{av}})(X_{12}/X_3)$$

for the constant $P_{13,12}$ (the coefficient of so-called 'internal' uptake during consumption, which is determined under laboratory conditions).

Numerous examples of the use of the described model are given in [76]; here some quantitative results of the calculations will be presented. For instance, a calculation of the behavior of mercury for certain values of k_{ij} in a water body shows a substantial annual decrease in the total mercury content of the sediments (about 15% per year); however, the rate of decrease in the mercury content of fish perceptibly lags behind the rate of its decrease in the sediments. It is of interest to note that spring floods cause large variations in all the quantities, but only during the flood period.

The author of [54] calculated that, for steady-state equilibrium, a permissible amount of methyl mercury in the blood not exceeding 0.1 μg/g will accumulate when consuming fish containing $55/X$ ppm of methyl mercury at a daily ration of X grams.

Emissions of halogenated hydrocarbons into the atmosphere serve as another example. The only process of removal of the halogenated carbons is their diffusion into the stratosphere, with a subsequent chemical transformation. This in turn leads to a decrease in the ozone content of the stratosphere.

There is, however, a certain condition which must be satisfied when using an approach in the 'dose-commitment' concept (this is shown in [51]). The transformation processes must obey a linear law, and in addition the 'dose-effect' relation must be linear, while the impact must be proportional to the dose or the integral level of the pollutant content and not a result of the effects of synergism. It must also be assumed that the transport processes are steady in time. It is more complicated to apply the above model to pollutants having considerable gradients in space and time.

The 'dose-commitment' model is of interest for the development of regional and global monitoring systems; by using this model when establishing ecological standards, we are able to distinguish the most important (high-priority) reservoirs and ingredients.

2.6. Establishment of Standards for the Peaceful Uses of Nuclear Energy

The diverse paths of an impact can also be taken into account when establishing standards for radioactive emissions arising from the peaceful use of nuclear energy (problems related to establishing regional ecological standards for the long-range transport of pollutants are discussed in [17, 20, 33] and they will be touched upon in Section 3.2). In this section a comparative assessment of the operation of nuclear and fossil-fuel power stations will be carried out, together with a calculation of the permissible radioactive emissions from peaceful underground nuclear explosions, with complete public radiation safety, including safety at long distances from the explosion site.

The need for a very rapid development of nuclear power engineering has become obvious. In the USSR this type of power production is particularly promising [31]. The global electrical power production by nuclear stations is expected to rise from several percent (the present state of the art) to approximately 25% of the total production by the year 2030 [57]. Moreover, in the near future we can expect the development of some new uses of nuclear power in industry, the public economy, and construction.

Nuclear power stations utilizing ^{235}U fission by thermal neutrons form the basis of nuclear power production at present. The future development of nuclear power will be associated with fast-neutron reactors (using ^{238}U), which will substantially increase the potential of nuclear power stations. It should also be stressed that nuclear fuel does not consume oxygen from the air or lead to an ever-increasing carbon dioxide content of the atmosphere. However, an important condition for the development of nuclear power is the absence of any adverse effects that can occur due to environmental pollution by radioactive waste and emissions.

Taking into account these things, as well as the rapid development of nuclear power throughout the world, we thought it appropriate to compare the loads on the population and on ecological systems due to the operation of nuclear and fossil-fuel power stations, not only of individual stations but of a large number of stations, over vast areas covering entire countries and continents. Such a comparative assessment was carried out by us in a paper presented at the fourth International Scientific Conference on the Peaceful Uses of Atomic Energy (Geneva, 1971) [12].

This assessment is based mainly on a comparison of the damage (or potential hazard) from products emitted into the air. Discharges of solid waste, small quantities of which are produced during the operation of nuclear power stations and which can be localized by means of burial, will not be discussed in this section.

Naturally the statement concerns power plants operating under safe conditions. However, significant power plant accidents, such as at Three Mile Island (USA), and particularly at the Chernobyl power plant (USSR) require that the safe operation of all power plants be guaranteed.

The hazard for man and for the environment presented by power stations operating on fossil fuels (coal, oil, natural gas) is associated primarily with the emission into the atmosphere of combustion products: sulfur dioxide, ash, nitrogen oxides, heavy metals, hydrocarbons, etc.

In recent years the total annual sulfur dioxide emission into the world atmosphere has reached 60 to 80 million tons, and the SO_2 content of the air over a number of large industrial cities exceeds the maximum permissible values (MPCs). The average SO_2 emission per unit of energy produced amounts to about 0.8 ton/(MW yr). This means that, unless measures (rather costly ones) are taken to clean fuel gases of the SO_2 (or else effective desulfurization of the fuel), the SO_2 emission into the atmosphere may reach by 2000 the astronomical figure of 300 million tons [12, 68].

At present a tremendous amount of dust (200 to 250 million tons) is also emitted into the atmosphere, but methods for cleaning the air of dust are being developed more effectively.

A potential hazard of the use of nuclear power is associated with the emission into the surroundings of a certain amount of radioactive products (of fission and fusion and of isotopes with induced activity).

Under normal conditions of operation, fission products accumulate in the core of the reactor, and there is practically no possibility that they will enter the surroundings. Only small amounts of gaseous isotopes (^{41}Ar, ^{14}C, ^{3}H, ^{133}Xe and traces of ^{131}I) can escape into the atmosphere. The dose of radiation from these products does not exceed 1% of the permissible value, even in the vicinity of the station [28].

The entry of long-lived gaseous ^{85}Kr (about 5×10^2 Ci/Mw yr) into the atmosphere is practically inevitable (for instance, when spent fuel elements are dissolved).

During accidents, the possibility of which cannot be ruled out completely (at least for probability calculations of the damage), the release of large amounts of volatile radioactive isotopes, such as ^{131}I, ^{132}I, ^{137}Cs, ^{89}Sr and ^{90}Sr, is possible. The amount of ^{131}I released during a major accident (1 accident per year for 10^3 reactors) can reach 10^3 Ci, while for a medium-sized accident (1 accident per year for 10^2 reactors) the corresponding figure is 10^2 Ci [37, 59].

The global radiation situation caused by the development of nuclear power engineering was estimated [12] on the basis of the assumptions that, first, about 5000 reactors each having an average power of 10^3 Mw would be put into operation by the year 2000 (their distribution over the land being relatively uniform), and, second, the possible number of major and medium-sized accidents would be about 5 and about 30 per year, respectively. Then, up to 5×10^3 Ci ^{131}I, 30 to 140 Ci ^{137}Cs, 4 to 7 Ci ^{89}Sr and 0.1 to 1.0 Ci ^{90}Sr will be released into the atmosphere annually (depending on the operation time of the reactors) [84]. It was calculated that the above isotopes travel over long distances, falling out with an effective velocity of 0.1 to 1.0 cm/s.

When reactors operate, ^3H and ^{14}C can enter the atmosphere [the amount of tritium entering the atmosphere varies from 10 to 30 Ci/(Mw yr)] [81]. Dissolution of the fuel elements will be accompanied by the emission of approximately 3×10^9 Ci ^{85}Kr (annually), into the atmosphere; the dose from tritium will be substantially less than the dose from krypton and will amount to 10^{-3} mrad/year for the entire population [52].

The doses for relatively short-lived isotopes were calculated on the assumption that they mix in the lower 200-m layer of the atmosphere, while krypton and tritium mix in the 500-m layer. The 200-m layer was also assumed for ash and sulfur dioxide emissions; the lifetime of SO_2 and of its conversion products in the atmosphere was assumed to be about 4 days.

Table 2.2 [12] presents some calculated concentrations of different substances in the atmosphere and at the Earth's surface during the operation of nuclear power stations [in fractions of the MPC (doses) for the atmosphere or of the maximum permissible amounts for the population contained in foods in the case of land surface contamination].

Interestingly enough, if the SO_2 emissions anticipated by the year 2000 amount to 3.0×10^8 t/yr, then its concentration in the air (over land) will be, on the average, only a few times less than the mean daily MPC of SO_2 (0.05 mg/m^3) and of its conversion products and in a number of regions will even exceed substantially the MPC. [In these calculations the height of the mixing layer was taken to be 1000 m, and the lifetime of SO_2 and its conversion products was taken to be 4 days.]

It should be noted, too, that in some cases the dose caused by natural radioactivity when traditional fuels are used may be comparable, or even

Table 2.2. Maximum average concentrations (or pollutant levels) in air of different radioactive and toxic substances during operation of nuclear power stations (in fractions of limiting amounts: dose limits, maximum permissible inputs for population or maximum permissible concentrations averaged over a year).

	Nuclear power stations				
	Operation without		Emergency situation		
	^{3}H	^{85}Kr	^{131}I	^{90}Sr	^{137}Cs
At ground level				10^{-6}	10^{-5}
In the air	5×10^{-6}	3×10^{-3}	7×10^{-4}		
Maximum value for given form of power		3×10^{-3}	$\sim 10^{-3}$		

greater than, the dose from radioactive emissions of the nuclear power stations [73].

An analysis of the data presented in Table 2.2 shows that 10^3 to 10^4 times as much air will be required to dilute the harmful substances contained in the emissions of fossil-fuel power stations down to the permissible concentrations, compared to the emissions of nuclear stations at the level of large-scale development of nuclear power and the adopted characteristics of the emissions (and even allowing for emergency situations at nuclear power stations).

In conclusion, it may be said that nuclear power can be developed and still ensure a sufficiently clean environment, and that it is less hazardous than the development of powerful fossil-fuel plants.

Both the direct problem and the inverse problem have to be considered for an underground nuclear explosion. In the case of the direct problem, the dose loads on the population living at considerable distances are to be calculated from the known parameters of the explosion. In the case of the inverse problem, using certain criteria for the permissible load on man (e.g., a small fraction of the permissible dose for the population or a dose containing a small fraction of the natural background radiation), the possible permissible emission of radioactive products into the atmosphere are calculated, i.e., standards (criteria) of safe nuclear explosions for peaceful purposes are established. Precisely such an approach was described in [8, 13, 31].

The radiation hazard from contained underground explosions (without breakthrough of the explosion cavity into the atmosphere) is caused by the seepage through cracked rocks of radioactive inert gases, and possibly also by tritium. In the case of excavation underground explosions (with a breakthrough of the explosion cavity into the atmosphere), the radiation hazard is caused by practically the entire complex of explosion radioactive products. Some of these can travel over very great distances: many hundreds or thousands of kilometers.

The following are the most hazardous radioactive products: fission radioactive isotopes ^{90}Sr, ^{89}Sr, ^{137}Cs, ^{131}I, $^{140}Ba + {}^{140}La$, some isotopes with induced activity (isotopes of tungsten, ^{203}Hg, ^{203}Pb), tritium, and some other (highly volatile) isotopes. Below we will consider the possible loads only on the population, but allowing for all possible routes of impact, i.e., taking into account passage along different trophic chains. A remote zone (located hundreds or thousands of kilometers from the explosion site) is selected, since in such a zone the radioactive fallout can cover vast territories, and large contingents of the population may be exposed to the radiation.

The dose loads in a remote zone are caused by external radiation from passage of the explosion cloud (or stream of gases), external radiation from the fallout, internal radiation from the inhalation intake of radioactive products from the cloud, and internal intake of isotopes with water and food from a contaminated region.

The fallout of radioactive products in the case of an excavation underground explosion of unit power (of the order of 1 kton due to the fission reaction) at a distance of more than 1000 km can reach 50 to 100 MCi/km^2 (at the time of fallout).

The greatest contribution to the dose load on the population is made by the external and internal radiation from radioactive products deposited on the territory.

Table 2.3 presents the doses of external radiation for 30 years from such a fallout, the inhalation doses during the passage of the cloud, and the doses of internal radiation when radioactive products enter human organisms with food during summer and winter fallout (in agricultural regions). As shown by Table 2.3, the highest doses occur when fallout is formed in summer.

The additional dose from tritium for a thermonuclear excavation medium-yield (170 kton) explosion has a value approximately an order of magnitude less than that from the ^{137}Cs fallout [6] after a 1kton fission explosion.

Calculations carried out taking into account all routes of the impact on man of radioactive products resulting from underground explosions for peaceful purposes (taking into consideration penetration into ecological systems) have shown that at a given stage the use of charges, at the technological level which existed in 1975, makes it possible to implement such major projects as the construction of channels, dams, and water reservoirs in arid regions, etc., all the while observing conditions of complete radiation safety for the population [8], i.e., using as a criterion (standard) only a small fraction of the permissible dose for the population or of the dose due to natural radiation.

Table 2.3. Doses of internal and external radiation from fission radioactive products (over 30 years, measured at distant point $R = 1000$ km) of underground fission explosion of 1 kton power.

Isotopes	Average composition of fallout on D + 1 (%)	External radiation dose (rad)	Inhalation dose (rem)	Internal radiation dose entering with food (rem)		Critical organ exposed to internal radiation
				Fallout in summer	Fallout in winter	
$^{140}Ba + ^{140}La$	$^{140}Ba - 30$	2.8×10^{-4}	3×10^{-5}	–	–	Bone tissue
^{89}Sr	50	–	2.4×10^{-4}	6.5×10^{-4}	2.5×10^{-5}	Bone tissue
$^{90}Sr + (^{90}Y)$	0.3	–	1.1×10^{-4}	1.1×10^{-2}	5.8×10^{-3}	Bone tissue
^{137}Cs	0.6	6.7×10^{-4}	2.2×10^{-7}	2×10^{-4}	6×10^{-5}	The entire body
^{131}I	1	1×10^{-5}	1.7×10^{-5}	2×10^{-2}(children) 1×10^{-3}(adult)	–	Thyroid gland
Total	$9.5 \times 10^{-4} \approx 10^{-3}$					

CHAPTER 3

Regulation of Environmental Quality

3.1. General Approaches. Ecological and Economic Aspects of Regulation

As noted above in the *Introduction*, while interacting with the natural environment, human society acquires all the material comforts necessary for life. Nature is also a source of satisfaction of man's esthetic needs.

However, as they interact with the natural environment, humans modify it, i.e., they remove vast territories from the natural biological cycle and they alter the character of the cycles of the most important elements.

Practically all anthropogenic impacts (both intentional and unintentional) affect and alter the natural balance of nature. It is difficult to detect some of these alterations against the background of natural fluctuations, whereas others have by now reached such major proportions that some are already irreversible.

These modifications can be either positive (necessary) for both man and nature itself (biological productivity increases and rejuvenation of biocenoses occurs) or negative (environmental pollution, exhaustion of nonrenewable resources, and a decrease in nature's ability to regenerate renewable resources).

From the author's viewpoint, it is correct to conclude that manmade changes in nature are necessary and that these changes and modifications of nature resulting from scientific and technical progress should be considered as positive events for mankind, while the 'deterioration' of the natural environment and the disturbance of the ecological balance do not stem from the general strategy of the industrial society, but rather are due to errors in the ecological technical policies. This was also the conclusion of Academicians E.C. Fedorov [24] and S.S. Shvarts [27]. However, with regard to the negative impact, it is clear that very large-scale expensive measures are needed to reduce (or prevent) this impact of man on nature.

On the other hand, an analysis of the anthropogenic impact on nature from the viewpoint of the kind of impact which adversely affects natural systems, and the kind which is positive, led Shvarts [27] to state that some general modifications introduced by industrialization can be considered to be potentially favorable for the development of the biosphere.

Shvarts is convinced that such factors as the simplification of biocenoses, the rejuvenation of these associated with industrialization and urbanization, the alteration of individual links of food chains, and the increasing role of animals as destroyers of primary organic matter do not represent a degradation of the biosphere, but rather its evolution under new conditions.

The joint efforts of man and nature as they interact can facilitate the development of stable biogeocenoses in an environment modified by man. The basic task of those interested in nature protection will be to foster the development of biogeocenoses able to regulate themselves under the conditions altered by man [27]. For this purpose it will be necessary to introduce certain elements of knowledge and culture into industrial and agricultural production.

However, we cannot overlook the fact that many anthropogenic impacts, and pollution in particular, have a very negative effect on nature. Consequently, active ecological and technical measures should be taken to prevent such harmful effects.

It is obvious that the large-scale introduction of new technological processes producing no waste would be extremely costly and time-consuming. If the material resources are limited, then it is necessary to justify the priority of the measures taken, and the priority should in turn be determined by the critical character of the impact on human health and ecological systems, taking into account the socioeconomic consequences.

Therefore, the first strategic problem of the regulation of environmental quality is the establishment of a system for observing, assessing, and forecasting the state of the environment, thereby providing adequate data for the determination of priorities [11] and the regulation of environmental quality.

The problem of regulating environmental quality should be solved concurrently with the problem of providing man with everything necessary for his vital activity, i.e., with normal nourishment, habitation, articles of long-term use, and public health services.

The trend and the sequence of the activities related to environmental regulation should be as follows [14]:

(1) Establishment of a pollution monitoring system; determination of pollution levels and places of higher risk;
(2) Establishment of a system monitoring the biota's responses to the anthropogenic impacts;
(3) Assessment of the load created on man and ecosystems (with respect to the permissible and critical loads); preliminary development of criteria for the permissible and critical impacts on different organisms and elements of the biosphere;

(4) Assessment of the ecological, economic, and esthetic damage to the natural environment;
(5) Forecasting the anthropogenic impacts on the biopshere (and on the weather and climate), associated with pollution; assessing the consequent loads and their consequences;
(6) Determination of the priorities when taking measures with a consideration of the actual state of the environment and of the ecological, economic, and esthetic damage (for maximization of the yield);
(7) Technical and technological decision-making; optimum distribution of enterprises, so as to decrease the impact on the environment;
(8) Taking technical measures to prevent pollution, i.e., developing methods for the treatment of discharges and emissions, as well as introducing technological processes producing little or no waste;
(9) Planning the development of economy branches in different regions, taking into account their ecological reserve and possible economic considerations;
(10) Determining the developmental scales of some trends in human activity, allowing for the ecological and economic possibilities and the limitations of their impact on the environment.

Economic, social, technological, and biological processes are so closely interrelated in the modern world that it becomes necessary to consider current production as the functioning of a complex ecological-economic system [19]. Here the ecological-economic system refers to the interrelation and interdependence of the functions of social production and of the processes occurring in nature.

Academician N.P. Fedorenko considers that we can now speak about the formulation of a new sphere of social production, namely the reproduction of natural resources [28].

So far the problem of the economic efficiency of the utilization of raw materials has been solved by calculating how much lower the expenses on the additional production S_1 in the given process are than the expenses for a similar production output of the competing technological processes S_i, i.e., on condition that

$$(S_i - S_1) > 0 \tag{3.1}$$

for any ith competing process.

It was shown in [8] that the condition for the expediency of fuller utilization is expressed by the following inequality:

$$(S_i - S_1 + \Delta R) > 0 \tag{3.2}$$

where ΔR is the decrease in economic damage from pollution of the environment due to the given technological process.

From this point of view, a relatively expensive technological process without waste may become more profitable than a process accompanied by environmental pollution (if, of course, condition (3.2) is satisfied).

Therefore, protecting the environment from impacts leading to negative consequences is economically justified and becomes an efficient sphere of application of socially useful work. Economists and ecologists long ago concluded that they have to join forces in order to make ecological-economic assessments of the interaction between human society and nature as a whole and of the changes in the state of the environment caused by economic activity, as well as to plan and carry out nature-protection measures, in particular. For this purpose, ecological-economic modeling is being expanded [1–4], and the economic damage is being taken into account when assessing the state of the natural environment [5, 10, 21, 23]. More and more attention is being given to the socioeconomic assessment of the effectiveness of environmental protection [7, 8, 25, 26]. The author of [19] suggested an ecological-economic model of nature utilization; it reflects simultaneously the processes occurring in both the economic substem and the ecological subsystem.

This model indicates the possibility of ensuring that there is no detrimental impact on the natural environment while at the same time maximizing thė economic effect; thus, the model makes it possible to select an optimum solution, from the point of view of both ecology and economy.

In [20] an example is presented whereby agricultural production is planned, together with protection of the environment against pesticide pollution and ecological-economic factors are considered. After specifying certain criteria, the authors of [20] obtained a quantitative answer to the question of what relation is appropriate when using pesticides of different stability and cost.

Taking economic factors into account may change the priorities of the measures taken to decrease the detrimental anthropogenic impact. Another important multiplier, determining the specific economic significance of a disturbed biospheric element (damage) or the level of pollution costs, appears in the formula for determining the ecological damage (see Chapter 4), in addition to the three basic components: the intensity of the impact factors, their toxicity (or biological efficiency), and the number of individuals of different populations or biospheric elements subjected to the impact.

Taking into consideration this multiplier may prove to be a decisive factor. For instance, if the economic damage from the ith impact on the m population

$$B_{im} = k_{im} A_{im}, \tag{3.3}$$

where A_{im} is the degree of the ecological impact, and k_{im} is a coefficient characterizing the ecological damage, then it is obvious that we can have $B_{im} > B_{jm}$ for $A_{im} \leq A_{jm}$ and $k_{im} > k_{jm}$.

Note that for unique ecosystems it is necessary to assume $k_{im} = \infty$ and $B_{im} = \infty$ at $A_{im} \geq A_c$, where A_c is the critical value of the ecological impact [3].

Taking the foregoing into account, we see that with respect to many ingredients it is unrealistic to allow the volumes of industrial waste which would already now provide for a desirable environmental quality everywhere, i.e., which would fully meet all the requirements of hygiene and ecology. In this case it is necessary to gradually decrease the total amount of discharged pollutants and to gradually improve the environmental quality, up to the levels complying with the hygienic and ecological standards for the permissible impacts on living organisms, and on natural and anthropogenic ecological systems. It should be emphasized that, if the ultimate goal of the regulation of environmental quality is completely determined by considerations of the perfect safety of the impact of the environmental factors on man (i.e., the absence of any detrimental effects on his health) and on natural systems, then the qualitative and quantitative characteristics of the steps or stages on the way to reaching the desirable environmental quality should also be determined taking into account economic considerations [15] and scientific and technical possibilities.

The authors of [20] present some considerations associated with the strategy of a transition from the present state to a more advanced state, with the aid of promising methods and technological processes, in conjunction with an economic criterion for the effectiveness of the reduction of pollution. The concept of 'pollution costs' is introduced in this work (as a part of the production costs).

The level of the pollution costs corresponds to the vector of the technogenic discharges, which determine their volume, their properties, and their propagation conditions.

The pollution costs include the following:

(1) expenses needed to bring the discharges into conformity with a certain vector of the technogenic discharges (prevention costs);
(2) expenses for compensation of the consequences of the discharges (compensation costs);
(3) expenses for indemnity of the losses of valuable products with the discharges (raw-material losses).

(The sum of the two latter kinds of expenses corresponds to the economic damage from pollution).

In addition, the concept of an economic optimum of pollution is introduced. This is a pollution level at which the minimum of the regional pollution costs (or of the total production costs, including the pollution costs) is achieved. Such an optimum occurs when the increase in prevention costs, for only an insignificant change in it, becomes equal to the decrease

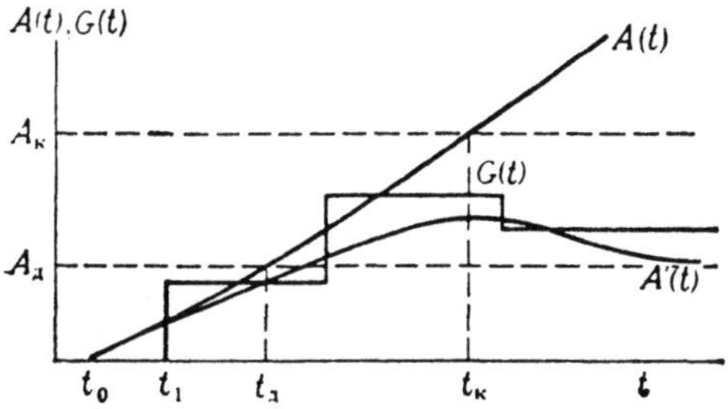

Figure 3.1. Change in degree of impact on biospheric element for different expenses in order to prevent this impact.

in compensation costs and raw-material losses [6].

This is equivalent to the statement in [11] that it is possible to consider as a justified value of the expenses G those values, for instance, of the expenses (or of the function $G(t)$, since the pollution level is a function of time) for which (with respect to the modulus)

$$\frac{\mathrm{d}G(t)}{\mathrm{d}(t)} = \frac{\mathrm{d}B(t)}{\mathrm{d}(t)} \tag{3.4}$$

(the complex and combined impacts of various factors are already taken into account here).

However, the most general approach to determining the economic optimum would apparently assume the sums of the expenses for nature protection and for damage from any impacts on it to be a minimum (for the maximum economic effect). (Let us recall the condition described during the formulation of approaches to the establishment of ecological standards: maximization of 'usefulness'). This approach is described by us in [11].

In this case $G^*(t)$ and t^* should be sought taking existing possibilities into account, when the sum

$$\int_{t_1^*}^{\infty} G^*(t)c(t)\,\mathrm{d}t + \int_{t_0}^{\infty} B(t)g(t)\,\mathrm{d}t = \min\left\{\int_{t_1}^{\infty} G(t)c(t)\,\mathrm{d}t + \int_{t_0}^{\infty} B(t)g(t)\,\mathrm{d}t\right\}, \tag{3.5}$$

where t and $G^*(t)$ are the values of t_1 and function $G(t)$ for which the sums of the expenses and damage are a minimum; $c(t)$ and $g(t)$ are functions characterizing the change with time in the 'significance' of the expenses and damage; $c(t)$ and $g(t)$ are monotonically diminishing functions when the possibilities of mankind and nature increase with time; t_0 is the time of beginning of the impact on a biospheric element. This situation is illustrated in Figure 3.1.

Suppose that an impact on a biospheric element starts at time t_0 and that the effect of the impact (the degree of the impact) changes as $A(t)$ (see Figure 3.1).

In this case the impact reaches the permissible value A_p at time t_p and the critical value A_c at time t_c. If we want to transform $A(t)$ into $A'(t)$ and to reach the value $A'(t) \leq A_p$ on condition that $A'(t)$ nowhere exceeds A_c, then we must put in certain expenses (in Figure 3.1 the expenses begin at time t_1, and curve $g(t)$ has three steps).

It is obvious that, other conditions being the same (especially if preventive measures are taken, i.e., $t_1 > t_0$), the earlier the expenses begin (at time t_1), the less they will be (integrally).

When the ultimate effect is determined, it is necessary to take into account the changes with time in the significance of the expenses and in the damage; these are determined in Equation (3.5) by the monotonically diminishing functions $c(t)$ and $g(t)$, respectively.

Returning to Equation (3.5), we note that the optimum economic conditions should be sought in conformity with this condition, naturally taking into account the real possibilities, that determine the possible time of the beginning of the expenses and their total sum $\int_{t_1}^{\infty} G(t)\,dt$ or $\int_{c_1}^{\infty} G(t)c(t)\,dt$.

In [7] the economic approaches to the solution of ecological problems are generalized. It is stressed there that in the USSR three basic principles of the formulation of planned requirements for the future state of the environment are used:

(1) the principle of not worsening the ecological situation;
(2) the principle of not permitting irreversible disturbances of the natural environment;
(3) the principle of maximum proximity to the optimum state of the environment.

It is obvious that a realization of these principles is only possible under conditions of development and implementation of an effective monitoring system, together with the development of criteria for the permissible ecological load (and public-health indices), as well as the establishment of hygienic and ecological standards, i.e., using approaches, most of which are discussed in detail in this book.

In order to determine the economic optimum of environmental quality, the concept of ecological expenses (as an element of socially necessary expenses) is of importance. The ecological expenses are the sum of the expenses for:

(1) prevention of ecological disturbances (changes) via measures of nature protection;
(2) prevention of the effect of the ecological changes on people and on property (recipients);

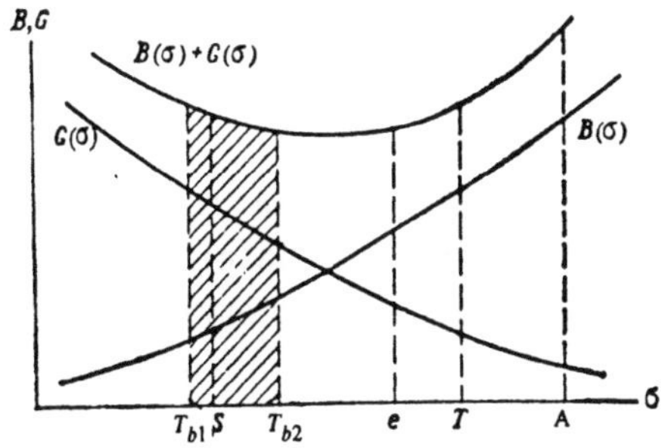

Figure 3.2. Damage B and expenses G (prevention costs) as functions of pollution level for fixed technology.
Pollution levels: A is actual level, T is lowest level for achieved technology, e is level corresponding to economic optimum, T_{b1}–T_{b2} (shaded area) is level corresponding to best achievable technology, S is level corresponding to national or world standards of environmental quality.

(3) elimination of the consequences mentioned in (2).

Expenses (1) are called 'nature-protective', and the sum of (2) and (3) constitutes the economic damage due to the ecological disturbances.

Prevention of the impact of the pollutants on the recipients and compensation for the negative consequences are often considered by economists to be alternative versions of the same thing. It is obvious that taking measures to prevent the effects will cost money, but that taking no measures will also require expenses. The economic damage from ecological disturbances is also determined by the expenses arising due to the rejection of nature-protection measures [7].

To make alternative ecologically and economically justified decisions, it is necessary to know how to estimate correctly the damage from existing and predicted anthropogenic impacts. Methods of carrying out such estimates are described in [5, 22, *et al.*].

The economic optimum of the environmental quality corresponds to a level of ecological disturbances such that the total ecological expenses are a minimum (see Figure 3.2.). In turn, the minimum of the ecological expenses is achieved when that variant of the nature-protective measures is chosen which ensures a maximum excess of the prevented economic damage over the expenses for nature preservation [7]. Note, too, that a still greater effect can be achieved by introducing production of the necessary product such that, with regard to the technology or the selection of the production mode, substantial expenses will not be needed to get this maximum excess (at the same time minimizing the load on the natural environment).

Also important is the statement that the economic optimum of the ecological disturbances should be determined not on the basis of the set of all possible disturbances, but rather only on the basis of the set of permissible (according to socioeconomic criteria) ecological disturbances [22].

When selecting measures for improving the use of natural resources, we can use the maximization of the profitability of the corresponding resources as a criterion of their effectiveness [6].

It is seen from the above how important ecological-economic assessments are for planning and for carrying out nature management. They make it possible to develop a correct nature-preservation strategy, aimed at minimizing the ecological expenses and maximizing the final product (excluding the ecological damage). An unjustified reduction of the resources intended for nature protection, while creating an apparent growth of resources intended for consumption and accumulation, will in the final analysis lead to even greater losses [25].

As already noted, regulation of the environmental quality should include the establishment of a monitoring system, implementation of a comprehensive analysis of the state of the environment, and an economic assessment of the potential damage from the anthropogenic impact, in comparison with the cost of nature-protection measures.

While selecting the different measures to be taken (determining the priorities), that is, adapting existing enterprises to meet the necessary nature-protection requirements (construction of treatment facilities or devising different filters) and working out or developing technological processes producing little or no waste, we also have to take into account such comprehensive assessments. Here it is important to pay attention both to the effectiveness of the nature protection against impacts and to the undesirability (or unacceptability) of incurring expenses without having an adequate effect on human society and nature.

For example, while selecting a strategy for economic development in an unpolluted area or in an area already subjected to a load, we can use effectively the methods of comprehensive environmental analysis, with due regard for possible impact thresholds and the probability of various effects of the impact.

If the concept of impact thresholds is adopted, then development in an unpolluted area, up to some impact level lower than the threshold, will be absolutely harmless and should not be questioned; development in areas where the permitted limit (reserve) has already been used, up to the established level (threshold), should, however, be prohibited. If the no-threshold concept is adopted, on the other hand, then a probability calculation will be required, taking into consideration the scale of the pollution or the impact in space and the scope of the contingent of exposed organisms.

The concept of absolute prohibition of economic development in unpolluted areas (often mentioned in different papers and reports) is not scientifically justified and does not, in the author's opinion, deserve comment (except for zones with endangered objects, nature reserves, etc.).

As already mentioned, the final choice of a strategy will depend on economic considerations as well.

Therefore, regulation of the environmental quality is aimed at limiting the anthropogenic impacts (primarily pollution) leading to negative consequences. Such a limitation should be supported by strong arguments, ecological as well as economic.

The following approaches (which have in part already found practical application in some countries and which have been discussed in different works, for instance, in [12, 29]) may be considered as possible variants of the criteria for limitations:

(a) Limitation of pollution by maximum permissible concentrations (MPCs) (public-health and fishery requirements);
(b) Limitation of pollution by maximum permissible loads (MPELs) on an ecosystem, on an entire area or region, or by critical ecological loads;
(c) Limitation of industrial and other emissions, based on the MPC;
(d) The same limitation, based on the MPEL;
(e) Complete prohibition of emissions (discharges) into unique (endangered) objects in unpolluted areas;
(f) Limitations motivated by the most sophisticated engineering and technology;
(g) Limitations in accordance with the optimum decision, made on the basis of an analysis of the expenses and the damage from pollution, taking into consideration the yield;
(h) Limitations in accordance with the preceding point, but under conditions of limited economic possibilities (the optimum is shifted);
(i) An attempt to obtain the least possible emissions (below the permissible limit) taking into consideration priorities determined with the aid of comprehensive analysis.

The strategy of environmental quality regulation involves an effective use of the above criteria. The use of a combination of two or several criteria is undoubtedly the most effective. Without going into details concerning the above approaches, let us just say that a combination of the approaches described in points (c) and (d), taking into consideration point (f) and of necessity observing point (h), seems to be the most suitable. This approach leads to a gradual step-by-step lowering of the standards limiting the discharges, down to observance of the requirements in points (c) and (d). Variant (i) is very attractive, but it is probably destined for the future.

A necessary condition for the use of the approach according to point (f) (and some other approaches) is an assessment of the probable occurrence of adverse effects for man, as well as for populations of animals and plants at each stage, where the environmental quality has not yet reached the desirable level. This assessment should distinguish an area where there exists a significant danger of adverse effects, within the range of which economic factors cannot play a decisive role.

In an area of lower values of such a probability, a comparison of the expenses for reducing pollution with the benefit of preventing the damage caused by pollution will make it possible to clarify the quantitative and qualitative parameters of the steps (stages) in the attainment of the desirable environmental quality.

A limitation of pollution according to the above criteria and a reduction of the adverse effects of pollution can be achieved in different ways.

The most efficient way is to reduce the total amount of pollutants discharged into the environment (by introducing new technological processes producing little waste, by improving treatment facilities, etc.).

Intensified dispersion and dilution of discharged pollutants in the atmospheric air and natural waters is also widely used. This means is, of course, less effective. It does not reduce the total amount of pollutants discharged into nature; it is designed to decrease pollutant concentrations in certain places, and it can bring into the polluted zone vast new territories and natural environments. In this case the assimilating capacity of the biosphere is used on an ever increasing scale. For this purpose, high chimneys are constructed for emissions, and the water-dilution coefficients are increased.

Regulation of the use of the environment, and of an air basin in particular, includes the above approach. However, such a solution does not reduce the load on the natural environment as a whole, and the expenses associated with this solution are rightly classified in [7] as economic damage from anthropogenic disturbances of the natural environment (since in this case the recipients receive protection from only some of the impacts).

Sometimes, when technical measures turn out to be difficult, it is probably possible to impose taxes or fines for discharges of pollutants (or for allowing the arrival of these substances to some territory) in any amounts or in amounts exceeding a certain limit. It is likely that this approach will draw attention in the future, when the transport of harmful substances from country to country will come under international control.

However, 'selling' the right to pollute, either inside an individual country or to other countries, is, in our opinion, unacceptable.

With regard to a comparison assessment of the different variants of programs for environmental quality regulation, it should be noted that, contrary to widespread opinions, the optimum variant is not always the one that minimizes the sum of the damage from pollution and the expenses to reduce pollution; it is probably necessary here to formulate some reasonable limitations.

Note that here we define the optimum variant as one which is optimum for human society, and which is more universal than the variant which is optimum with regard to an analysis of the direct expenses and the damage from pollution [15].

Similarly, in [20] it is assumed that, when the economic optimum is achieved, the problem consists in changing perspective to the social optimum corresponding to a pollution level such that all the established environmental quality standards are observed. However, according to the authors of [20], a necessary condition for the transition from the economic optimum to the social optimum is the requirement that as simultaneously as possible scientifically justified standards will be reached for feeding the population, for providing housing, for high environmental quality, etc.

Sometimes, the economic optimum is reached at the same time as the social optimum.

In general, if the damage B and the expenses G are denoted by B_T and G_T, respectively, for the achieved level of technology, by B_e and G_e for the optimum solution with an analysis of the expenses and damage, and by B_S and G_S when the national standards are achieved (B_S can be equal to 0), then various relations between these quantities are possible, for instance,

$$B_T > B_e > B_S. \tag{3.6}$$

This situation is illustrated in Figure 3.2 which shows the economic damage $B(\sigma)$ and the expenses (or prevention costs) $G(\sigma)$ as functions of the pollution level, as well as the sum of these functions.

States corresponding to the actual environmental pollution A are indicated on the ordinate axis.

With the introduction of the most sophisticated technology at all plants, the total pollution decreases (to point T). Suppose that the economic optimum is not yet reached. (In Figure 3.2, $B_e < B_T$, which, as a rule, corresponds to the actual situation.) It is the opinion of the author that correctly devised environmental quality standards should ensure either the condition $B_S = B_e$ or else a maximum closeness of these quantities.

Under conditions of improved technology, a step-by-step decrease of B_T down to $B_T \leq B$ and $B_T = B_S$ is possible. However, the requirement that $B_T \to B_S$ practically always entails quite considerable expenses, and in some cases (for different ingredients, production types, regions) it is unrealistic at present. It is possible to satisfy the above condition with the best achievable technology (i.e., in Figure 3.2 corresponding to the interval $T_{b1} - T_{b2}$, when $T_{b2} \to 0$).

Standardization plays an important role in determining the strategy of environmental quality regulation; taking economic factors into account when introducing standards is of indisputable importance. On the other hand, it is obvious that the economic consequences of the introduction of standards directly depend on the validity of the standards. The above-described comprehensive analysis of the state of the environment and of the man/environment interaction should play an extremely important role in the verification of the standards.

3.2. Establishing Standards – the Most Important Element of Regulation

In the preceding section some approaches to the strategy of limiting environmental pollution were formulated.

These approaches may be classified into several large groups as follows:

(1) Limitations based on the compulsory observance of environmental quality standards;
(2) Limitations based on the establishment of limitations for discharges into the environment;
(3) Limitations associated with selecting and observing the economic optimum when analyzing expenses and damage;
(4) Limitations based on the results of a comprehensive analysis of the environment.

Clearly, the strategy of a limitation based on the results of a comprehensive environmental analysis is the best (optimum) approach. However, this way is the most complicated one, and at present the first group, and to a certain degree the second group, of approaches is used more often.

An approach based on public-health requirements for environmental quality played a very progressive role during the early stage of establishment of standards for environmental pollution, which was started in the USSR in the 1930s. In many countries it is the basic approach used now. In essence, such an approach is in conformity with the principle of 'zero damage'. However, when postulating the impact threshold (and such a postulate predominates when working out public-health standards), we find that this approach permits discharges into the environment of considerable amounts of pollutants σV, where $\sigma > \sigma_0$ (MPC), and V is the volume within which the discharged pollutant spreads. This approach is not always permissible from the ecological viewpoint (see Chapter 2), since it can lead to substantial hazardous loads on biological systems, due to the migration and accumulation of harmful substances.

On the other hand, the requirement for the observance of such standards in zones where there is practically no population or exposed biocenoses (e.g. coniferous and deciduous forests) is too stringent.

The concept of 'zero damage' associated with a definite impact threshold is questionable itself, since, as we broaden our knowledge about impacts and improve the measurement techniques, the impact threshold is, as it were, lowered, and thus the border of 'zero damage' is shifted.

Monitoring and interpreting the environmental pollution are rather complicated, and the results of this monitoring make it possible to judge only whether the general situation is favorable or unfavorable. It is nearly impossible to obtain from the available data any information about the pollution sources and their intensity. This information is thus inadequate for environmental regulation.

The approach associated with the limitation of discharges into the environment has some significant advantages. Such limitations could be realized in the following ways:

(1) by reducing some irrational technical procedures or technological regimes (e.g., by reducing the idling time of automobiles, by prohibiting open waste burning and opaque or black stack emissions into the atmosphere, limitation of the content of sulfur and other harmful substances contained in fuels);
(2) introducing, if possible, state-of-the-art technological processes at all plants and the best technological processes that could be achieved in the near future at some of them;
(3) imposing taxes or fines for pollution (it should be noted that this is the most irrational way; it is thought that in the Soviet Union, where there are no contradictions between personal interests and the interests of society, this means is unacceptable; in non-Soviet literature there are also rather explicit statements that the right to contaminate the environment should never be sold [29]);
(4) by introducing scientifically justified standards for discharges by plants and by sources of any kind into the environment.

These standards should be based on the observance of certain environmental quality standards (e.g., MPCs and MPELs, taking into consideration the above (see Chapter 2) approaches to working out the MPELs and the above remarks with respect to the MPC values).

Therefore, regulation of the environmental quality from the hygienic and ecological viewpoints is only possible via the introduction and strict monitoring of the maximum permissible discharges (MPDs) by pollution sources (establishment of standards for discharges and for the total impact).

The specific climate characteristics, as well as the already existing ecological load in the given region (the existing background), should also be taken into consideration when introducing MPD standards.

It is necessary to take into account, too, the above remark concerning the presence (and amount) of biological objects exposed to the impact in the zone, for which the MPD standards are introduced.

In many cases the required reduction to the permissible load values is possible by setting up temporarily compatible values (TCVs) of the discharges, with a gradual step-by-step decrease of these. To accomplish this, the TCVs can be determined on the basis of the achieved or achievable technical level (it being necessary, of course, to take account of the safety of both man and ecosystems).

Regulation of the environmental quality involves the use of mechanisms limiting environmental pollution by decreasing the emissions (discharges) of pollutants into the atmosphere, the hydrosphere, and the biosphere.

Sources of high power (high intensity) and substances representing the greatest hazard to the biosphere should be the first to be restricted. These substances and their sources are identified with the aid of a monitoring system: a system for the observation and assessment (at present and in the future) of such pollutants and their sources.

Potential future pollution sources (at projected industrial enterprises, municipal economy enterprises, etc.) are also subject to preventive limitation.

The criterion for the limitation of pollutant discharges into the environment is the need to achieve high environmental quality, ensuring human health and well-being, as well as normal functioning of the ecosystems without any negative consequences. Observance of the maximum permissible concentrations (MPCs), and of any amounts of substances in a localized area or in a biosphere element such that the maximum permissible ecological loads (MPELs) in this area are not exceeded, can serve as such criteria. In the latter case it is a question of setting up standards for the load (e.g., the discharge of some pollutant into an individual natural object (a lake or a river), a biospheric element, or even the biosphere as a whole.

Therefore, the concentrations σ_i in the environment should satisfy the following conditions:

$$\sigma_i \leq \sigma_{io} \quad \text{or} \quad \int_R \int_t \sigma_i(R,t)\,\mathrm{d}R\,\mathrm{d}t \leq q_{io}$$

where σ_{io} is the MPC of the ith ingredient in the environment; q_{io} is the amount of the substance in the given part of the biosphere, corresponding to the maximum permissible ecological load, and the concentrations $\sigma_i(R,t)$ are integrated over the time and space limiting the given element of the biosphere. The value of σ_i is not to exceed a certain threshold value at which significant changes in the biosphere are possible, while the effect of the given pollutant is linear within the entire variation range (see Section 2.4 for more details).

For instance, if it is found that the maximum permissible load on the ozonosphere corresponds to a decrease in the total amount of ozone molecules by no more than 1.0% due to the impact of different chemical substances, then q_{io} is the amount of the given substances in the troposphere that, when diffused into the stratosphere over a long period will lead to a decrease in the total ozone in the ozonosphere by no more than 1.0%.

When limiting the pollution sources, it is necessary to be guided by the above criteria, here taking into account the dispersion of the substance in the environment, and the transport and transformation in the different parts of the environment.

To restrict the discharges (emissions) of pollutants, the values (standards) of the maximum permissible discharges (MPDs) into the environment are introduced.

In any case, when working out the MPDs, it is necessary to know (to study) the possible form of the relation between the concentration of the substance in the natural environment σ_{il} (for the ith substance in environment l) and the discharge Q_{il}:

$$\sigma_{il}(R,t) = F[Q_{il}(R,t), v_R, w_R \ldots], \tag{3.7}$$

where F is a function determining the dependence of σ_{il} on the discharge Q_{il} and on the meteorological parameters and characteristics (wind speed, eddy diffusion, precipitation, etc.), as well as the dependence of the effect on a biosphere element on the concentration or total amount of the pollutant in the environment and on the time of impact:

$$\eta_{il}(R,t) = f(\kappa_{il}), \tag{3.8}$$

where in the simplest case $\kappa_{il} \to \sigma_{il}$, and in the general case

$$\kappa_{il} = \int_R \int_t \sigma_{il}(R,t)\, \mathrm{d}R\, \mathrm{d}t. \tag{3.9}$$

The quantities σ_{il} and η_{il} can be correlated with the MPC and MPEL values, thereby establishing, in view of all the above considerations, the amounts of permissible discharges Q_{il}.

The described approach can be used to establish standards for the maximum permissible discharges in any environment, when pollutants spread through air and water, using the MPC and MPEL criteria for the individual media and for several media (allowing for pollutant transformation and migration).

When discussing all the above approaches to pollution limitation and when considering the requirements of obligatory (or nonobligatory) observance of the MPC and MPEL values, we have to know the ecological reserve of the given ecosystem (reserves of human health, and of the health of any other organism, population, or ecosystem as a whole). Observance of the MPCs and MPELs guarantees the use of the potential of man and the ecosystem within the framework of the ecological reserve, at a level far from the critical values of the state of the organism (or organisms).

However, there apparently exists for any organism or ecosystem some wider (extended) ecological reserve which makes it possible to work out more flexible MPCs, not including the critical states, but still satisfying the technical and economic requirements. The supposition of the existence of such an extended ecological reserve for any organism or ecosystem makes it possible to formulate the concept of a step-by-step decrease of the MPDs, taking into consideration the technical and economic feasibilities, down to values that would satisfy the MPCs and MPELs. The MPD values may refer to individual pollution sources (gaseous, liquid, and solid discharges or waste), a group of sources, individual zones, areas, or even regions. For instance, a field can become a pollution source after the application of fertilizers, when the latter are washed away or when some of the chemical substances get into agricultural products, etc.

When considering the load on a forest or a lake, we could even take an entire city, district, region, or even country as a pollution source. For instance, in the case of long-range pollutant transport (over hundreds or thousands of kilometers), a country, or even a continent, might be considered as a pollution source.

The establishment and introduction into practice of MPDs is a complicated project, which must take into account the physical-graphical peculiarities of the given area, its ecological state, the size and distribution of population, the technological level of the given production, and many other factors. Naturally, the MPC values may be different, even for the same ingredients in different regions and at different times.

The establishment of discharge standards per unit production also seems very useful, keeping in mind how important it is, when working out the MPDs, to take account of the achieved (or achievable) technological level of some production process. These standards are not basic (the basic standard should be the standardized MPD for the pollution source, ensuring that there is no danger to the health of the population and ecosystems, as well as high environmental quality), but they may be extremely useful for imposing restrictions on the pollutant discharges in one branch; such a mean standard for a branch can be established for reasons of orientation for newly constructed industrial enterprises, for already existing plants of different categories, etc. When, during the establishment of MPD standards for a pollution source, it is necessary to take account of the achieved (or achievable) technological level, this level can be expressed quantitatively by the establishment of standards for discharge into (or input to) the environment per unit production, for industrial enterprises, per unit traveled for motor transport, etc. This approach has already found practical application in some countries (USA, Sweden, etc.).

Especially important for the limitation of pollutants and for the regulation of environmental quality is the ecological-economic approach, together with the use of comprehensive environmental analysis.

Similarly, an analysis of possible anthropogenic climate changes and fluctuations, as well as the causes and sources of anthropogenic impacts associated with climate changes, makes it possible to formulate the limiting values of these impacts at which the climate and the climate system will lie within the range of natural fluctuations (e.g., within the time interval of the last century) on global and local scales [6].

It is obvious that such criteria may be formulated for a complex of impacts and individual most effective factors (the ratios of the intensities of individual factors may be different for an unchanged effectiveness of a complex impact). For instance, these criteria may refer to the limiting concentration of carbon dioxide in the atmospheric air, to the rates of its arrival in the atmosphere, to some release of anthropogenic heat and various pollutants (e.g., halogenated hydrocarbons) into the environment, or to a certain dust content of the troposphere and stratosphere, taking into account the actual complex impact of various factors on the climate system.

The formulation of these criteria involves a determination of the critical (potentially most hazardous) factors of the impact on the climate system and a determination of the permissible impacts on the most vulnerable biosphere elements and on the ecosystem as a whole.

Obviously, in order to take effective measures for limiting environmental pollution on global and regional scales, it is necessary to establish a system for setting up ecological standards on an international level. Such actions have already been outlined at the Convention on Transboundary Long-Range Air Pollution (1979). They were further developed at a multilateral conference (Munich, 1984) on determining the causes of and preventing damage to forests and water bodies resulting from air pollution in Europe. At this conference delegates from many countries expressed their willingness during the next 10 years to reduce by 30% the emissions or transboundary flows of sulfates from their territories.

As already noted, in order to establish the standards, it is necessary to distinguish the critical factors of the impact and the critical elements of the biosphere and biota with respect to this impact. On a local scale, this is first of all human health, while on global and regional scales the critical element of the biosphere is often the natural ecosystems (e.g., in the case of acid rain) or even abiotic elements (e.g., the atmosphere itself in the case of elevated ionization due to the impact of radioactive emissions).

For instance, in Section 8.3 the impact of the emission of radioactive isotope ^{85}Kr on the atmospheric air is described. The ^{85}Kr concentration in the atmosphere, equal to 3 nCi/m^3, is shown to alter the total resistance of a column of atmosphere by 10%, which is a 'disturbing situation'. It is also estimated there that such a global concentration is established when 750 Ci of ^{85}Kr per year are emitted into the atmosphere. Probably this value should also be assumed as a criterion for establishing ecological standards for the above-given most important pair, namely a critical factor (a critical biosphere element): radioactive emission of ^{85}Kr, the variation of the electrical properties of the atmosphere (but not the radiation effect on people and ecosystems).

In the event of long-range transport of conversion products of sulfur (or nitrogen) oxides, and acid rain on an international scale, it is necessary to establish standards only for external emissions (i.e., beyond the limits of the source territory), since the establishment of standards for 'internal' pollution is a probem of national competence [18].

In order to determine the external pollutant flows, a matrix is constructed (see Chapter 9), with the aid of which the flow from any country of the region to any other is determined. Such matrices for the European region are prepared at the East-European and West-European synthesizing meteorological centers. If we define, in accordance with the adopted model, the operator S_i enabling calculation of the load L_{ij} in the ith country, induced by the amount of the substance q_j^{ext} going beyond the boundaries of the jth donor-country (and reaching the ith country):

$$L_{ij} = S_i(q_j^{ext}), \tag{3.10}$$

then we obtain the criterion for establishment of standards

$$\sum_{j=1}^{n-1} J_{ij} = \sum_{j=1}^{n-1} S_i(q_j^{ext}) \leq \beta L_i^{lim}$$

where L_i^{lim} are the limiting loads on the critical environment in the ith country and β is determined by international agreements ($0 \leq \beta \leq 1$).

Thus, to set up ecological standards for discharges in a region, it is necessary [17, 18] to work out common ecological standards of the MPEL (L_i^{lim}) and quotas of pollution (loads) (β), and to have unified calculation methods and a common regional network of observations of transboundary flows (the two latter components have already been developed and are available now).

Consequently, the ability to correctly distinguish critical effects and impacts using a monitoring system (in combination with modeling) and to take timely measures to limit these, within the framework of the indicated criteria, will eliminate the very possibility of undesirable, or even hazardous, anthropogenic changes in the state of the environment.

PART II

The Principles of Monitoring

Monitoring anthropogenic changes

Ecological monitoring

Climate monitoring

Implementation of monitoring systems

CHAPTER 4

Monitoring Anthropogenic Changes

4.1. Definitions. The Principal Goals and Methods of Monitoring

Information about the state of the natural environment, and about changes in it, has been used by man for a long time. Regular observations have been carried out for more than a hundred years now, and these include meteorological, phenological, and other observations.

With the development of industry and engineering, when man became more able to exert an influence on nature, to transform it, and to use natural resources more intensively, geophysical information became ever more important and necessary. With the aid of this kind of information it is possible to determine the optimum natural conditions for the implementation of different measures, to predict both favorable and unfavorable factors for economic development, and to take measures to reduce the effect of unfavorable conditions on human life and activities. This information should include observational data on the actual state of the environment and forecasts of changes in the natural conditons.

For a long time the observations were only of variations in the state of the environment resulting from natural causes. In recent decades, however, man's impact on the environment has sharply increased. It has become obvious that the uncontrolled exploitation of nature can lead to very serious negative consequences. In this connection, detailed information on the state of the biosphere has become even more necessary.

It is known that the state of the biosphere changes due to natural and anthropogenic impacts. However, there is a significant difference between the results of these impacts; when the state of the biosphere continuously changes due to natural factors, it returns, as a rule, to its initial state. The temperature, pressure, air humidity, and soil moisture tend to mainly hover around some relatively constant mean values. Seasonal changes in the biomass of plants and animals are examples of these natural variations. The mean values characterizing the state of the biosphere (its climatic characteristics in any region of the world, the natural compositions of different parts of the environment, the cycles of water, carbon, and other substances, and the global biological productivity) change appreciably only over a very long time (over thousands of years, and sometimes even over hundreds of thousands or millions of years). Large equilibrium ecological systems and geosystems also change extremely slowly under the influence of natural processes. These gradual evolutionary changes occur only during time periods measured by historical epochs.

Unlike changes in the state of the biosphere induced by natural causes, the changes due to anthropogenic factors can occur very fast. For instance, the changes due to such causes during the last few decades are comparable with some of the natural changes which occurred during thousands, or even millions, of years.

Both short-term and long-term natural changes in the state of the environment are to a certain degree observed and studied by the geophysical services of many countries (hydrometeorological, seismic, ionospheric, gravimetric, magnetometric, and other services).

In order to distinguish anthropogenic changes against the background of natural changes, it became necessary to set up special observations of the changes in the state of the biosphere occurring under the influence of human activities.

A ***monitoring system*** is defined as a system of repeated observations of one or more elements of the natural environment in space and time, for definite purposes in accordance with a previously prepared program [43].

The term 'monitoring system' appeared prior to the UN Conference on the Environment (Stockholm, 5–16 June 1972). The first proposals concerning such a system were set forth by a special SCOPE commission (Scientific Committee on Problems of the Environment) in 1971, and this system was mentioned in the recommendations of the Stockholm Conference. The basic elements of monitoring are described in [43]. The term 'monitoring' was probably introduced to counterbalance (or supplement) the term 'control', which implies not only observation and data gathering, but also an active regulation of the environment.

At that time the monitoring of pollution was discussed; the monitoring of natural resources was given little attention. Later, a number of working groups discussed the problem of monitoring as a whole. However, some serious contradictions could be seen in the conclusions and proposals of these groups. Many of them included the duties of already existing geophysical services in the duties to be carried out by the monitoring system. It was suggested to establish a universal Earth Service, that would supply information on any changes in the state of the natural environment in the world, but then the duties of this Service and those of the monitoring system were confused.

In the USSR, discussions of a monitoring system livened up before the First Intergovernmental Meeting on Monitoring, convened in Nairobi (Kenya, February 1974) by the Governing Council of the United Nations

Environment Program (UNEP); it should be stressed, however, that observations of many changes in the biosphere caused by anthropogenic factors had already been made earlier, in particular by the Hydrometeorological Service of the USSR.

The author of the present book described the differences between the above concept of a global observation system and the scientific principles of monitoring in a paper presented at the above-mentioned Nairobi meeting (some items of the paper were included in the proceedings of the meeting) and in papers presented at the First Soviet-American Symposium on the Comprehensive Analysis of the Environment (Tbilisi, March 1974) and the first meeting of the group of experts on Project No. 14 of the Man and the Biosphere program (Moscow, April 1974). The main contents of these papers were published in *Meteorologiya i gidrologiya* [8] and in the proceedings and materials of the meetings [9, 49]. Later, these problems were described in more detail [21, 37].

In 1975 an article by Academician I. P. Gerasimov on the scientific basis of present-day environmental monitoring was published [4].

Although the authors of [8, 10, 43] used the definition of monitoring mainly as applied to anthropogenic changes in the natural environment, this term was interpreted by some participants of the international meetings in a very broad sense. For instance, it was once again suggested that the existing geophysical services (their duties) be integrated into a monitoring system; at the First Intergovernmental Meeting on Monitoring (1974) the problem of 'economic', 'scientific' monitoring was posed (in this case not an economic or scientific basis for monitoring was meant, but rather the integration by the monitoring system of systems of observation or gathering of information on economic progress, scientific potential, and current research in different countries). These proposals were not accepted at the meeting.

The basic ideas and objectives of the Global Environmental Monitoring System (GEMS) were described at this same meeting [51], where the participants focussed their attention, on the one hand, on warning about changes in the state of the environment associated with pollution, and, on the other hand, on warning about threats to human health and the threat of natural calimities; ecological problems were also considered.

At the intergovernmental meeting on monitoring a list of major pollutants was prepared, so that they could be taken into consideration when establishing a monitoring system; it was also decided to monitor the parameters needed for the interpretation of pollution data. The participants agreed that international cooperation in the development of global monitoring should be based on the existing national and international systems, and that specialized United Nations agencies be used intensively for the coordination and implementation of monitoring programs. The majority of the resolutions adopted at this meeting were approved at the Second Session of the UNEP

Governing Council and obtained recognition [44].

However, a number of the goals of a global monitoring system, which are in conformity with the interests of both developed and developing countries, nevertheless introduce a certain vagueness into the distribution of functions among the already existing systems (e.g., the World Weather Watch of the World Meteorological Organization) and the suggested monitoring system. Perhaps it was hoped to show, as a result of the necessary work, the systems developed earlier and functioning already for a long time, instead of directing efforts to fill in gaps in the existing information systems, associated with the absence of data on changes in the state of the natural environment due to anthropogenic impacts. Of course, this work should be based on the experience, structure, and network of the existing geophysical services.

In the author's opinion, the meaning of the suggested term 'monitoring system' should be specified in detail.

In [8, p. 4] it was stated that "It would be more correct to define a monitoring system as an observation system making it possible to distinguish changes in the state of the biosphere caused by human activities." This definition of a monitoring system was also supported by Academician I. P. Gerasimov [4] and by Munn [43].

To be more exact, such a system should be called *a system monitoring anthropogenic changes in the natural environment.* In a joint work by the author and R. E. Munn [39], some compatible views on monitoring anthropogenic changes in the state of the environment and renewable resources are presented.

Thus, the term 'monitoring system' is not just a new designation of already existing geophysical services, but rather will refer to a system which is synthesized to distinguish anthropogenic effects in the environment using the information and some elements of existing geophysical services.

When considering the main goals formulated for the Global Environmental Monitoring System (GEMS), with regard to isolating changes in the environment induced by anthropogenic impacts, it is impossible to discover any contradictions in the stated goals of a monitoring system in the above definition. However, the recognition and prevention of natural meteorological and hydrological calamities are the job of existing meteorological and hydrological services, while disease prevention lies within the province of the appropriate public-health services. These services should, of course, be improved and developed, but it is doubtful whether they should be integrated into a unified monitoring system.

It should be noted that a system for monitoring anthropogenic changes in the environment is not a fundamentally new system, requiring the establishment of a new observational network, telecommunication lines, data processing centers, etc. Rather, it forms a part of the universal system of observations and control of the environment, which has already been developing in the Soviet Union and other countries for many years (hydrometeorological or meteorological services). It might also become a part of, or be based to a certain degree on, the World Weather Watch of the World Meteorological Organization and other services.

Thus, a pollution monitoring system can be, and must be, a part of the already existing service of environmental observation and control, and it must use the experience, observation network (including measurements of new elements, of course), telecommunication lines, and data-processing centers of the existing service, with the development of some elements.

As already mentioned, in order to maintain a system of environmental observation and control, whereby changes induced by anthropogenic factors are distinguished, we must have detailed data on the natural changes and fluctuations of the state of the environment; the implementation of a monitoring system envisages the acquisition (or availability) of such information.

Monitoring comprises the following basic activities [8]:

(1) observation of factors influencing the environment, as well as observation of the state of the environment;
(2) assessment of the actual state of the environment;
(3) forecasting the state of the environment and assessing this state.

Consequently, a monitoring system is a system of *observations, assessment, and prediction* of the state of the natural environment, not including regulation of the environmental quality. It is obvious, however, that a necessary condition for a correctly organized management of the environmental quality is the establishment of a monitoring system.

A monitoring system may cover either local areas or the world as a whole (global monitoring). A basic feature of a global monitoring system is the possibility of assessing the state of the biosphere on a global scale, on the basis of the data of this system.

A national monitoring system commonly refers to a monitoring system within the framework of one state; such a system differs from global monitoring not only by the scale but also by the fact that the main goals of national monitoring are to gather data and to assess the state of the environment with national interests in view. For instance, an elevated level of atmospheric pollution in individual cities or industrial centers may be insignificant with regard to the state of the biosphere on a global scale, but it may be of importance with regard to taking measures in the given area, at a national level.

A global monitoring system should naturally be based on the national monitoring subsystems, and it should include their elements (there is no need to fully include these subsystems in the global system, however, since their sphere of activity covers only strictly national problems).

Sometimes 'transboundary' monitoring or 'international' monitoring is referred to. It would probably be more correct to use this term for monitoring systems serving the interests of several countries (in order to solve problems of transboundary transport of pollutants from country to country, etc.).

Thus, a monitoring system is a multipurpose information system. Its main goals are: observation of the state of the biosphere, assessing and forecasting its state, determination of the degree of the anthropogenic impact on the environment, and determination of the factors and sources of this impact as well as the degree of impact.

The main goals of a monitoring system, and various aspects associated with its planning and implementation, were considered in detail at the international symposia on the integrated global monitoring of environmental pollution, held in Riga (December 1978) [18] and in Tbilisi (October 1981) [20].

Let us consider a universal plan of an information system for monitoring the state of the natural environment, suitable both for the system as a whole and for any geophysical service forming a part of this sytem (a hydrometeorological service or a pollution-observation system, monitoring either pollution or the anthropogenic changes in the biosphere). Since pollution monitoring is a relatively new element in the system for observing the state of the natural environment, let us consider it in more detail; we will explain the universal plan mainly as applied to monitoring environmental pollution.

The most universal approach to determining the structure of the system for monitoring anthropogenic changes in the environment is to divide it into the following blocks: 'Observations'; 'Assessment of the actual state'; 'Prediction of the state'; and 'Assessment of the predicted state' (the existing geophysical services were designed according to this same scheme).

The scheme given comprises all the above blocks (subsystems). This approach was described in [8, 11, 13]. The components and blocks of the system for monitoring anthropogenic changes in the environment will be considered below in more detail.

Figure 4.1 shows the individual blocks of the described system, as well as the feedforward and feedback relations between these blocks.

The 'Observations' and 'Prediction of state' blocks are closely interrelated, since it is possible to forecast the state of the environment only if adequately representative information on the actual state is available (feedforward relation). The preparation of forecasts, on the one hand, implies a knowledge of the laws governing changes in the state of the environment, availability of a scheme, and the possibility of numerical calculations, while, on the other hand, the orientation of the prediction should to a great extent determine the structure and composition of the observation network (feedback relation).

Data characterizing the state of the environment, obtained from obser-

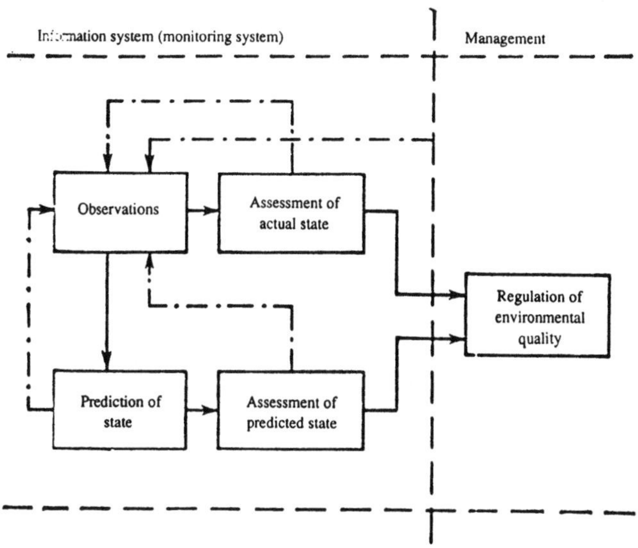

Figure 4.1. **Block diagram of monitoring system.**

vations or forecasts, should be assessed depending on the field of human activity in which they are used (with the aid of specially chosen or developed criteria). Assessment implies, on the one hand, determination of the damage from the impact and, on the other hand, selection of optimum conditions for human activities and determination of existing ecological reserves. In assessments of this kind it is understood that the permissible loads on the environment are known.

Information geophysical systems, as well as the information system for monitoring anthropogenic changes, are a component of the system of environmental management, of man/environment interaction, since information on the current state of the environment and on the trends of its changes should form a basis for the program of environmental protection and should be taken into account during economic planning. The results of the assessment of the current and forecasted state of the biosphere, in turn, make it possible to define more exactly the requirements for the subsystem of observations (this forms the scientific basis of the monitoring system, determining the composition and structure of the network and the methods of observation).

Figure 4.2 shows the place of the monitoring system in the system of environmental management (control) [10]. The energy and information flows in the diagram are arbitrarily aligned with one another.

An element of the biosphere with a level of state B, subjected to an anthropogenic impact (A), changes its state ($B \to B'$).

With the aid of the monitoring system (M), a 'picture' of this altered state (and, if possible, also the initial state) is obtained, the data are generalized,

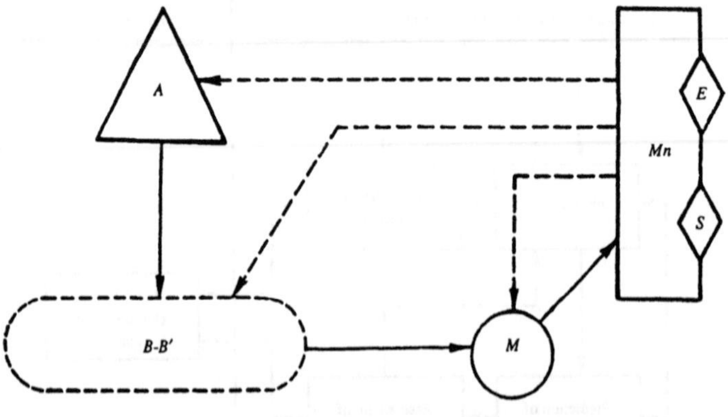

Figure 4.2. Place of monitoring system in system of environmental management.

and the actual and forecast states are analysed and assessed. This information is transmitted to the management (decision making) block (Mn); see also Figure 4.1. On the basis of this information, depending on the level of scientific and technical development (S) and economic possibilities (taking into consideration the ecological and economic assessments) (E), measures are taken to limit or curtail the anthropogenic impacts and to effect a preventive 'strengthening' or subsequent 'curing' of the element of the biosphere. Undoubtedly, a combination of the above approaches is possible, and the monitoring system is also being improved (the indicated actions are shown in the diagram by dashed lines).

Note that, since the assessment of the actual and forecast states of the environment is a component of the monitoring system (see Figure 4.1), some investigators identify this part of the monitoring with an element of the environmental management.

Observations of the state of the environment should include observations of the sources and factors of the anthropogenic impact (including sources of pollution, radiation, etc.) and observations of the state of biosphere elements (including the responses of living organisms to the impact, as shown by changes in their structural and functional indices; see Figure 1.3). This implies, of course, the acquisition of data on the initial (or background) state of the elements of the biosphere.

The above approach provides for tracking of the entire cycle of anthropogenic impacts: from the impact sources right up to the influence on and reaction of individual parts of the environment and complex ecological systems (Figure 4.3). Figure 4.3 shows a classification of the successive monitoring 'stages' [10]. Classification of the monitoring system and all its possible directions constitutes a complicated and cumbersome problem; this problem will be considered in Section 4.5.

The assessment of the state of the environment implies a comprehensive

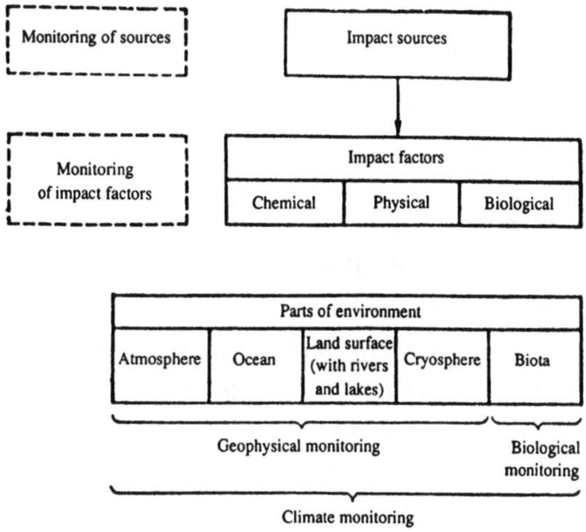

Figure 4.3. Diagram and classification of monitoring system.

analysis of the state induced by the impact of various factors in different parts of the environment (these factors often act simultaneously and intensify the effect of the impact). For instance, if we consider the impact on the biotic component of the biosphere, then the ecological damage will obviously depend on the kind of impact, its intensity, and the quantity and types of organisms subjected to the impact. For an integrated assessment of the state of the environment and a determination of the dynamics of this state, simultaneous meteorological, hydrological, and some biological observations should be carried out, making it possible to correctly interpret and distinguish the anthropogenic changes against the background of natural processes.

Some works dealing with the development of systems for monitoring anthropogenic changes have appeared in recent years. Several important papers were presented at the Second International Symposium on Integrated Global Monitoring of Environmental Pollution (Tbilisi, October 1981) [19]. At the ninth session of the UNEP Governing Council (May 1981), it was noted that the objective of the International Symposium on Integrated Monitoring should consist in the development of proposals aimed at imparting a truly global character to the Global Environmental Monitoring (GEMS) activities.

Presented at this symposium were papers on the establishment of a system for integrated environmental monitoring [1, 6, 14, 16, 25]; these elaborated ideas set forth in the papers presented at the First Symposium (Riga, December 1978 [18]) [3, 10, 20, 22, 26].

At the symposium papers were also presented on monitoring the background environmental pollution [17, 29], on monitoring the impact of pollution on land and marine ecosystems and climate, on the establishment of standards for environmental quality and anthropogenic loads [24], on modeling the spreading of pollution and the behavior of ecosystems, as well as papers dealing with the assessment and prediction of the impact of pollution on the state of ecosystems and the use of different observation methods.

It was stressed in the recommendations of the symposium [23] that, for the establishment of the GEMS network, it will be necessary to use as much as possible the already existing BAPMoN (WMO) stations [54] and biosphere reserves, and to organize observations according to the GEMS Program in conjunction with existing regional background monitoring systems.

A European monitoring system is an important component of the GEMS. The need for its establishment was indicated in the Convention on transboundary air pollution over great distances. The first results of integrated observations in regional GEMS subsystems have already been obtained, especially in the region of the European countries (Comecon countries), the USSR, and the USA.

In a special report *The World Environment 1972–1982* [55], prepared by the UNEP for the decade, it is stated that "By the end of the decade in the USSR particular attention was given to the development of integrated monitoring systems, in which observations are carried out in all parts of the environment and are coordinated in space and time. This approach was recognized as an effective basis for the harmonization of different monitoring programs".

In [20] Munn notes that joint efforts by specialists in hydrometeorology, plant physiology, limnology, soil science, etc., are necessary for the development of an integrated monitoring system. It is gratifying to note that this process of joining forces has already started in the USSR.

The correctness of such methods was also confirmed in the papers dealing with the development of monitoring systems, presented at the First International Congress on Biosphere Reserves (Minsk, September 1983) and at the First International Symposium on Integrated Global Ocean Monitoring (Tallin, October 1983).

In the following sections the above subsystems, i.e., the observation, prediction, and assessment of the state of the environment, will be described in detail; special attention will be given to monitoring anthropogenic pollution.

4.2. Observations of the Change in the Biosphere and of Impact Sources and Factors

In order to analyze and forecast the ecological situation on both a global and a regional scale, it is necessary to know about many geophysical processes, together with the different anthropogenic effects and situations responsible for them. This pertains first of all to detecting and studying the anthropogenic factors of the impact.

The acquisition of initial data and the primary analysis of these are provided by a background monitoring system, which is being estabished on a national level in a number of countries, as well as by the global monitoring system under the auspices of UNEP.

The complex of anthropogenic factors affecting the state of the biosphere and human health are extremely multifarious. They include pollution of the environment by various chemical substances, emission of inert material, physical (including mechanical) impacts disturbing the natural cover of the planet, heating of the biosphere, extraction of renewable and nonrenewable resources, etc. [39].

The effects and results of anthropogenic impacts leading to a change in the basic elements of the biosphere, and to large-scale geophysical and ecological changes, are even more multifarious. Anthropogenic impacts, either directly or via the above changes, affect human health and well-being and they result in various socioeconomic changes. A list of the effects of the basic anthropogenic factors on the state of the biosphere and on human health and well-being was presented in Table 1 of the Introduction.

Below we will consider the main factors, elements, and processes requiring a thorough observation and study. A classification of the factors, states, and processes (which should be observed within the framework of the monitoring system) is presented in Table 4.1.

Observations of local sources of impacts and pollution, as well as of impact factors, are placed into a special section of the observations (section A). Such sources can be natural (volcanic eruptions, spontaneous discharge of gases, oil, etc.) or anthropogenic (industrial discharges; agricultural sources, such as livestock farms and fields after treatment with fertilizers and pesticides; air, water, and land transport; municipal pollution sources, etc.).

Observations of the factors of impacts (mainly anthropogenic) should receive primary attention. It is difficult to analyze the state of the environment and to distinguish the causes of its changes without a thorough study of such impact factors as different pollutants, radiation, etc. Observations of impact factors are also included in the section of observations of the state of the environment (section B), since they sometimes characterize the state of the environment (e.g., according to geochemical data).

Observations of extraterrestrial sources of different impacts (solar flares, solar and galactic fluxes of particles), which should be taken into account in the assessment and prediction of the state of the environment, are made by heliogeophysical and astronomical services.

Section B comprises observations of the state and changes of the environment, using geophysical data obtained from successive or continuous measurements of the parameters characterizing the instantaneous state of the

Table 4.1. Classification of states of environment and public health, reactions of natural systems, sources, and factors of impact, covered by monitoring system.

Observations	Classification
A Impact sources and factors	A.1. Pollution and impact sources
	A.2. Impact factors (pollutants, radiation, etc.)
B State of environment	B.1. State of environment characterized by physical and physicogeographical data
	B.2. State of environment characterized by geochemical data and data on composition and character of pollution
C State of biotic component of biosphere	C.1. Reaction of biota: responses and consequences for: (a) individual organism; (b) population; (c) community or ecosystem
D Reaction of large systems and biosphere as a whole	D.1. Reactions of large systems (weather and climate)
	D.2. Reaction of biosphere as a whole
E State of human health and well-being*	

environment. Such observations are already being carried out by a number of geophysical services. Observations of natural disastrous phenomena (volcanism, earthquakes, tsunami, droughts, hurricanes, floods, mud slides, snow avalanches, soil erosion, etc.) are included mainly in the sphere of activity of the above services. *

Physicogeographical data, including data on the distribution of land and water, on the relief of the Earth's surface, on natural resources (mineral, land, plant, water, and animal resources), on population, on urbanization, etc., also present very important information about the state of the environment.

Section B covers observations of the state of the environment (and its changes) characterized by geochemical data, i.e. observations of the cycles of substances in nature, of the content of foreign admixtures in the biosphere (including radioactive substances), of different specific physical characteristics of the environment, including observations of noise and heat pollution (the concept of pollution in the latter two cases is rather arbitrary, but commonly used), and of different kinds of radiation (ionizing and nonionizing radiations).

* Monitoring human health constitutes a special section of research. In this Monograph mainly sections A–D are considered.

The following observations also belong to section B: observations of the chemical composition (of natural and anthropogenic origin) of the atmosphere, precipitation, surface and ground water, ocean and sea water, soil, bottom sediments, vegetation, and animals, as well as observations of the principal paths of pollution spreading. These observations are usually considered the most important ones made in the monitoring system. Note that a special report on the Global Environmental System [43] discusses the monitoring of different ingredients in the natural environment. It should be noted that similar pollutants may become the sources of pollution in other parts of the environment. However, solely local sources are placed in a separate section (section A).

Section C includes observations of the reactions of the biota (the living component of the biosphere) to different factors of impacts and changes in the state of the environment. Some such are observations of the responses (reversible changes) and consequences (irreversible changes) observed in the biota. Observations of the functional and structural biological indices are possible. The increase of biomass per unit time and the uptake rate of different substances by plants and animals are examples of functional indices, while the abundance of plant and animal species and the total biomass are structural indices. These observations should be made at different levels: of an individual organism, of a community and of an ecosystem.

Man occupies a particular place in nature. Observations of man's responses to different impacts should be placed into a separate section, in view of its importance and its specific nature. Monitoring of factors associated with human health also includes observations and assessments of environmental conditions affecting human health, the spreading of different diseases, etc.

Section D – observations of responses of large systems (weather, climate) and of the biosphere as a whole – includes the entire system of observations listed in Sections B and C, as well as observations of the state of the climate system, and these require special generalizations and assessments.

Observations arranged according to a certain system and aimed at tackling the above-mentioned problems can be carried out with respect to physical, chemical, and biological indices. In order to determine the dynamics of the state of the biosphere, measurements should be repeated at regular time intervals, and with respect to the most important indices they should be continuous. An observation system can be arranged on the basis of point measurements (at stations), including remote observations, or on the basis of areal surveys and the use of integral indices; a combined use of these principles is possible and convenient. Data of areal surveys and integral indices should play a very important role in a monitoring system. Strong emphasis should be placed on the use of aircraft and satellite observations. When considering and analyzing the results, it is important to distinguish changes in the state of the environment, and the response of the biota to these changes, occurring due to the anthropogenic impact. For this purpose, it is important to know the initial state of the environment, i.e., the state prior to man's substantial interference. This initial state can be partially reconstructed using the results of observations carried out over an extended time period, as well as from data on the composition of bottom sediments, glacial layers, or tree rings, referring to a period prior to the beginning of man's pronounced influence on the environment.

While studying the anthropogenic effect on the biosphere, we should determine the global background state of the biosphere at present in places remote from any local sources of the impact (pollution), as well as the background state characteristic for each region (these levels can differ significantly).

4.3. The Assessment and Prediction of Anthropogenic Changes in the Biosphere

An assessment of changes, and tendencies of changes, in the state of the environment should describe an unfavorable situation, should indicate the cause of this situation, and should help determine the actions needed to restore or normalize the situation or, on the contrary, should indicate particularly favorable short-term or long-term situations and the existence of natural possibilities for the effective use of the available ecological natural reserves in the interests of man.

Thus, to assess the state of the environment taking into consideration anthropogenic changes, it is necessary to know how, on the one hand, to determine the potential damage from natural and anthropogenic impacts and, on the other hand, to recognize the additional natural possibilities that could be used in the interest of man. For this purpose, it is necessary to know the maximum permissible loads on the environment and the ecological reserve of the given ecosystem. Let us consider these issues in succession.

It is possible to distinguish arbitrarily between the ecological, economic, and esthetic damage [9]. It is not difficult to see that the ecological damage must depend directly on the degree of impact of different factors on the biosphere; the economic and esthetic damage will to a great extent be related to the ecological damage.

First let us consider the problems of ecological damage and ecosystem reserves. The ecological damage from an impact is determined by the deviations from a certain state (characterizing the limit of the permissible impact) of an ecosystem, a community, or a population exposed to the given impact.

Unfavorable phenomena can already occur for a slight deviation from the optimum state, but with different probabilities (the more the deviation, the higher the probability); if the critical level is exceeded, the probability is very large (beyond the critical level, 50% or 100% probability of the occurrence of irreversible changes can be assumed).

In order to better understand the problem of deviations of the state from the optimum, let us introduce a certain function of the state of an ecosystem or of some other element of the biosphere (see Section 2.1), characterizing, for instance, the productivity, the rate of exchange of matter and energy in the system, or a combination of these or other similar factors corresponding to the structural and functional indices of the ecosystem [12]. This function $\eta(\mathbf{R}, t)$, which changes in space and time, may be written for an ecosystem as a whole or for any population, and then it can be considered within its permissible limits.

Problems of stability of the variables are considered in [8]. In order to assess the stability of an ecosystem, a universal function is introduced; this function, called the measure of homeostasis, is equal to the ratio of the functional indices of the ecosystem to the structural indices.

It is obviously possible to determine the critical and maximum permissible values of the function η_c and η_p, which should be different (although sometimes coincidence of these levels is possible). In most cases, function η_c (as well as η_p) has two sets of values, one maximum and one minimum, and the interval of permissible (and optimum) values lies between them; a change of η with time is caused by a change in the external natural conditions (temperature, moisture supply, etc.).

Clearly, any deviation of the ecosystem from the actual state, due to the impact of external factors, will be determined by the degree of the impact of these factors on the n-ecosystem.

Then it is possible to write (similarly to formula (2.4)) that for the given impact A_n the state of the ecosystem is transformed from the initial state (the background state or one already somewhat altered compared with the background) characterized by a function η_0 to the actual (changed) state characterized by a function

$$\eta_a = A_n \eta_0. \tag{4.1}$$

Ecological damage is associated with a deviation which causes reversible or irreversible damage to the ecosystem. In this case possible 'breakdowns', or changes at the populations level or the community level, serve as an index of the ecological damage.

The difference between the actual and maximum permissible (and sometimes critical) states of the ecosystem can be called the ecological reserve of this system.

If we consider the impacts on the biotic component of the biosphere, it is obvious that the ecological damage will depend on the answers to the following questions: How many (and which) organisms will be subjected to the impacts? What will be the doses and intensity of the impacts?

It can be stated that the degree of the impact on the living component of the biosphere depends on the intensity I of the factor affecting the biosphere (pollutant concentration, radiation intensity, etc.), on the biologically harmful effect ϵ (e.g., the toxicity for the given population and ecosystem), which may depend on I nonlinearly, on the number of organisms $N_{n,m}$ of the m population of the n ecosystem subjected to the impact, and on the sensitivity K_m of the m population in the ecosystem to the given impact [11]. When assessing the impact, we have to take into consideration the simultaneous impact of all the factors acting in different parts of the environment.

The pollutant concentration, which varies in space and time, $I_{il}(\mathbf{R},t)$ (or the intensity of the impact of some factor) is a function of the sources Q_i $(\mathbf{R},t)$ and depends on the dispersal and the transfer, i.e., on the hydrometeorological or physical characteristics of the environment

$$I_{il}(\mathbf{R},t) = F(Q_i, v_{\mathbf{R}}, w_{\mathbf{R}}, v_z, \ldots) \quad \text{(for the atmosphere)} \tag{4.2}$$

where $v_{\mathbf{R}}$, $w_{\mathbf{R}}$, v_z are the wind speed, the eddy diffusion, and the rate of gravitational settling, respectively.

In turn

$$Q_i(\mathbf{R},t) = \sum_j q_{ij}(\mathbf{R},t), \tag{4.3}$$

where q_{ij} is the amount of the ith substance discharged by individual j-sources. The distribution of $I_i(\mathbf{R})$ depends to a great extent on the spatial distribution of the sources.

When assessing the degree of the impact, we assume that we know the laws governing the transformation of the given ingredient of the ith class to another $(i+1)$ class and the laws of the transition from one part of the environment to another $(l \to l+1)$

$$I_{i+1,l} = \omega_i(t) I_{il} \tag{4.4}$$

$$I_{i,l+1} = \mu_l(t) I_{il} \tag{4.5}$$

where ω_i and μ_l are the coefficients of the corresponding transformation or transitions.

Thus, the degree of the impact A_n can be written in general form for the n ecosystem (community) in a certain region (area):

$$A_n = \int_R \int_t \sum_m \sum_i \sum_l I_{ij}(\mathbf{R},t)\omega_i(t)\mu_l(t) C_{ilm} \epsilon_{ilm} N_m K_m$$

$$\times \nu_{i,j+k} \xi_{m,m+k} \, \mathrm{d}\mathbf{R} \mathrm{d}t \tag{4.6}$$

where $\nu_{i,j+k}$ characterizes the effect of the simultaneous impacts of the ith $(i+k)$, and any other ingredients; $\xi_{m,m+k}$ is a coefficient which takes into account the effect of the simultaneous impacts on the mth, $(m+k)$ (and any other) population of the given ecosystem; C_{ilm} is a geometrical factor which takes into account the impact on the given organism (element of the biosphere) of the intensity I actually distributed in time and space. This factor, for instance, takes into account the ratio between the distributions of the concentration of the ith substance in the environment and in the organisms of the m-population.

If A_n is estimated in absolute values of ecological damage for the given ecosystem or community, then a coefficient K_m is introduced in the formula. This coefficient characterizes the significance (or, in some cases, the sensitivity or criticality) of the m population in the given ecosystem; in this case $0 \leq K_m \leq 1$; when $K_m = 0$, the given population is insignificant for the given ecosystem and can be lost to the system or replaced by some other population. For endangered ecosystems, for any population $K_m > 0$.

Coefficient K_m is equal to unity for the most important populations, primarily for the human race. The most important populations can be distinguished for each ecosystem. For example, the lemming population, according to Academician Shvarts, is the most important population for polar ecosystems; for the ecosystem of Lake Baikal it is the zooplankton epishura.

Equation (4.6) is also true for abiotic elements of the biosphere (in this case m stands for the number of any element of the biosphere). Moreover, in a general universal assessment of the impact degree and the ecological damage, both the biotic and the abiotic components of the biosphere should be taken into account in Equation (4.6).

Formula (4.6) can be used to assess the state of the biosphere and the ecological damage. It can also be used when estimating the economic damage and when planning a system for monitoring the state of the environment. This formula is too complex to be used in practice, but it can be simplified considerably if we separate its most significant components.

In many cases it can be assumed that $\nu_{i,i+k}, \xi_{m,m+k} = 1$, i.e., we neglect the additional intensification or attenuation of the total impact when several ingredients are acting together (absence of synergism or antagonism) or when some ingredient is acting simultaneously on several populations in the ecosystem.

Only those populations can be considered for which we can assume that $K_m \approx 1$.

By considering the degree of the impact at each time for some part of the environment l, we can write, with the above assumptions, the equation

$$A_{n,l}(t) = \int_{\mathbf{R}} \sum_m \sum_i I_i(\mathbf{R},t) C_i \epsilon_{im} N_m(\mathbf{R},t)\, \mathrm{d}\mathbf{R}. \tag{4.7}$$

This formula is simpler for practical use.

In order to estimate in practice the total impact of all the diverse and dissimilar factors, and thus to determine the ecological damage caused to the given ecosystem, the quantity $b_{ilm} = I_{il} C_{ilm} \epsilon_{ilm}$ in the integrand of formula

(4.6) or (4.7) should be expressed in comparable units, e.g., in the maximum permissible values $b_{ilm}(0)$ or critical values $b_{ilm}(\max)$ of the given impact factor; the rest of the quantities should be expressed in relative units or in normalized form. Let us denote A_n as $A_n(0)$ and $A_n(\max)$, respectively. The distribution of the number of the given population should be expressed in normalized form:

$$n_m(\mathbf{R}, t) = \frac{N_m(\mathbf{R}, t)}{\int_{\mathbf{R}} \int_t N_m(\mathbf{R}, t)\, \mathrm{d}\mathbf{R}\, \mathrm{d}t}. \tag{4.8}$$

A comparison of the values of A_n for different time intervals makes it possible to determine the trends of the change in state of individual ecosystems (in cases where the state of the ecosystem is to a great extent determined by the impact degree and depends in the same way on the impact in time).

An impact A_n can tentatively be considered permissible for an ecosystem if, for instance, the following conditions are satisfied:

$$\int_{\mathbf{R}} \int_t n_m(\mathbf{R}, t) b_{ilm}(\mathbf{R}, t)\, \mathrm{d}\mathbf{R}\, \mathrm{d}t < b_{ilm}(\max) \tag{4.9}$$

for any factor of the impact i and for all l and m;

$$A_n(\max) < 1. \tag{4.10}$$

The latter can be replaced by the more rigid condition

$$A_n(0) \leq 1. \tag{4.11}$$

When assessing the state of the environment and the potential ecological damage, we are supposed to use certain criteria of permissibility of the impact, criteria of environmental quality, and criteria of permissibility of the impact source intensity under the existing actual conditions.

Here we are interested in the maximum permissible concentrations (MPCs) of pollutants for assessment of the permissible amount of the substance acting in the environment, the maximum permissible doses for assessment of the permissible effect of the impact, the maximum permissible discharges (MPDs) for assessment of the permissible intensity of the pollution source, and the maximum permissible load (MPEL) for assessment of the permissible ecological load on an individual ecosystem or within the framework of the entire region. In the latter case, we can conditionally refer to the MPEL for the given region as a whole.

In order to establish the MPEL, we have to take into consideration possible combined and complex impacts on the ecosystem, as was shown in Chapter 2. For example, when mercury arrives in water bodies via washout from the soil, it changes to a highly toxic form, which makes the problem of mercury pollution extremely serious (even when the initial mercury contents of the air do not exceed the MPC).

Establishment of the MPEL should be based on the concept of the stability or critical state of the ecosystem or of its individual components and levels.

As a rule, there is a quite significant difference between the maximum permissible values κ_p and the critical values κ_c of the impact factors, established for populations.

However, for some ecosystems, an impact value is taken to be the maximum permissible value if, even for considerable modification, it does not lead to destruction of the ecosystem. In this case κ_p is obviously equal to κ_c.

In order to assess the economic damage, we take all quantities in the integrands of formulas (4.6) and (4.7) in absolute values, and we introduce a coefficient L_{im} characterizing the economic damage per unit ecological damage for each population from the ith ingredient; this approach is only applicable for the permissible values of A_n.

To take into account the total damage, it is necessary to carry out an integrated (comprehensive) analysis of all the kinds of impact on all the components of the biosphere.

An evaluation of the state of the biosphere requires a study of the relationships $\omega_i(t)$, $\mu_l(t)$ and a knowledge of the quantities $\nu_{i,i+k}$, $\xi_{m,m+k}$ and $b_{ilm}(0)$, $b_{ilm}(\max)$ and L_{im} for the basic (or all) impact factors.

Assessment of the esthetic damage is the most complicated task, since we can speak about esthetic damage only with regard to man's perception of the natural environment. Damage associated with a change in the habitual situation or landscape for animals refers to ecological damage. A quantitative relationship between the esthetic damage and the economic damage can be established, e.g., in the case when it is possible to determine the damage caused by a lesser number of tourists due to a decreased esthetic value of the environment (the so-called recreational damage [9]) or when deterioration of public health results from people's refusal to go for a holiday or from worsened conditions for relaxation.

However, the concept of 'esthetic damage' is broader and it reaches beyond such cases; for instance, the decreased esthetic value of unique and protected ecosystems defies any economic assessment (in some cases, the economic damage (C) may be considered irreparable, i.e. $C \to \infty$).

Prediction of the state of the biosphere and assessment of the predicted state constitute a component of the monitoring process.

Forecasts are based on current and historical data on the state of the environment (these data are obtained from observations and from analyses of their findings).

The study of series of observations, and a determination of regularities in environmental changes, make it possible to determine the trends of these changes.

Forecasts of the state of the biosphere should be based on the results of investigations determining the regularities of natural processes, and regularities of the spreading and migration of pollutants, of their transformations, of their influence on the state of the environment, and of the reactions of different organisms to a change in this state.

During the first stage it is necessary to forecast changes in the intensity of the sources of different impacts and pollutants, and to predict the factors of the impact in the natural environment, for instance, the total amount of pollutants in different parts of the environment, their spatial distribution, and changes in their properties and concentrations with time.

Information concerning plans for human activity is necessary for the preparation of such forecasts.

When preparing forecasts of the state of the biosphere, we can assume that human activities do not change (e.g., we can suppose that pollution sources are constant) or we can take into consideration plans for an increase in the volume of economic development (which can lead to an increase in the level of environmental pollution), as well as measures to reduce this level (which restrains or decreases pollution). Forecasts of pollution in the biosphere may be considered as a first (or even zeroth) approximation of such prediction, assuming that, starting at a certain moment, pollution sources will be absent. This makes it possible to take into account more accurately all the geophysical, geochemical, and biogeophysical processes associated with the 'life', transformations, and migration of pollutants.

Here particular attention should be paid to a possible increase in the toxicity and mobility of various substances.

The next stage consists in forecasting possible changes in the biosphere, and in its biotic component, under the influence of pollutants already existing in nature (and other impact factors), as well as newly entering or appearing pollutants.

It should be noted that man-induced changes which have already occurred in the biosphere can still affect the state of its individual elements for many years to come (particularly when we consider the genetic consequences), even if the additional external impact decreases or ceases completely.

The assessment and analysis of the predicted state of the biosphere make it possible to select certain directions requiring priority measures to reduce their adverse effects; forecasts make it possible to plan and implement not only concrete measures aimed at attenuating already existing (and possibly increasing) impacts, but also measures aimed at protecting against adverse effects which have not yet appeared. Besides regions where the state of the biosphere is unfavorable because of the fast development there of industry and the economy, forecasts make it possible to recognize problems requiring particular attention on a global and regional scale.

A comprehensive analysis of the existing situation and an analysis of various effects of the impact, allowing for trends in the development of human activities, make it possible to consider that in the coming years (and possibly decades) problems related to the spreading of the following pollutants in different parts of the environment will be the most serious:

- sulfur dioxide and products of its transformations (sulfuric acid and sulphates);
- heavy metals (mercury, lead, and cadmium), particularly mercury, taking into consideration its migration and transformation;
- carcinogenic substances, in particular benzo(a)pyrene;
- petroleum and petroleum products in seas and oceans;
- organochlorine pesticides, and in urban centers also carbon dioxide and nitrogen oxides.

The appearance of these pollutants will be observed on a large scale either due to coverage of a great number of towns and settlements or else due to transport over long distances of sulfur transformation products, heavy metals, and petroleum products.

Among the global problems which should be distinguished, since the anthropogenic impact, and primarily the discharges of various pollutants into the environment, will play an important role, we list the following:

- a possible climatic change due to the anthropogenic impact on the atmosphere and other anthropogenic changes in nature;
- a possible disturbance of the ozone layer due to the influence of nitrogen oxides and halogenated hydrocarbons of anthropogenic origin (such disturbances can also induce a climatic change);
- pollution of the World Ocean by petroleum products.

An analysis of the prognostic data will make it possible to introduce certain amendments to human economic activity and to correct in an optimum manner the interaction between human society and nature. Therefore, forecasting the state of the biosphere, as we also demonstrated earlier, represents a necessary link in the management of environmental quality.

4.4. Probability Approaches to Risk Assessment of the Potential Hazard

Procedures for the assessment and calculation of different kinds of damage due to the anthropogenic impact on the natural environment were described earlier.

In some cases, however, it is necessary to estimate the possible degree of the damage, that is, the potential damage associated with some probability of impact and depending on many factors whose appearance is also a probability event. In such cases the concept of risk, i.e., the probability of the appearance of a certain damage, is often introduced. The risk is thus a probability characteristic of the hazard presented in the given case to the environment (and to man) due to possible anthropogenic impacts or other phenomena and events.

It was shown in Section 4.3 that the damage (primarily the ecological damage) depends on the degree of the impact on the given system (or element of the biosphere). It should be emphasized, however, that we are referring here to the possible damage, since the expression for the impact degree (see Equation (4.6)) includes quantities whose values are associated with a definite probability. For instance, the number of exposed organisms depends on the direction of spreading of the given pollutant; when assessing the impact degree, and then the potential damage, though, a certain probability that the substance will spread in this direction is assumed (e.g., a wind-rose for the given territory is used for propagation through the atmosphere); the distribution in time and space of exposed organisms is also a probability quantity.

In this sense the assessment of the impact degree is equivalent to estimating and calculating the risk of a given impact on an element of the biosphere. For instance, Lawrence [42] states that for risk assessment it is necessary to determine the conditions of the impact (radiation), to identify undesirable effects, to find the dose/effect ratio, and then to estimate the total risk. This is similar to, but somewhat simpler than, the assessment of the degree of impact on a biosphere element suggested by us back in 1974 [9].

Risk assessment implies an assessment of the form and degree of the hazard, and an understanding of the hazard or potential damage associated with the impact [41].

Risk calculation provides a quantitative expression of the probability of some consequence, whereas risk assessment provides a comparison of the risk of some concrete consequence with other kinds of risk (taking into account existing approaches to the assessment of these other kinds of risk) [47]. This definition of risk assessment is close to the concept of environmental damage assessment introduced by us (occurring or anticipated damage); in both cases some specially developed comparison criteria are used for the assessment.

It is obvious that the data on risk can be used differently for the decision-making process; in some cases all possible effort should be made to avoid unfavorable risk or to balance a certain risk with some other kind of risk (balanced risk) or to compare the given risk in terms of benefit-risk and damage-benefit. Here a social assessment is carried out, the social effect of the risk being considered.

The latter can be illustrated easily. The construction of supports for high-voltage power lines furnishes a simple example. When choosing the strength of the supports, we have to analyze the costs of supports of different strength, comparing the strength with the probability of occurrence of excessively strong winds (or icing up) in the given region as well as with the risk of destruction of such supports in some extreme cases. A variant is considered optimum if the total social or ecological damage from destruction of the high-voltage lines plus the cost of their construction is a minimum.

Therefore, in some extreme cases replacing these lines is cheaper (and more reasonable) than constructing superstrong supports guaranteeing their preservation under any conditions. The object of such investigations is to establish strength criteria in accordance with the calculated risk of destruction under extreme meteorological conditions.

The social significance of assessments of the state of the environment is considered in Section 3.1, where it is stressed that a necessary conditon for the transition to the social optimum is the requirement of a more or less simultaneous provision of the population with food, habitation, and high standards of environmental quality.

Probability approaches and a risk assessment of the appearance of some undesirable phenomena or consequences are also applicable during the assessment of complex situations, when it is possible to analyse and evaluate (quantitatively or even qualitatively) individual aspects (and their interaction) of such situations or programs.

The concept of risk and its quantitative assessment were widely applied and are being applied in the implementation of space and nuclear programs [46]. Recently a similar approach has also been developed and applied to evaluate the possible consequences of the anthropogenic impact on nature. For instance, in [41] some approaches to this problem are discussed in detail. Even just the detection of a potential hazard is often an important achievement. For example, the attention of scientists and of society has been attracted to a possible climatic change, to a disturbance of the ozone layer, to an increase (taking accumulation and transformations into account) in the toxicity and pollution hazard of heavy metals, as well as to unfavorable consequences associated with these phenomena, and these are all serious signals. Of course, further efforts are necessary to determine and assess the degree of hazard associated with this threat and its consequences.

In some cases the assessments of the probability of some unfavorable event are so reliable that they can be used to calculate the planning and implementation of practical measures (to calculate different variants of the measures, to take into consideration the possible economic damage, etc.).

Thus, the first step of the risk assessment is identification of the hazard, i.e., determination of the real danger for man and for the environment. Research plays an important role here. Representatives of fundamental science have drawn attention to such problems as possible climatic change, the threat to the ozone layer, and the significant genetic consequences of the anthropogenic impact on nature and man. The task of identifying the hazard obviously reduces to a search for danger signs and distinguishing such signs against the existing background.

To identify any hazard, we make use of methods of approbation, selection (e.g., of different samples), modeling of the behavior of different substances in the environment, monitoring, and diagnostics (assessment of symptoms and impact consequences). Note that problems related to assessment, diagnostics, and forecasting shoud pertain to the monitoring system. Diagnostics begins with observations of anomalies, and on the basis of these anomalies the 'disease' has to be determined correctly. Practically all the data obtained through monitoring require diagnostic assessments. The first steps in the identification of the hazard and the assessment of the risk are to decide what the hazard is and how great the risk is, i.e., it is necessary to determine the probability of occurence of the given hazardous event and the probability of adverse effects. It is shown in [41] that for risk determination and assessment foresight, intuition, and extrapolation are useful tools.

The concepts of intuition and foresight are close to each other, in that they are both based on implicit information [56] (a vague information source or a psychological conviction that information is either inadequate or absent). Foresight often refers to events which either have not occurred at all or else occur extremely seldom (e.g., major disasters, large-scale natural changes in the global climate); intuition may refer to the forecasting or assessment of more commonplace events. The extrapolation of experience gained is very close to intuition, and if the extrapolation is based on vaguely defined experience, then it is identical to intuition.

Forecasts of future events are often based on the method of extrapolation. Weather forecasting with the aid of analogs is an example of this. The statistical prediction of floods and droughts is also an example of the calculation of risk by the extrapolation method.

For these purposes specially developed models are used, into which other, earlier obtained regularities can also be introduced.

Into this same category we can place the spreading to closely related species of the appearance of anthropogenic impacts (e.g., pollution) and different effects studied earlier in other species. It has been proposed to use this technique for the organization of ecological monitoring at the background level (see Chapter 5).

Qualitative considerations and comparative estimates are often used in risk calculations; there exist methods for transforming these into quantitative data (see, e.g. [50]). This is, however, by no means simple and is subjective. For instance, more than one hundred interrogated residents of Saint Louis (USA) interpreted a probability forecast of rainfall (with 60% probability) in the following way: 29% considered rain to be 'almost certain', 33% said 'probably', 28% said 'some chance', and 10% said 'slight chance' or even 'probably not' (similar difficulties were encountered for the opposite interpretation) [30].

Qualitative considerations and methods of estimation by experts have some major weak points, due to their subjectivity; subsequent checks have shown that such data often exaggerate the significance of recent and frequently recurring events, as well as more impressive effects (the anticipated probability of the event is distorted). Considerable difficulties are encountered in assessing infrequent events. All these difficulties refer to the assessment of both the risk of anthropogenic impacts and their negative consequences.

As already noted, the social determination (assessment) of risk completes the chain of hazard identification + risk assessment (estimation) + social assessment.

The significance of the risk in the latter case is assessed by different methods, such as absolute assessment of the given risk or comparison with other kinds of risk and with the expenses needed to avoid it, taking into consideration the possible 'benefit', the positive effects.

In this connection we can consider the forbidden (unfavorable) risk. The balanced risk, and the benefit-risk and cost-benefit ratios.

The forbidden risk obviously refers to the category of events which are absolutely inadmissible from the point of view of society. For instance, a forbidden risk in ancient society was formulated as a taboo, while in modern society, it takes the form of laws, rules of etiquette, and basic norms or standards.

There are, of course, some contradictions. For instance, in the USA carcinogenic substances are forbidden in foodstuffs but not forbidden in water; in the USSR some pesticides are forbidden in foodstuffs but not forbidden in water bodies, etc.

The same problems may be treated differently in different countries. In spite of the fact that all developments are based on the same laws of physics, there exist, for instance, some differences between the standards of different countries, such as with respect to the permissible quality of the environment. Interrogations have shown [41] that different people have quite diverse ideas of the degree of hazard or risk from natural disasters or from adverse events (auto accidents, etc.).

In non-Soviet literature describing probability approaches and the concept of risk, we often come across the concept of the permissible, or acceptable, risk. For instance, the monograph by Lawrence [42] contains this term in the title. It is our opinion, however, that this term can be interpreted ambiguously. It is natural to ask: For whom is the given risk acceptable? Is it acceptable for an individual person or for society? And the answers to these questions may well be contradictory.

It is obvious that, in order to fulfill nearly any task or to implement many programs, it is necessary to think over ways of implementing the programs with different degrees of risk or with different probabilities of solving (or not solving) the stated problem. Even in such a simple problem as how to get from point A to point B, when choosing the means of transport it is necessary to analyze a time-comfort-safety system. For this interpretation the goal is to make a scientifically grounded, optimum decision, i.e., a decision associated with the least risk when implementing the program and with the maximum probability of solving the problem (with the least damage). In this case we should probably refer not to the acceptable risk but rather to the minimum balanced risk for the maximum success. The use of medicines which cause side effects is an example of a balanced risk; their use is absolutely necessary in some cases, taking into consideration the overall assessment of the state (health) of the organism (or system) as a whole.

When using an approach determining the balanced risk, we must thoroughly examine the degree of risk of each individual event, as well as the changes under different conditions. For instance, people with different professions are subjected to different dangers. In a number of countries pilots, miners, or steel workers run risks on the job which are 10 times greater than those of people engaged in other production spheres [36]. It has been estimated that for English workers during the course of a day the probability of running a risk of injuring their health varies by a factor of 600 (the most probable risk is on the way to or from work by car, and especially by motorcycle) [32].

The natural limits of risk for man range from 10^{-2} (per capita probability of sickness) to 10^{-6} (the lower level of risk from a natural disaster or other serious danger). In some countries this level corresponds to 10^{-5} or even 10^{-4} (Bangladesh [41]).

Protective measures against natural calamities and anthropogenic impacts are usually worked out with the aid of the balanced risk method, taking into consideration natural calamities or dangers encountered by citydwellers every day (while using cars, etc.). This approach is realized when protecting against earthquakes (e.g., in Long Beach, California, a death risk of 10^{-5} to 10^{-7} per capita per year is used as a criterion for protective measures, a value comparable with the risk of accident which a person faces in everyday life [57].)

To establish the permissible radiation dose for the population, the natural radiation background is taken as a basis, i.e., the additional risk of radiation is comparable with the slight risk from the background radiation encountered constantly by humanity.

When assessing the existing risk, society makes different efforts and incurs expenses to reduce this risk. There exists a certain correlation between different risk levels and attempts to reduce this risk. It can be assumed that this relationship is different in countries with different social systems. The huge amounts of money allotted to social security in the USSR and to the improvement of safety techniques at industrial enterprises speak for themselves. Obviously, in this case a shift toward a maximum possible reduction of risk (increased safety) at work and in everyday life takes place.

Otway and Erdman [45] present data on the quantitative relationship between the risk values and the reaction of society to it, as embodied in attempts to reduce risk (the figures presented are for the USA). According to the data in [45], if the probability of accidents is 10^{-3} (per capita per year), then urgent measures are taken to reduce the danger; if the probability of accidents is 10^{-4} (per capita per year), then money is spent on signal lights, traffic regulation, and firemen. A risk of the order of 10^{-5} is discussed; children are cautioned against the danger of playing with fire and drowning; medicines are kept away from children; some people do not travel by plane. A risk of 10^{-6} probability does not trouble the 'average' man. Different kinds of assessments and a degree of risk are used to arrange life and domestic property insurance.

Experience shows that people run a voluntary risk considerably more calmly, or even willingly, if it is associated with some benefit (time saving, comfort, or entertainment). For instance, many people who know that driving their own car is associated with a certain danger (significantly more danger than traveling by public transport) still prefer to drive themselves. Similar examples are smoking, hunting, etc. A plot of the benefit (B) versus the risk (R) is shown in Figure 4.4 [53]. It is seen from Figure 4.4 that there is an exponential function $B \sim R^n$, with $n = 3.0$ for voluntary and forced risks (in [48] $n = 1.8$ for voluntary risks and $n = 6.3$ for forced risks). It is also seen from the figure that some representatives of society would rather run voluntary risks 100 times more frequently (or to run greater risks) than to run a forced risk. Naturally, plotting a quantitative benefit – risk relation is a very difficult and questionable task, but plotting the benefit or positive effect (in relative units) as a function of the risk (or vice versa) is much easier (Figure 4.5).

When analyzing the benefit-risk relation, it is very important to analyze the distributions of benefit and risk. These distributions may not be identical, especially in countries with different social systems and different systems for the distribution of material goods. This is also recognized by western specialists. For instance, according to Kates [41]: "Both benefit and risk can be dispersed (diffused) or concentrated in space, time, or social classes, which makes it difficult to arrive at a correct decision". In practice, the greatest conflict arises precisely because of differences in such distributions. A different distribution of benefit and risk in time and space makes their assessment difficult.

With regard to dispersed and concentrated benefit and risk, we can take the use of pesticides as an example, since the effect of pesticides is felt in specific fields while the risk is scattered. On the other hand, under occupationally unhealthy conditions of work, the risk is concentrated at the

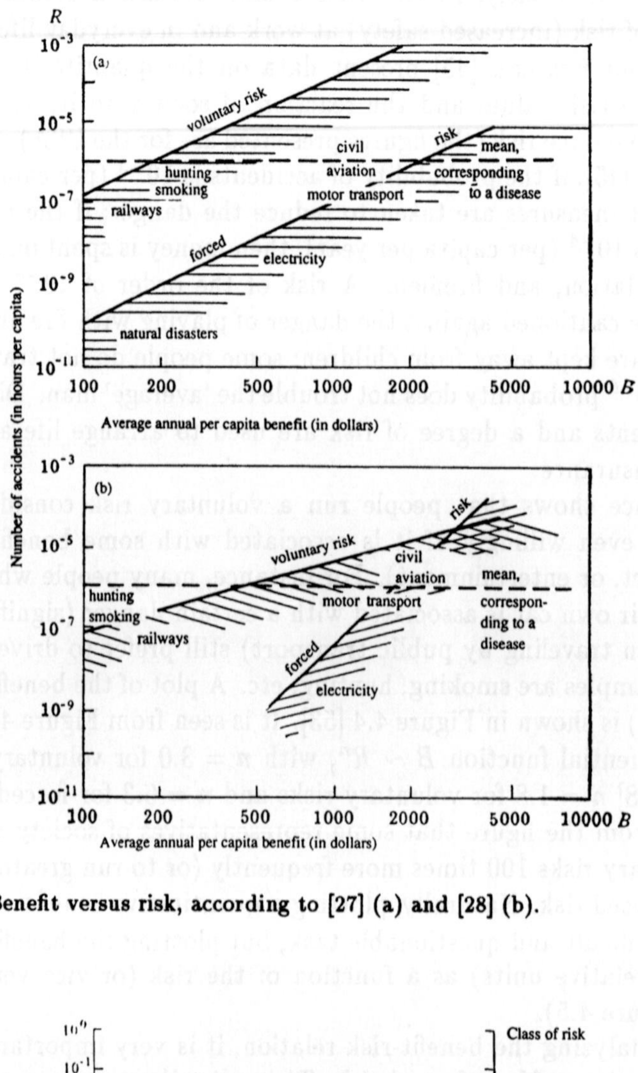

Figure 4.4. Benefit versus risk, according to [27] (a) and [28] (b).

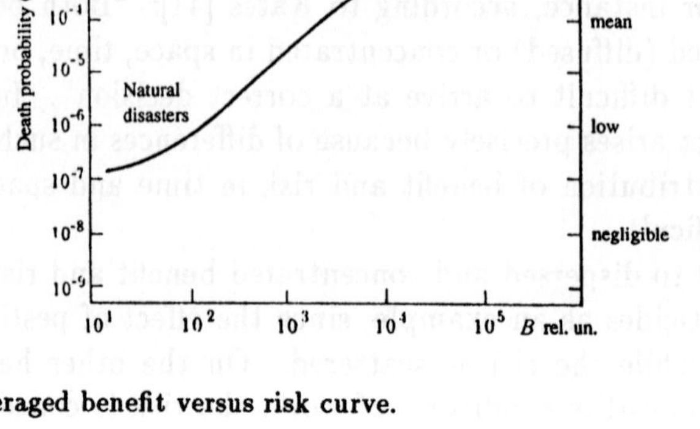

Figure 4.5. Averaged benefit versus risk curve.

Table 4.2. International comparison of permissible standards for chemical substances in a technological environment.

Ratio of national standard to lowest standard	USA		West Germany		GDR	
	Number of standards	%	Number of standards	%	Number of standards	%
1.00	19	11.2	22	13.8	21	19
1–2	26	15.4	32	20	32	29.1
2–5	42	24.8	30	18.8	27	24.6
5–10	27	16	24	15	7	6.4
10–50	38	22.4	34	21.8	20	18.2
50–200	17	10.2	18	11.2	3	2.7
Total number of standards	169	100	160	100	110	100

Ratio of national standard to lowest standard	Sweden		Czechoslovakia		USSR	
	Number of standards	%	Number of standards	%	Number of standards	%
1.00	21	26.6	19	27.5	147	87
1–2	15	18.9	21	30.4	13	7.7
2–5	19	24.1	16	23.2	4	2.3
5–10	10	12.6	6	8.7	4	2.3
10–50	12	15.2	6	8.7	1	0.7
50–200	2	2.6	1	1.5	0	0
Total number	79	100	69	100	169	100

Note. The lowest standard (the most rigid for all countries) for each chemical substance was assumed to be equal to 1.

given enterprise, while the positive effect, that is, the benefit for society, is sometimes tacit in nature.

In the study of problems asssociated with environmental pollution, much importance is attached to cases when a positive effect is achieved rapidly while the damage from the pollution manifests itself only very gradually, i.e., the risk is distributed over time (the opposite situation is also possible).

Different approaches to the social assessment of risk (taking into consideration the cost and benefit) in different countries can be demonstrated by comparing the norms or standards of permissible concentrations for a great number of chemical substances (up to 169) in the technological environment (Table 4.2 [41]).

It is precisely these standards which characterize the preference of society during its assessment of the safety, benefit, and risk, although the above quantities can reflect different methodologies of determination and different degrees of study of the appearing effects (the less the event has been studied, the higher should be the safety factor).

Table 4.2 shows that the most rigid standards are those in the Soviet Union. If sufficient money is allocated, this provides safety for workers, while a realization of the MPCs in the natural environment guarantees the safety of the population as a whole.

Risk estimation and the social assessment of risk are of importance for management operations, as well as for the determination of priorities when selecting alternatives in the battle against undesirable consequences.

For these purposes different models exist, which in [41] are divided into rational models (constructed on a scientific basis) and bureaucratic models. The latter take into account the role of different opinions of the opponents and supporters of the given measures, conflict situations, the need for decision making, and in some cases consideration of prejudice and preconceived opinions.

A check has shown that in many cases the benefit is overestimated and the cost is underestimated (as was the case in all checks during estimates of water-resource projects in the USA [31]). Thus, the benefit often seems to be quite definite, while the risk looks uncertain. Sometimes a prejudice toward adversity is observed. For instance, in some cases it is simpler to arrive at an agreement confirming the debatableness or undesirability of some statement than to agree on something acceptable to all [40].

It should be noted in conclusion that many of the described aspects go beyond the limits of physics and ecology, and lie rather within the province of psychology. However, all the above-indicated things should be carefully taken into consideration during a general risk assessment. The various elements of risk assessment (hazard identification, risk calculation, and social assessment) are closely connected with each other.

4.5. Establishment And Classification of a Monitoring System

An adequately representative network of observations (measurements) of the most important impact factors, the indices of the state of the environment, must be established in order to monitor properly the state of the biosphere. These factors and indices may be different, depending on the specific monitoring problem.

In accordance with the definition given earlier, the basic objectives of biosphere monitoring can be formulated as follows: on one hand, observations of the state of the biosphere, and assessment and prediction of changes in its state caused by the anthropogenic impact; on the other hand, determination of the degree of this impact (with detection of anthropogenic effects) and identification of its sources and factors.

Taking the above objectives into consideration, it is first of all necessary to find the factors leading to the most serious long-term changes in the environment (and the sources of such impacts), as well as to distinguish the biosphere elements which are most subjected to the impact (or are more vulnerable), or else are critical, key elements, whose damage can lead to ecosystem destruction. Formulas of the type of (4.6) and (4.7) can be used to find these factors and indices.

Since it is very difficult to organize observations of changes in all the parameters, we have to distinguish in the integrand of formula (4.6) the most significant components, which mainly determine the degree of the impact on the environment A_n (or the ecological damage). A classification of the impact factors and biosphere elements which have to be taken into consideration when determining the priorities is presented in Table 4.3.

Attention should first of all be given to the component $I_{il}(\mathbf{R}, t)$ for the most detrimental impact factors i in the environment l where these factors are most hazardous (or where their travel out of these parts is hazardous), the time t, and the place where I_i reaches the highest values. Since the hazard of the impact is to a great extent determined by the factor ϵ_{ilm}, therefore it is important to take this quantity into consideration. As already noted, the quantity I_i depends on the impact sources, which is why so much attention should be given to monitoring the most dangerous sources $q_{ij}(\mathbf{R}, t)$ and $Q_i(\mathbf{R}, t)$ (for notation, see Section 4.3).

It is very important to take into account the actual distribution $N_m(\mathbf{R}, t)$ or $n_m(\mathbf{R}, t)$ of the organisms of different populations, and the possibility of contact of certain organisms of the most significant or most vulnerable populations with pollutants or with other impact factors. This aspect of the problem is by no means always taken into consideration, and a higher priority is often given to more (potentially) hazardous impact factors which have practically no contact with large contingents of individuals, rather than to factors which are not very toxic but which affect a great number of organisms. Therefore, only a comprehensive study of the problem can yield a correct result.

It is imperative to organize observations of the reactions of the biotic and abiotic components of the biosphere to the impact. The results of these measurements should correspond exactly (quantitatively) to the factor b_{ilm} in the integrand of Equation (4.6), expressed in units of the maximum permissible loads. This correspondence is even more obvious for the abiotic component of the biosphere. As stated above, the highest priority should be given to organizing observations of the direct impact on man.

The described approach to the determination of priorities when selecting the impact factors and indices is the most important stage in the scientific establishment of a monitoring system. When selecting the observation points in space and determining their frequency of observation, we should try to arrive at an approximately equal change in the integrand of Equation (4.6) over these intervals.

Table 4.3. Classification of impact factors, biosphere elements, indices and indicators of their properties and state.

Impact factors and sources, biosphere elements, and their properties	Component (constituent) of formula (4.6)	Note
I. (1) Ingredients, impact factors, their intensity, distribution in space and time	i $I_{il}(\mathbf{R}, t)$	
(2) Pollution sources, impact factors.	$q_{ij}(\mathbf{R}, t)$, $Q_i(\mathbf{R}, t)$	$I_{ij}(\mathbf{R}, t) = f\left(\sum_i \sum_j q_{ij}, \dots\right)$
(3) Properties of substances and impact factors,		
their chemical and physical transformations,	$\omega_i(t)$	$I_{i+1,l}(\mathbf{R}, t) = \omega_i(t) I_{il}(\mathbf{R}, t)$
passage from one part of environment to another.	$\mu_l(t)$	$I_{i,l+1}(\mathbf{R}, t) = \mu_l(t) I_{il}(\mathbf{R}, t)$
(4) Properties of substances and impact factors, their biological efficiency (toxicity, etc.),	ϵ_{ilm}	$I_{ilm}(\mathbf{R}, t) = I_{ilm}(\mathbf{R}, t)\epsilon_{ilm} \times$ $\times \nu_{i,i+k}, \dots, \nu_{i,q}, \dots$
effect of synergism with homogeneous impact factor $i + k$,	$\nu_{i,i+k}$	
effect of synergism with accompanying factor q	$\nu_{i,q}$	
II. Part of environment (atmosphere, hydrosphere, soil, biota).	l	

Impact factors and sources, biosphere elements, and their properties	Component (constituent) of formula (4.6)	Note
III. (1) Biosphere element,		
population,	m	
ecosystem,	w	
distribution of organisms in space and time	$N_m(\mathbf{R}, t)$, $u_m(\mathbf{R}, t)$	
(2) Properties of organisms, their significance and vulnerability,	K	
exposure to impact of factor i,	ϵ_{ilm}	
effects for simultaneous impact on several populations	$\xi_{m,m+k}$	
IV. Accompanying factors:		
(1) environmental properties: traveling speed of conservative pollutants	$v_{\mathbf{R}}, w_{\mathbf{R}}, v_z, \ldots$	$I_{il} = F(Q_i, v_{\mathbf{R}}, w_{\mathbf{R}}, v_z, \ldots)$
(2) external factors influencing state of systems:		
temperature, humidity, solar radiation, atmospheric turbidity, etc.		

It should be noted that, in the solution of different problems, the determination of priorities for the monitoring subsystems can yield different results for the very same impact factor. For instance, on the one hand, the damage caused by an increased CO_2 content of the atmosphere is insignificant for some ecosystems, and in many cases increased CO_2 may even be useful, since it enhances the productivity of plants. On the other hand, considering its possible impact on some large systems, such as the climate of our planet, CO_2 accumulation leads to an increased greenhouse effect and possible climatic changes with different adverse effects for the biosphere.

Let us consider some examples of the determination of priorities in the establishment of a monitoring system [12]. With regard to territories, the highest priority should be given to urban centers and to zones containing drinking-water sources and spawning grounds for fish. The parts of the environment deserving the highest priority are the atmospheric air and the water in fresh-water bodies (especially low-flow bodies); the high-priority ingredients are: for air, dust, sulfur dioxide, heavy metals (mercury), carbon dioxide, nitrogen oxides, benzo(a)pyrene, and pesticides; for water, biogenic products, petroleum products, and phenols; the major pollution sources (in urban centers) are motor transport, fossil-fuel power stations, nonferrous metallurgical works, etc.

As an example, let us consider a method which was used to determine the priorities for pollution monitoring at the first Intergovernmental Meeting on Monitoring, held in Nairobi in 1974 (the author was a member of a working group assigned to the task of developing procedures and establishing priorities for different pollutants). First some criteria were selected for determining priorities, based on the properties of the pollutants and the possibility of organizing measurements [49, 51]. These criteria are presented in abridged form below.

(1) The scope of an actual or potential effect on human health and well-being, on the climate, or on (land and water) ecosystems.
(2) A tendency to degrade in the natural environment and to accumulate in man and in food chains.
(3) The possibility of a chemical conversion in physical and biological systems, as a result of which secondary (daughter) substances may prove to be more toxic or harmful.
(4) Mobility.
(5) An actual or possible tendency to become concentrated in the environment and/or in man.
(6) The frequency and/or scope of the impact.
(7) The feasibility of measurements at the given level in different countries.
(8) Values for assessment of the location in the natural environment.
(9) Suitability from the viewpoint of the overall spreading, for similar measurements within the framework of global and subregional programs.

Table 4.4. Classification of priority pollutants according to classes of priority.

Priority class	Pollutant	Medium	Type of measurement program
I	Sulfur dioxide plus suspended particles.	Air	I, R, B
	Radionuclides ($^{90}Sr+$ $+^{137}Cs$)	Food	I, R
II	Ozone	Air	I, B (in stratosphere)
	DDT and other organochlorines.	Biota, man	I, R
	Cadmium and its compounds.	Food, man, water	I
III	Nitrates, nitrites.	Drinking water, food	I
	Nitrogen oxides	Air	I
IV	Mercury and its compounds	Food, water	I, R
	Lead	Air, food	I
	Carbon dioxide	Air	B
V	Carbon oxide	Air	I
	Petroleum hydrocarbons	Sea water	R, B
VI	Fluorides	Fresh water	I
VII	Asbestos	Air	I
	Arsenic	Drinking water	I
VIII	Microtoxins	Food	I, R
	Microbiological infection	Food	I, R
	Reagent hydrocarbons	Air	I

Note. B = base (global), R = regional, I = impact.

A great number of pollutants were classified in grades (from 0 to 3), according to each of the developed criteria. The priorities were determined from the largest sums of the grades (the greater the sum, the higher the prioritiy). The priorities determined in this way were then divided into eight classes (the higher the class, i.e., the lower its number, the higher was the priority), the part of the environment and the type of measurement being indicated (impact, regional and 'base', global). Table 4.4 gives the media affected by some pollutants, together with their priorities and measurement programmes.

The types of measurements which should be carried out when it is difficult to measure the pollutant itself (indirect monitoring) were also listed. For this purpose, the following quantities should be measured:

(a) water-quality indicators (colibacteria, 5-day BOD, COD, blue-green al-

gae, and their primary productivity);
(b) soil-quality indicators (salinity, acidity/alkalinity ratio, nitrite and organic nitrogen concentrations, organic matter content of soil);
(c) health indicators for man and animals, indicators of plant damage (diseases, genetic consequences, sensitivity to medicines);
(d) plant indicators of pollution.

In addition, we have listed the corresponding observations necessary for interpreting the measurements of pollutants, including selective meteorological, hydrological, and geophysical parameters, kinds and compositions of diets, and some others.

At the Nairobi meeting some of the above-described approaches were used to determine the priority pollutants, but unfortunately not all of them (e.g., the distribution and abundance of organisms in contact with pollutants were only taken into consideration indirectly).

Thus, even the organization of a system monitoring anthropogenic pollutants seems to be an extremely complicated problem. Although a number of attempts have been made, no sufficiently complete or sufficiently clear scheme for the multipurpose monitoring of anthropogenic changes in the environment has as yet been developed.

Comprehensive proposals have been made on the development of a global monitoring system (see, e.g. [44, 51]), but in general these developments have two significant drawbacks: first, in the proposed schemes there is no demarcation between the activities and objectives of already existing geophysical services and those of the proposed monitoring system; second, these proposals involve a great number of well-known parameters which should be measured to analyze the state of the environment in general, but the priority of these parameters is seldom formulated in practice.

Because of all this, the proposed global-monitoring schemes are being put into practice very slowly. However, there do exist some developments (including some already implemented) of different special monitoring systems which were devised depending on the requirements and feasibilities. As examples, we can cite the systems for measurement of urban air pollution and the global system for observations of the fallout of tritium and other long-lived radioactive isotopes. The background monitoring subsystem which is being organized at present on the basis of the biosphere reserves should also be mentioned here (it will be discussed in more detail in Section 5.4). Since such subsystems are undoubtedly useful in practice and, as a rule, can be modules of a complete (global) monitoring system, their establishment and description are important, and they make a valuable contribution to the establishment of a multipurpose monitoring system.

We will try to present a classification of approaches to the development of more or less complete multipurpose monitoring systems, as well as of possible (or already existing) approaches to the creation of different monitoring subsystems [13].

As mentioned above, different systems for the observation, assessment, and prediction of the state of the environment are already operating (or may be organized) on a limited (local) scale. Such subsystems often solve individual specific problems. They may be considered to be part of a sufficiently complete monitoring system, designed to estimate anthropogenic changes in the biosphere and to distinguish the causes and sources of such changes (or of changes which are possible in future).

Before going on to classify existing or possible monitoring systems, let us consider some universal approaches.

The establishment of a global monitoring system with a simultaneous solution of all the problems arising is the most universal approach. Here, the monitoring of anthropogenic pollution and anthropogenic impacts, and changes not associated with pollution can undoubtedly be distinguished.

The second trend is usually included in ecological monitoring [9]; obviously, the monitoring of pollution and of related effects arising in the biosphere (and in particular, the responses of the biota) is also a component of ecological monitoring. The second trend is often associated with the use of natural resources and, in this connection, with the influence of man on nature, as well as with the influence of natural spontaneous processes on human activities (and with man's influence on the development of these processes).

This approach was reflected in the formulation of the Global Environmental Monitoring System (GEMS), which is being developed under the auspices of UNEP, and in the international action plan of this organization [44, 51]. However, at the first Intergovernmental Meeting on Monitoring (Nairobi, 1974), monitoring of environmental pollution and of the impact factors associated with it was considered to be a priority activity in the global monitoring system.

Observations are supposed to be made at the impact level (the level of a strong impact on a local scale), the regional level, and the background 'base' level.

As already noted, in the implementation of such a universal monitoring system it is absolutely necessary to know the background state of the biosphere (both at present and during the period prior to the appreciable influence of man).

It has been suggested that the background global state of the biosphere be studied at so-called background 'baseline' stations [43], which are being organized in a number of countries, e.g., in the USSR, on the basis of the biosphere reserves [5].

Figure 4.6 shows the locations of the WMO background 'baseline' and regional stations for measurements of atmospheric pollution and biosphere

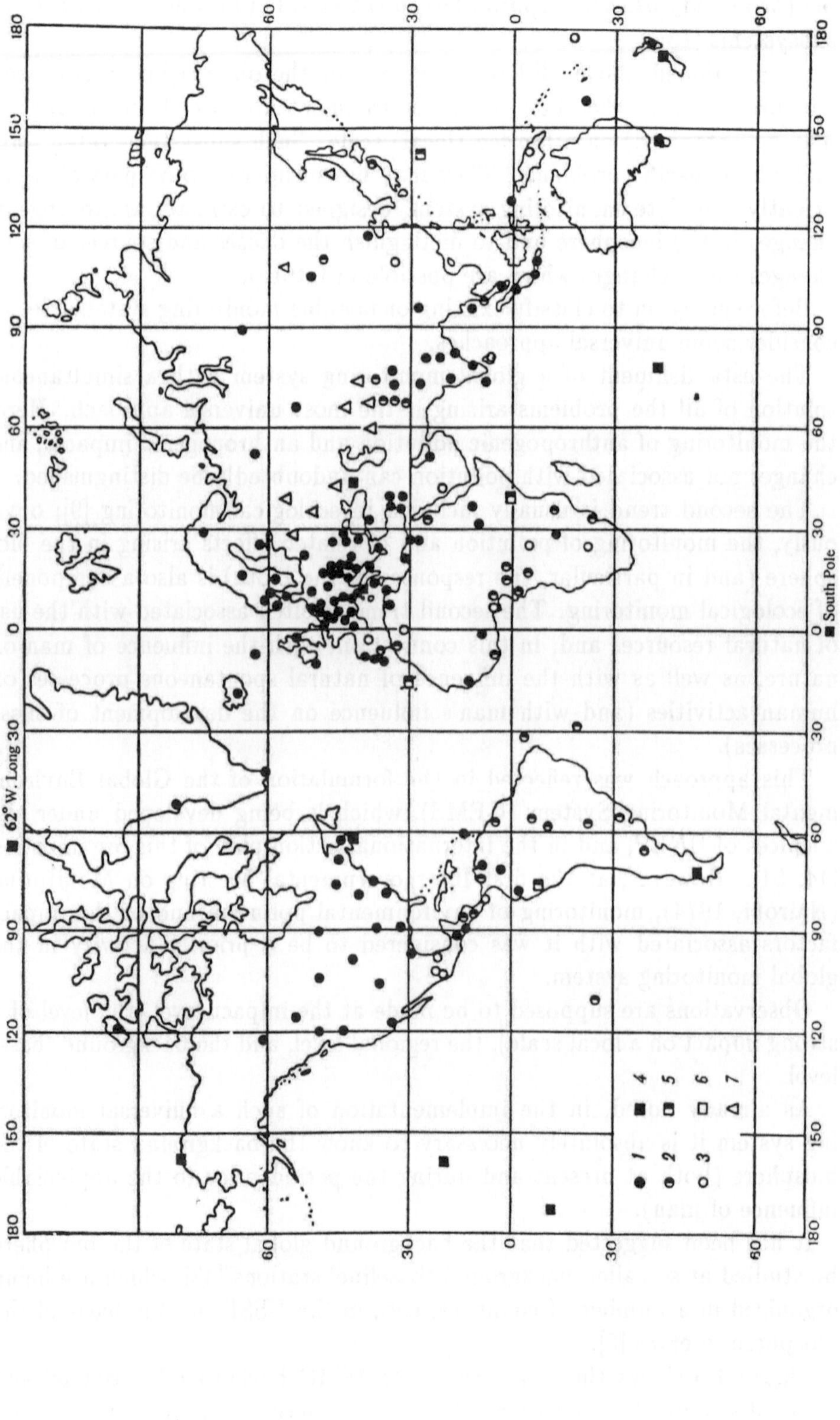

82° 30′ N.L.
62° W. Long 30
South Pole
1
2
3
4
5
6
7

reserves (the latter are shown only for the USSR).

The background state of the environment in the past, prior to man's impact, can be determined from an analysis of the rings of old or dead trees, and from samples of annual layers of glaciers and bottom sediments (so-called historical monitoring or paleomonitoring).

The universal approach is, of course, reflected in the formulation of the national system monitoring the state of the environment in an individual country. In the USSR this approach is implemented first of all within the framework of the State Service for Observing and Monitoring the Environmental Pollution Level [15, 21]. This service is designed to implement the monitoring of environmental pollution and is functioning successfully at present.

When classifying different approaches and when orienting the monitoring subsystems in order to achieve certain goals, it is convenient to distinguish a subsystem of observations (assessment and prediction) *of the responses of the basic components of the biosphere*: the abiotic component (geophysical monitoring) and the biotic component (biological monitoring).

Geophysical monitoring includes a determination of the response of the abiotic component, both on a microscale and on a macroscale, right up to the response and determination of the state of two large systems: climate and weather. It also includes the monitoring of selective meteorological and hydrological environmental characteristics necessary for the interpretation of data on atmospheric pollution and turbidity. Finally, the monitoring of different elements of the inanimate component of the biosphere, including manmade structures and buildings, also pertains to this monitoring subsystem.

The main objective of *biological monitoring* [27] is to determine the state of the biotic component of the biosphere and its response to the anthropogenic impact, and also to determine the state function and the deviation of this function from the normal natural state at molecular and cellular levels, as well as the levels of organisms, populations, and communities.

The study of the content of different ingredients in the biota can only conditionally be classed with biological monitoring. This problem refers to measurements of pollutants in different countries. Observations of the state of the biosphere with the aid of biological indicators can also be grouped with biological monitoring.

Figure 4.6. WMO network of baseline and regional stations measuring atmospheric pollution and network of biosphere reserves in USSR.
Regional stations (april 1982): (1) operating, (2) planned (with definite locations), (3) proposed (locations not specified).
Baseline stations: (4) operating, (5) planned (with definite locations), (6) proposed (locations not specified), (7) stations of background integrated monitoring system on territories of biosphere and other reserves in USSR.

Biological monitoring includes the monitoring of living organisms-populations (according to their abundance, biomass, density, and other functional and structural features) subjected to an impact. In this monitoring subsystem it is convenient to distinguish the following observations:

(a) observations of the state of human health and the effect of the environment on man (medical-biological monitoring).
(b) observations of the most important populations, both from the viewpoint of the existence of an ecosystem characterizing by its state the well-being of one or another ecosystem and from the viewpoint of great economic value (e.g., valuable species of fish);
(c) observations of populations which are most vulnerable to the given kind of impact (or else to the integrated impact) (e.g., responses of vegetation to the impact of sulfur dioxide) or observations of 'critical' populations with respect to the given impact (e.g., the response of zooplankton epishura to the discharges of paper and pulp mills into Lake Baikal);
(d) observations of populations-indicators (e.g., lichens).

A particular part in biological monitoring should be assigned to *genetic monitoring* (observations of possible changes in genetic features of different populations).

Ecological monitoring is apparently more universal, in that it covers the problems of both biological and geophysical monitoring in their intimate relationship. This is of particular importance when observations are carried out at the level of ecological systems.

Academician I.P. Gerasimov [4] suggests a somewhat different classification, according to which monitoring is subdivided into three stages: the first stage is bioecological (public-health) monitoring, including observations of the state of the environment from the viewpoint of its effect on human health; the second stage is geoecological, geosystemic, natural-economic monitoring, covering observations of changes in natural systems, and their transformation to natural-technical ecosystems; the third stage is biospheric monitoring, covering observations of the biosphere parameters on a global scale.

Monitoring of pollutants and other impact factors in different parts of the environment is very important, with regard to the practical establishment of a monitoring system on any scale and for any purposes (e.g., when establishing a national monitoring system).

Monitoring in different parts of the environments (different media) includes:

- monitoring of the surface layer of the atmosphere and the upper atmosphere; monitoring of atmospheric precipitation;
- monitoring of the hydrosphere, i.e., of surface water on land (rivers, lakes,

and reservoirs), ocean and sea water, and groundwater;
- monitoring of the lithosphere (primarily of the soil).

It is possible to distinguish monitoring of the atmosphere, the ocean, the land (with rivers and lakes), and the cryosphere (constituting the climate system).

Monitoring of the content of different substances in the living component of the biosphere (biota) can also be classed with this kind of monitoring.

Particular attention should be given to the passage of different substances from medium to medium, due to pollutant transport, spreading, and migration.

The classification of monitoring systems according to impact factors and sources appears to be no less important from the practical point of view.

Monitoring of impact factors, that is, the monitoring of different pollutants (monitoring of ingredients) and other factors, includes the monitoring of electromagnetic radiation, heat, and noise.

Here the monitoring should first of all include the most hazardous impact factors, and the most stable and mobile toxicants having toxic daughter products formed during breakdown and transformations, which are hazardous in combination with other substances. Obviously, it is necessary to consider separately the monitoring of radioactive isotopes in the environment. This kind of monitoring is called chemical monitoring.

Monitoring of pollution sources is carried out in a special subsystem, although it can in principle be covered by the preceding subsystem.

Among the sources of impacts, and especially of pollution, let us cite point stationary sources (factory chimneys, etc.), point mobile sources (transport), and space (areal) sources (fields sprayed with chemicals, cities, etc.).

Classification *according to the scale of impact* (in space, in time, at different biological levels) is also possible.

When classifying with respect to the spatial scale of the impact, we can distinguish observations referring to a point, to small areas, to districts, and to regions from observations on a global scale, processed, averaged, or generalized according to a certain system. Determination of the amount of pollutant, and of objects of the impact in one or another part of space (their distribution, density, and integral over a substantial area), can be placed into this category. Of interest, too, is the problem of the redistribution over different territories of the pollutants and the organisms subjected to the impact.

As already noted, it is sometimes convenient to devise ad-hoc observation systems to solve some specific problems.

At present it is necessary to distinguish several global problems which are of primary importance for all of humanity, in connection with their potential serious adverse effects. These are: pollution of the World Ocean, destruction

Table 4.5. Classification of monitoring systems (subsystems).

Principle of classification	Monitoring systems (subsystems) which exist or are being developed
Universal systems	Global monitoring (on baseline, regional, and impact levels), including background monitoring and paleomonitoring National monitoring (e.g., USSR State Service for Observation and Monitoring of Environmental Pollution Level) International monitoring (for instance, monitoring of transboundary pollutant transport)
Reaction of basic components of biosphere	Geophysical monitoring Biological monitoring, including genetic monitoring Ecological monitoring (including above-mentioned types)
Different parts of environment	Monitoring of anthropogenic changes (including pollution and reaction to it) in atmosphere, hydrosphere, soil, cryosphere, and biota
Impact factors and sources	Monitoring of pollution sources Monitoring of ingredients (e.g., individual pollutants, radioactive radiations, noise, etc.)
Urgency and global character of problem	Ocean monitoring Ozonosphere monitoring
Methods of observation	Monitoring according to physical, chemical, and biological indices Satellite monitoring (remote-sensing techniques)
Systemic approach	Medical-biological monitoring (of human health) Ecological monitoring Climate monitoring Variant: bioecological, geoecological, and biosphere monitoring

of the atmospheric ozone layer, and a possible change in the Earth's climate due to anthropogenic impacts.

In this connection, the organization of special monitoring systems to observe and study the effects of the anthropogenic impact on the ocean (*ocean monitoring*), the ozonosphere (*ozonosphere monitoring*), and the climate (*climate monitoring*) is completely justified and desirable.

It is also possible to classify monitoring systems according to a number of other features, e.g., ad-hoc monitoring in order to detect the long-range (including transboundary) spreading of some harmful substances (sulfur dioxide, radioactive products, etc.), or to classify monitoring according to the observation methods. However, the latter classification is supplementary, and it indicates the observation possibilities rather than the information-bearing nature of the monitoring system.

However, at the meeting on monitoring (Nairobi, 1974) this factor was seriously taken into consideration when the priorities of pollutants were determined [49].

Measurements can be distinguished by chemical, physical, and biological indices (and by hydrobiological indices for water objects).

The monitoring of different changes in the biosphere, determined by remote-sensing techniques (particularly with the aid of satellites), can be very interesting and timely.

Table 4.5 presents the above-described classification of possible monitoring systems (subsystems).

CHAPTER 5

Ecological Monitoring

5.1. Ecological Monitoring. Integrated Monitoring of the Biosphere

Ecological monitoring is an integrated subsystem of biosphere monitoring. It includes the observation, evaluation, and prediction of anthropogenic changes in the state of the abiotic component of the biosphere (including changes in the pollution levels of the environment), the reaction of ecosystems to these changes [20], and anthropogenic changes in ecosystems, changes related to pollutant impact, agricultural land use, forest cutting, urbanization, etc. Thus, ecological monitoring includes both biological and geophysical (physical) aspects. The functioning of the ecological monitoring is considered successful if it ultimately evaluates and predicts the state of the ecosystems and assesses the ecological equilibrium in the ecosystems.

This requirement distinguishes ecological monitoring from the other subsystems of biosphere monitoring or monitoring the natural environment, e.g., climate monitoring (see Chapter 6). Climate monitoring and ecological monitoring have much in common: both cover a wide range of elements of the biosphere. For instance, climate monitoring includes monitoring the state of the climate system (atmosphere – ocean – land surface – cryosphere – biota). Climate monitoring, however, provides such an acquisition of data so as to ultimately be able to estimate possible climatic changes (in this example we only have to know the state of the biota, which affects the albedo of the underlying surface).

Ecological monitoring is of great significance when assessing the state of the biosphere on a large scale, up to a global scale. This is precisely why the theory of monitoring (from the viewpoint of assessing the state of the biosphere at the background, global, and regional levels) has become so widely developed. These issues were raised in the report [9] discussing Project 14 of the UNESCO Program 'Man and the Biosphere' ('A study of the Pollution of the Environment and Its Impact on the Biosphere') [69], at the first Soviet–American Symposium on the joint project 'Biosphere Reserves' [1]; later the major trends of ecological monitoring and a program for their implementation were formulated in more detail, as well as the corresponding joint programs [10, 16, 37, 54]. A proposal in [15] would establish a special, constantly operating, ecological service, designed to satisfy all the requirements of ecological monitoring. Some attention is paid to ecological monitoring within the framework of the Global Environmental Monitoring System (GEMS). It was decided that ecological monitoring should be connected with the two major tasks of the GEMS [68];

(1) assessment of critical problems related to agricultural activities and land use;
(2) assessment of the reaction of terrestrial ecosystems to the influence of the natural environment.

In the global system specified by the UN Environmental Program (UNEP), ecological monitoring is defined as monitoring of the renewable resources of the biosphere. In [68] ecological monitoring also includes monitoring of the soil state, of the vegetation cover, of the water resources (hydrologic cycle), of the marine resources, and of the biosphere (biotic component).

In [58] the problem of evaluating the responses of marine and land ecosystems to environmental impacts (which is undoubtedly the most important element of ecological monitoring) is considered within the framework of global monitoring. Information on the change in area of tropical and deciduous forests can be used to estimate the impact on land ecosystems on large scales. Such data are evidently of particular significance for modeling the global carbon cycle.

Some quite representative observations of the circulation of substances emitted into the atmosphere and then deposited onto the underlying surface can be organized at a comparatively small number of stations (9 to 12).

In order to estimate the critical problems connected with agricultural practice and land use, it is necessary to carry out periodic mappings of urban areas, desertification areas, areas of forest cutting and forest planting, coastal areas, irrigated and nonirrigated agricultural areas, permafrost areas, foggy areas, and open-pit mines. Clearly, desertification areas can serve as indicators of climatic changes: changes in their areas lead to a change in the albedo of the Earth's surface. Forest cutting areas and forest planting areas are significant for assessing the albedo and the cycle of biogenous matter. The coastal areas are impacted substantially by man. Agricultural areas are directly connected with food production; the balance of irrigated and nonirrigated areas affects the hydrologic cycle, soil retention, and soil fertility. Permafrost areas are indicators of climatic change; foggy regions are of interest as indicators of shifts in the hydrologic cycle.

A program for ecological monitoring of the background pollution of the biosphere and the reaction of biological systems to this pollution is given in [20, 37]; this program is considered in detail in Section 5.3.

In [10] it is proposed to monitor the state of the reference natural ecosystems and their anthropogenic transformation, and also to compare the regional characteristics of the state of the environment, in order to generalize these, right up to the values characteristic of the global state.

When discussing the functional characteristics of the reference natural systems and their anthropogenic changes [10], the main attention is focused on the biotic and soil components of natural ecosystems. Observations of the ecosystem state should be followed by measurements of the radiation balance, of the thermal and hydrological regimes of the environment, and of other factors which characterize the physical state of the environment. We should also observe ecosystems subjected to different degrees of anthropogenic impact, ranging from protected ecosystems to ecosystems forming after a significant intentional or unintentional human impact.

As the integral indicators characterizing the changes in the ecological equilibrium, let us take the following: balance of the bioproductivity (the ratio of the primary bioproduction to the secondary), the rate of formation of bioproduction (the ratio of the productivity to the total biomass), and the intensity of the cycle of nutrients.

When organizing observations of the change in the state of an ecosystem, we should first of all concentrate on the possible anthropogenic disturbances and reconstruction associated with forestry, agriculture, and animal husbandry.

Besides a possible change in the state of the ecosystems as a whole (estimated according to the integral characteristics, for example, according to the change in forest area), it is very important to consider the response of biological systems to the anthropogenic impact at the level of the organism, the population, and the community. The levels of various impacts (first and foremost of pollutants), the migration, and the transformation of pollutants should be thoroughly studied, along with the biological effects of these pollutants. For instance, in order to solve the major problems of the abovementioned Project 14 of the UNESCO Program 'Man and the Biosphere', namely to evaluate the present state of the environment and the predicted ecological trends produced by pollutant impacts, the following approaches were formulated [69].

(1) The indicators of the environmental state which reflect ecosystem changes produced by pollutant impacts are identified. The development of criteria and a methodology for the selection of test organisms is included here.
(2) The 'dose-response' relationship is determined, via field observations and laboratory experiments.
(3) The paths and processes of transformation, and the incorporation of pollutants in the ecosystems, are analyzed.

The results of chemical and biological measurements are supposed to be used for these purposes.

Biological monitoring, which is closely related to ecological monitoring, being a component of it, is also discussed in this section. As already mentioned, the purpose of ecological monitoring is to detect anthropogenic changes in ecosystems (against the background of natural fluctuations). This problem can be solved using different methods, based both on separate measurements of some pollution characteristics of the biota and its response, and on continuous measurements of the integral characteristics over quite large areas. These methods will be described in this chapter. Some examples and procedures will be given here, in advance of principles of ecological monitoring, where the focus will be on the ecosystem response to the anthropogenic impacts, as the main basis of ecological monitoring.

For instance, one of the most widespread anthropogenic impacts on lake and water-reservoir ecosystems is the eutrophication process, which speeds up their aging. This process is the result of an increased amount of nutrients and organic substances (first of all, nitrogenous substances), which enter the water bodies with runoff from flooded land and agricultural fields, and with municipal sewage. With an increased 'blooming' of the water (an increased amount of blue-green algae), the oxygen content of the water drops; this results in a reduced number of some populations which are more sensitive to an insufficient amount of oxygen and to the appearance of toxins. Thus, observations of indicators characterizing eutrophication of the water body form a significant element of ecological monitoring (see [25]).

Another example of an anthropogenic impact which has been detected with the help of ecological monitoring is acidification of the surface water (and soil) produced by acid rain (when an enormous amount of SO_2 is emitted into the atmosphere (see Chapter 9)).

At pH < 5.8 many diatomaceous and green algae disappear; zooplankton representatives (daphnia) already disappear at pH ≤ 6.0. An increased content of the SO_3^{2-} ion in precipitation up to 0.6 ppm leads to pH $= 4.6$. As a result, the pH in many investigated Swedish lakes has been found to decrease to 6.0 or 5.0. Fish reproduction undergoes a substantial impact at pH < 5.5. Gudgeon, loach, and trout have disappeared completely in many lakes with acidified water. There have been cases of grave ecological consequences as a result of the presence of detergents in waste water.

Aquatic ecosystems, the living components of which are mainly algae and protozoa, usually react very quickly to pollutants. This reaction shows up either as a reduced number of species or as a change in the distribution of the number of individuals by species (in an undisturbed system such a distribution is often close to lognormal). The number of certain species can either decrease or increase (an increase is observed in the case of more tolerant species, because of the reduced competition). The mollusk number also decreases significantly in river and lake ecosystems as a result of their pollution by waste waters, while in estuaries a decrease in the number of arthropods is observed.

Such an ecosystem response can evidently also be used when organizing ecological monitoring and introducing biotesting [3].

For instance, in order to detect the impacts of nontoxic organic and inorganic matter in aquatic ecosystems we can use the eutrophication process, which is accompanied by an increased biomass of blue-green algae, a disappearance or reduced amount of different organisms because of insufficient oxygen and the appearance of plankton decomposition products, toxins of blue-green algae, and an increased heterotrophic part of the biocenosis.

In the case of thermal pollution of water bodies, we should consider the following reaction of the ecosystems: replacement of the diatomaceous communities by green and blue-green algae, and a reduced species diversity of protozoa. In the case of toxic substances and radioactive isotopes, we should take note of bioconcentration in single-celled organisms, a reduced species diversity, and a reduced number of individuals of many species.

Pollution of the marine environment, the World Ocean, is a special subject, to be covered in Chapter 11.

The pollutants entering the World Ocean are distributed nonuniformly in it, producing areas of increased pollution in ecological zones with maximum biomass. This can result in a transformation of natural biocenoses and a reduction in bioproductivity.

Such pollutants as petroleum and its products (polycyclic aromatic hydrocarbons), heavy metals (mercury, lead, cadmium, copper, zinc), and chlorinated hydrocarbons present the greatest potential danger. For example, the content of petroleum products in the Pacific amounts to 200 to 300 ppb, and an oil film or tar balls are observed over vast areas of the ocean.

The thin nearsurface microlayer of the ocean is a collector of many pollutants, and the films formed in it effect the gas exchange, heat exchange, and light propagation. This can result in ecological and climate changes.

Heavy metals primarily spread through the coastal and shelf zones of the ocean.

The accumulation factors of heavy metals in plankton vary from 10^2 to 10^4. The concentrations in commercial fish of the Pacific are up to 60 ppb, and in the Baltic Sea they are an order of magnitude higher. A substantial amount of DDT and its derivatives, as well as polychlorinated biphenyls (PCBs), enter the ocean (mainly due to atmospheric transport).

The PCB content of sea water (in the eastern Atlantic) is 20 to 30 times higher than the DDT content. PCBs have a higher stability than DDT. The coefficient of accumulation of PCBs by microplankton is 1.7×10^5; marine flora appear to be able to develop in the presence of PCBs under conditions of coacidification. There is a chance for microflora to adapt to chlorinated hydrocarbons [1], and, similarly, some marine micro-organisms have adapted to another very dangerous pollutant, namely carcinogenic benz(a)pyrine (BAP), and they have acquired the ability to decompose it, by using it as a carbon source.

The coefficient of accumulation of BAP in hydrobionts of the Bering Sea is 2.9×10^3, while in the Mediterranean it is 5 times higher.

The problem of microbiological pollution and the eutrophication of sea water, is a very serious one, particularly in coastal zones.

The peculiarities of the pollution of the marine environment can be used when organizing a marine monitoring system (see Section 5.4).

The impact of pollution on land ecosystems is also accompanied by changes which can be used as specific indicators for ecological monitoring. For instance, a reduction in the species diversity is observed, as well as a change in the distribution of individuals by species. Acid precipitation increases the soil acidity (see Chapter 9), and the increased mobility of a number of ions (aluminum, manganese, iron, etc.) produced thereby results in their redistribution. The decreased pH can alter the soil microflora and cause damage even to higher plants. Soil pollution has a considerable effect on the physicochemical characteristics, structure and productivity of the soil, and much damage is caused to soils by acid precipitation. An increased soil acidity decreases the rate of decomposition of organic compounds and reduces the productivity of nitrogen-fixing bacteria. Cases where plants have been injured directly by acid precipitation are well known. Acid precipitation falling on tree bark damages lichen species growing on the bark. Since lichens absorb nitrogen and pass it on to plants, this effect may well change the nitrogen cycle and the nitrogen balance of ecosystems, while the accumulation of sulfur can disturb the biogeochemical cycle.

The spreading of heavy metals has also a negative effect on the state of the vegetation (both due to the deposition of harmful substances on leaves and due to root uptake).

When pollutants enter into plant organisms, they affect the development of their vegetative and reproductive organs, as well as the tissues and the cellular apparatus which form them [40].

The arrival of a substantial amount of pollutants in the environment (for example, sulfur oxides and nitrogen oxides) can have some very serious ecological consequences.

Since the early 1970s scientists have observed damage to trees (fir, spruce, pine, etc.) in Europe, caused by the substantial air pollution and acid rain there. Investigations in West Germany [29] indicated that in 1982 8% of the forest area could be considered damaged. In 1983 (allowing for an extremely dry summer) 34% of the forests were considered damaged. Yellowing and dropping of needles, underdeveloped shoots, thinned-out crowns, and mycosis injury of the plants were observed.

The above effects can be used as biological indicators when establishing a monitoring system [27, 40, 42]. Often effects which are hardly noticeable during studies of a separate organism or species are revealed when we consider the state of the system as a whole. It is precisely this feature that emphasizes their importance for ecological monitoring.

When establishing an ecological monitoring system, we should carry out observations of the possible changes for different intensities of the impact; ecological monitoring at the background level should be of special concern. Changes in the state of the biosphere over large territories (even insignificant levels of change) can be observed using remote techniques, including satellite systems (see Section 5.5).

5.2. Ecological Monitoring at Different Impact Levels

Ecological monitoring of the biosphere provides monitoring of anthropogenic changes in the natural environment, together with monitoring of the effects produced by these changes and monitoring of the impact factors. Ecological monitoring should take into account all the major changes produced by any anthropogenic impacts, against the background of the natural variability (as already mentioned above).

Anthropogenic impacts on the biosphere are distributed very irregularly throughout the world; on the one hand, there are zones where such impacts are almost absent (central Antarctica, the southern part of the Southern Ocean, vast nature reserves on different continents), while, on the other hand, in some parts of the world the anthropogenic impact has reached such a level that the original ecosystems, relief, and even the type of landscape, have been changed completely (indrustrial regions, especially open-pit mines, areas of intensive agriculture, urbanized territories); the environment of such an area contains admixtures (chemical substances) which are completely absent in nature.

The objective is to organize ecological monitoring with due regard for the alternate arrangement of different areas (sometimes a large distance apart), where the loads due to anthropogenic impacts (measured by the intensity of various impact factors) change by many orders of magnitude. In areas of intensive anthropogenic impact, the appropriate reactions of the ecosystems correspond to the above factors.

Figure 5.1 shows a plot of the intensity of the impact factor κ (for example, the concentration of the air pollutant) as a function of the impact effect η (ecosystem response) (for instance, the reaction of coniferous and leaf-bearing trees to the impact of sulfur dioxide and sulfuric acid in precipitation) at different distances from the impact source (in any direction).

Figure 5.1 shows that a very low level of biological effects (for instance, slight damage to the leaves of leaf-bearing trees) corresponds to the background pollution level κ_b (which is very low for SO_2).

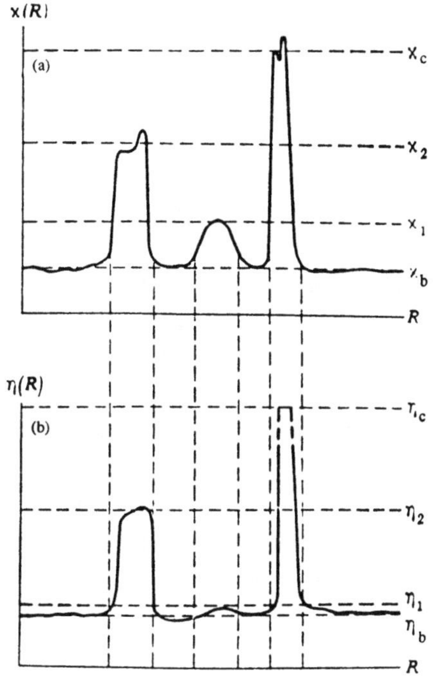

Figure 5.1. Variations of intensity of impact factor κ (a) and of reactions of ecological system η (b) with distance.

High values η_1 and η_2 correspond to high values κ_1 and κ_2 (the total effect can also be nonlinear). Impact κ_3 exceeds the critical level (such impact is impermissible); it corresponds to $\eta(\mathbf{R}) \to \infty$ in Figure 5.1.b, i.e., to destruction of the ecosystem, in this case to dessication of coniferous or leaf-bearing forests.

The objective of ecological monitoring is to observe the impacts on an ecosystem and its response to these impacts, as well as to evaluate the observed changes in the biosphere, at present and in the future, at different levels of impact:

- against the background global and regional levels (the quantity $\kappa_b \to \eta_b$ can be found as the mean value of the minimum level, prevailing in wild or relatively wild areas);
- at intermediate levels ($\kappa_1, \kappa_2 \to \eta_1, \eta_2$);
- in places of 'high' impact levels ($\kappa_c \to \eta_c$)

It is important to assess the possible impact at the intermediate and critical levels, not only on the ecosystem in regions of direct impact, but also on the entire biosphere. (For example, depleted elements of the biosphere would be restored or replaced at the expense of some of the reserves of ecosystems in the vicinity of the damaged areas). Zones with a damaged or destroyed ecosystem can gradually have a negative effect on ecosystems of adjacent areas (e.g., desertification or secondary pollution produced by pollution in neighboring regions).

Thus, if we consider the change of the ecosystem response η both in space and in time, then we should determine η_b on the global and regional scales (in the future, strictly speaking, up to infinity).

The establishment of an ecological monitoring system for all levels of impact, on both global and regional scales, is very complicated.

For the time being, some factors of the impact and the states of individual subsystems of the biosphere are being monitored only over limited areas. Some background observations are carried out only for some impact factors and for a small number of biosphere responses. The obtained data on background levels are not always correctly extrapolated to large areas or to the entire world. Observations in areas with a high load on the environment are also carried out according to separate indicators (mainly according to the pollution level), in order to be able to take prompt action in case of obvious impermissibility and in order to take certain planned measures to reduce 'intermediate' loads.

Observations of the state of ecosystems in areas of high load are carried out on a comparatively small number of natural objects, mainly for research purposes. The organization of regular ecological monitoring in a zone of high anthropogenic impacts is an independent problem of great importance.

The organization of ecological monitoring at the background level (global and regional) is a task of no less importance. As already noted, this requires not only the establishment of quantitative relations between the impact factors and the individual reactions of ecosystems, but also a thorough study of the change of these relations in time due to the above effects (effects of neighboring and more remote zones with substantially changed conditions produced by intensive impacts). It can happen that under such conditions the impact of some pollutant with the previous concentration will cause a quite different response of the biological system (because of changes in temperature, humidity, degree of pollution, or other environmental characteristics).

These effects can be calculated using an approach which is very close to that described in Section 2.5, where the concept of the dose-'commitment' was discussed. In the general case we have to find the impact function of a single source on any element of the biosphere on a global scale, and then integrate this function over space and time.

As already noted, within the framework of ecological monitoring, the observation (and study) of changes in the state of the abiotic and biotic components of the biosphere should be implemented at different levels: the impact level (when the impacts are of significant value), the regional (intermediate) level, and the background (global and regional) level.

In order to implement the monitoring of anthropogenic changes in the environment, it is necessary to select the most representative (from the viewpoint of the ongoing changes and the susceptibility to changes) elements in the ecosystem, the characteristics and responses. To that end we must study the character of the reactions of the biosphere elements to the anthropogenic impacts, with the help of in-situ and laboratory experiments, mathematic modeling, and a thorough analysis of the results of the field observations.

Field observations. Observations of the factors of the anthropogenic impact and of the biological responses, together with an analysis of the results of these observations, make it possible to distinguish (according to the indicators selected) major trends of the change in the state of the ecosystems. Long series of observations enable us to determine the rates of such changes and to select those indicators for which the rate of change is very high, which can indicate that the situation created is undesirable or even critical.

It is very complicated to determine the character of the relation between the various impact factors and the biological responses of the system. If there is a large complex of impact factors, an analysis of the results of only the field observations is very cumbersome, and it does not always lead to an unambiguous solution: finding out the real causes of the changes. This is evidently explained by the complexity of the combined and complex impact on the biological system of many simultaneously acting factors and by the low background levels of the impact.

Experimental investigations. As a strategic basis when studying and analyzing the impact of various factors on complicated biological systems, it was proposed in [50, 51, 64] to carry out a many-factor experiment (mathematically and economically established), using methods and techniques developed for systems 'with a bad structure' [35]. Such an experiment makes it possible to provide a simultaneous and independent analysis of a large number of factors affecting biological systems. The objective of the experiment is to study probable (anticipated) situations, which are predicted on the basis of the current trends toward environmental change. In order to determine the levels of the selected variables when setting up the experiment, it is advisable to specify some criterion, for example, constancy of the rate of change of the selected variables (pollutant concentration, certain indicators of the state of the environment).

With the help of such an experiment, we can model real situations with a changed state of the environment (situations which can take place in the future), and we can identify certain factors responsible for such changes, for instance, eutrophication of water bodies produced by the arrival of phosphorus and nitrogen compounds.

Mathematical modeling. The 'impact-response' relationship in complicated ecosystems can be established, and the extent of the anthropogenic impact (including the anthropogenic impact on the climate) can be determined with the aid of mathematical modeling. Models enable us to study the sensitivity of an ecosystem to a variation in some impact factor.

The above approach, as applied to the effect of pollution on an ecosystem (including the background level), is dealt with in [20, 54]. The numerical values of the coefficients in such models are assumed to be determinable with the aid of laboratory experiments.

Simulation-balance mathematical models of natural ecosystems can be used to predict the future of these systems, although only for periods less than the succession periods.

However, since the above simulation models do not take into account the responses of the biota to anthropogenic impacts for a long-term period, they can lead to significant errors.

A more universal model with an exogeneous succession block (in order to predict the responses of an ecosystem to an impact for time intervals longer than the succession periods) is suggested in [20, 24].

For such an approach the dynamics of the system of biological species is simulated by a system of difference equations

$$x_i(t+1) = K_i[x_1(t), \ldots, x_N(t);\ \mathbf{a}_1, \ldots, \mathbf{a}_N;\ \mathbf{u}]x_i(t), \tag{5.1}$$

where $x_i(t)$, $x_i(t+1)$ is the number (or other quantity characterizing the biomass) of the ith population at moments t and $t+1$, respectively; $\mathbf{a}_j = (a_j^1, \ldots, a_j^k)$ is a vector ecological parameter characterizing the state of the jth species; $\mathbf{u} = (u_1, \ldots, u_k)$ is a vector parameter characterizing the state of the elements of the abiotic component of the environment (a complex of hydrometeorological, geochemical, and other characteristics) or the impact factor; K_i is the reproduction coefficient of the ith species.

The ecological parameters characterize such properties of individuals as fertility and the resistance to various biotic and abiotic factors of mortality. In the modeling of natural ecosystems, the trajectories of system of Equations (5.1) should possess a certain stability: the fluctuations of each species number (density) should be limited. Otherwise an unacceptable situation occurs: the loss of any species from the ecosystem (which means formation of a new, transformed ecosystem).

The climax state is stable relative to a possible rearrangement of the genetic pool: each individual is maximally adapted to the ecosystem conditions. The geometric-mean net reproduction of such individuals is equal to unity, while that of any others (e.g., new individuals), being less well adapted, is less than unity.

This property of the climax ecosystems is formalized as an optimality principle, whereby the real parameters a_j^n lead to a maximum for the functional

$$\varphi_i(\mathbf{a}_i) = \widetilde{K}_i[x_1(t), \ldots, x_N \mathbf{a}_1^r, \ldots, \mathbf{a}_{i-1}^r, \mathbf{a}_{i+1}^r, \mathbf{a}_N^r;\ \mathbf{u}] \tag{5.2}$$

where K_i is the geometric mean (in time) of the net reproduction of the ith

species.

The ecological parameters are known to be interconnected in a certain way (they form sets $A_i(\mathbf{u})$, known as Matthew–Kermak sets), but they differ when the environmental conditions are different. Some parameters of the abiotic component of the environment are also interconnected (soil moisture and temperature, concentration of matter and wind velocity, etc.), although in many cases the dependence is one-sided; for instance, the concentration of atmospheric pollutant practically does not affect the wind velocity.

A change in the state of the abiotic part of the environment and/or a change in the factors which affect a biological system. i.e., a change, $\mathbf{u}_k \rightarrow \mathbf{u}'_k$, leads to a change in the state of the ecosystem; an exogenous succession appears, and the rearrangement of the ecosystem leads to a new climax state, which corresponds to values $\mathbf{u}_k$.

Generally the problem of ecological monitoring includes a determination of sets $\mathbf{u}$ and $A_i(\mathbf{u})$. However, it is advisable to select as few parameters as possible (while retaining the most significant). According to the authors of [20], the optimum organization of an ecological monitoring system should ensure that the integral indicators are included in the prognostic models, i.e., net reproduction of species under various environmental conditions. It is precisely this factor which reflects well the species responses to the impact. The ecological parameters which determine the net reproduction include the fertility of the species and the resistance to mortality factors.

The next, equally complicated, problem related to the establishment of ecological monitoring is selection of the most representative and sufficiently sensitive species, out of a large number of biological species.

Since it is very laborious to investigate the responses of all species of the ecosystem to the environmental impact, therefore the main thing is to select representative species, with a subsequent interpolation of the observation results of the selected species to the others. Minimum variance of the measured quantity within the interpolation region can be used as the interpolation principle.

It is noteworthy that linear interpolation will be the most accurate in a coordinate system in which the autocorrelation function of the biota response field will decrease the most slowly; a phylogenetic tree represents such a coordinate system [37]. In such a system at the transition from higher taxons to lower taxons the position of the main points is not as important (and a change when selecting a species within the genus does not affect the accuracy as much). The additional possibilities appearing thereby should be used to select the most sensitive species and the species which are most important ecologically.

Such an approach seems optimum when establishing ecological monitoring subsystems under various anthropogenic impacts at all possible levels and for a different intensity.

Biological monitoring is closely connected with ecological monitoring. Its aim is to identify and evaluate anthropogenic changes related to a change in the biota and biological systems, and also to estimate the state of these systems. When implementing ecological monitoring, the state of the biological systems can be determined by assessing the correspondence between the given biogeocenosis and the criterion of a 'good' biogeocenosis [53]. To do this, we have to evaluate:

- the production of all the main links in the food chain;
- the correspondence between high productivity and high production (which determines the compensatory activity of the biological systems);
- the structural stability and diversity of separate trophic levels;
- the rate of exchange of matter and energy in an ecosystem which is characteristic of a possible biological selfpurification of the system.

In an ecological monitoring system determinations of the number of individual species and their state can be used as indicators of the environmental conditions and their effect on biological systems. Shvarts [53] considers the following to be suitable specific indicators of the environmental conditions: the content of chemical substances in various tissues of organisms at various levels of the food chains; the growth rate of trees; the energy of photosynthesis; the microbiological activity of soil; the growth of lichens; and the development of various hydrobionts. These data can be supplemented by data on the change in the structure of biogeocenoses and on their spatial and functional interrelationships. Of particular importance is an analysis of the biocenotic homeostasis, when simplifying individual food chains.

The described procedure forms the basis of biological monitoring (within the framework of ecological monitoring). Most authors [20, 46, 51, 60] believe that biological monitoring includes observations of impact factors, but that the main focus of biological monitoring should be observations of the biological consequences, namely the responses of biological systems to external impacts, to changes in the state of the environment.

According to [47], biological indicators can be divided into two categories: functional indicators, which are expressed as the time derivatives of certain functions (e.g. productivity indicators, respiration, metabolism rate, photosynthesis rate), and structural indicators, which can be expressed by an integral over time as a total result (e.g., an indicator characterizing the number of species and individuals, the amount of biomass, the change in the size and mass of individuals, the content of substances in the ecosystem).

With reference to Academician V.E. Sokolov, the author of [10] gives some more categories of biological indicators (signs). In that work structural indicators are called structural-morphological indicators. In addition, we have ethological signs (change in activity, migration, change of food regimen) and pathological signs, which correspond to disturbances of the homeostasis (various diseases, decreased resistance to cold and heat).

The following factors are used as functional indicators for biological monitoring [48]:

- growth (productivity) indicators;
- loss indicators (respiration, lifetime dying of organic matter because of shedding and moulting);
- state indicators (food consumption and digestion, rate of circulation of various elements in ecosystem), etc.

The selection of the structural indicators depends on the organization level of the biological system (see [45]). In [10] it is noted* that, when monitoring the state of the biota, we should pay particular attention to:

- fluctuations of the total population number (and the causes of these fluctuations);
- changes in the age and sex composition of the population;
- a change in the reproductive processes and the reproduction intensity;
- a change in the reproductive cycle;
- a change in the embryonic and postnatal development.

Let us stress once again that a simultaneous recording of the indicators which refer to the impact factor and the biological response are of great importance for biological monitoring; this makes it possible to interconnect all the observations characterizing the state of the biosphere as a complete system [51]. A specific program for studying the real rates of successions and diagnostics of anthropogenic and other exogenous impacts is presented in [43]. Biological monitoring at different levels is discussed in [28].

Observations (particularly long series) make it possible to identify the major trends of changes in the state of the biosphere, to determine the rate of changes of individual indicators characterizing the state of the environment, and to specify indicators with a very high rate of change. Attention should be given not only to an obvious change in the structural signs, but also to a change in individual biological responses (e.g., changes in the productivity of some populations). We should establish relations (correlations) between indicators reflecting, on the one hand, the intensity of the impact factors and, on the other hand, the change in the biological responses in the ecosystems.

Biotesting and the use of vulnerable species are of ever-increasing importance in biological monitoring, in order to detect anthropogenic effects [3, 25, 27, 36, 40, 42].

Fedorov [50] considers how to analyze the state of a biological system on

---

* This portion of the program is based on proposals made by coworkers at the Severtsev Institute of Evolutionary Morphology and Animal Ecology of the the USSR Academy of Sciences (note taken from [10]).

Table 5.1

Level of biological organization	Suitable indicator
Subcellular	Membrane potential of cells
Cellular (organismic)	Increased respiration rate
Population	Assessment for individuals of population with bilateral symmetry in sign distribution, generation time, sex ratio, age-group ratio
Trophic grouping (simple community)	Species diversity, chlorophyll content
Complex community	Ratio of outputs of various trophic groupings

the basis of an assessment of indicators which have been selected in a special way.

Homeostasis is considered to be a mechanism which controls and regulates the change in the ecosystem characteristics. It is postulated that the complex of estimates characterizing the processes which are regulated in the system by homeostatic mechanisms conform (under normal conditions) to a Gaussian distribution [48]. Such a situation, which has been experimentally confirmed, enables us to separate the normal states from states outside the norms. For example, estimates within the interval $y \pm \sigma$ indicate a good state of the system, while estimates beyond this interval indicate a bad state.

Substantial difficulties may be encountered when studying biological responses: if a relatively large number of variables are considered in the experiment (the study of the state of the system), then the number of recorded responses in the biological system can be practically infinite. Consequently, the method of data discrimination, and the principle used to select these data, should be ascertained.

In [48] the following (systemic) approach is suggested: five levels of biological organization with suitable indicators selected according to some criterion are specified. The selected levels and the corresponding suitable indicators are given in Table 5.1 (the number of indicators can be limited substantially). These indicators lie outside the framework of biological monitoring per se, and they constitute an 'ecological spectrum'.

Fedorov [50] gives some rules for selecting variables when monitoring the state of biological systems:

(1) select indicators which refer only to processes with homeostatic mechanisms;
(2) give preference to indicators characterizing a nonspecific response to the impact factor;
(3) give preference to integral indicators.

It is assumed that the presence of a norm according to some indicators and pathology according to other indicators, as well as of their ratio, provides a single-valued determination (and identification) of the 'disease' of the biological system.

If there are no specific indicators when various factors affect the environment (e.g., when numerous indicators determine the flow of energy and matter in the system), then the solution is to unite indicators assessing the environmental quality as either good or bad. To do this, we have to introduce a quality assessment for each variable d_i and a generalized quality assessment D, taking into account the results of the assessments d_i.

According to Fedorov [49, 50], the effect of a harmful impact can be evaluated by selecting a singular generalized indicator of the state of the ecosystem, which incorporates the individual responses of the system via a so-called desirability function.

Fedorov [49] presents the results of experiments in which the effects of various pollutants (their mixtures) on a biological system were reliably assessed (and compared) with the help of this criterion. He also suggests using the above approach to determine the priority of the pollutants, from the viewpoint of their possible negative effects on biological systems.

5.3. Ecological Monitoring on Land (Based on Biosphere Reserves)

The study of ecological variations and the organization of ecological monitoring at the background level seem to constitute the most complicated problem.

The organization of orderly ecological monitoring at the background level began in the Soviet Union with the creation of such a system on the basis of the biosphere reserves. The discussion of the problem (in this formulation) began in 1974. This aspect was included in the sphere of Soviet-American cooperation in the field of environmental protection. The organization of the biosphere reserves (stations) in order to study, monitor, and predict anthropogenic changes in the state of the biosphere was discussed in the first symposium report, presented in 1975 by the Soviet participants I. P. Gerasimov, Yu. A. Izrael, and V. E. Sokolov [9]. Thus, the main tasks of the biosphere reserves (in accordance with Project 8 of the UNESCO program 'Man and the Biosphere') to preserve natural ecosystems and the gene pool of flora and fauna were supplemented by an integrated ecological monitoring program, providing for continuous observations of the background indicators of the state of the biosphere and man-made changes in the biosphere, as well as investigations needed for the scientifically based selection of the appropriate parameters for monitoring the state of the environment.

It was intended to study a number of geophysical characteristics such as the solar radiation flux, to organize observations and studies of anthropogenic transformations of the cycle of major chemical elements, and to investigate the soil cover, the water balance and the hydrologic cycle.

It was also planned to study thoroughly pollutants of the environment and to measure a wide range of ingredients in different parts of the environment.

The biota investigations were supposed to be carried out at the biocenotic level (biota composition, variations in it, functional vital activity, and biological productivity), the species level (dynamics of populations of indicator species), the physiological level (photosynthesis, respiration, growth, reproduction) and the molecular-genetic level (mutagenesis, teratogenesis). These investigations should be closely connected with observations of the abiotic component of the biosphere.

Therefore, it was decided to carry out comprehensive investigations in the biosphere reserves both of the external factors of the environment and of internal processes and phenomena in ecosystems, at the background level and in a buffer (transition) zone, where different types of land use could be studied and results of anthropogenic effects could be compared with the background state of the ecosystems.

The project of organizing the first five biosphere reserves in the USSR was considered in [9].

Later, issues of ecological monitoring on the basis of biosphere reserves were developed in reports presented at the First Soviet-American Symposium devoted to that very problem [1, 2], in Project 14 of the UNESCO Program 'Man and the Biosphere', in reports at the First International Congress on Biosphere Reserves (Minsk, September 1983) [11, 12, 23, 44], as well as in [4, 28, 56, etc.].

The integrated program of ecological monitoring at the background level based on biosphere reserves was developed in 1977–1979 by a number of Soviet research institutions: the Laboratory for Monitoring Environment and Climate of the USSR State Committee for Hydrometeorology and Environmental Monitoring of the USSR Academy of Sciences (subprogram of integrated background monitoring, including monitoring of environmental pollution and monitoring of biological responses to the effect of pollutants); the Institute of Geography of the USSR Academy of Sciences (subprogram of studies of functional characteristics of reference natural ecosystems and their anthropogenic modifications); Severtsov Institute of Evolutionary Morphology and Animal Ecology of the USSR Academy of Sciences (subprogram of monitoring the state of the biota). The program was discussed in [10, 20, 66]. This program also included proposals made by nature protection organizations of the USSR Ministry of Agriculture (see, for instance, [32]). A brief summary of this program is given below.

The program of ecological background monitoring based on the biosphere reserves includes the following sections:

(1) monitoring environmental pollution and other factors of the anthro-

pogenic impact;

(2) monitoring biota responses to the anthropogenic impact, primarily background pollution levels;

(3) observations of a change in the functional and structural characteristics of virgin ('reference') natural ecosystems and their anthropogenic modifications.

The results of the above observations can be interpreted only by carefully studying the background geophysical characteristics of the environment (hydrometeorological observations, including observations of the radiation balance, etc.). Measurements of the hydrometeorological quantities pertain to the first section of the background ecological monitoring, that is, to 'abiotic' monitoring [37].

The observations pertaining to the first section (monitoring of environmental pollution) should be so organized that the results obtained provide sufficient information about the concentrations of different pollutants (including the biota), about migration processes and the circulation of these pollutants, and about their accumulation and transformation [19].

Measurements are made of the pollutant fluxes from the atmosphere to the underlying surface, and of the migration with surface and underground waters and via food chains in ecosystems.

When selecting the pollutants to be included in the program of measurements in biosphere reserves, criteria like the following (see [7, 39, 42] as well as Chapter 4, above) were taken into account: the abundance of pollutants, their stability and mobility in the environment, their ability to affect biological and geophysical systems (including the climate and climate system). A list of the pollutants so selected (for different parts of the environment) is given in Table 5.2. This list is similar to one compiled in the USA [34].

The periods between measurements (the measurement frequency) differ for different media: for air and atmospheric precipitation these range from 10–20 days, and for other media they range from 2 to 6 months.

Also to be carried out are measurements characterizing the state of the environment (atmosphere turbidity, pH of water) and observations of a number of hydrometeorological quantities for interpreting problems of transport, scattering, and migration of pollutants, and of solar radiation (including UV radiation).

As noted, when considering a given section of the program of ecological monitoring at the background level, much attention has been given to the priority of the measured impact factors and effects in the biosphere produced by the most significant anthropogenic impacts. The impacts causing the strongest responses of ecosystems, which are therefore the most ecologically dangerous, have been given high priority.

However, this problem should also be considered from the viewpoint of

Table 5.2. List of chemical substances to be determined in parts of environment at background stations in biosphere reserves.

Measured pollutants	Part of environment				
	atmosphere	atmospheric precipitation	surface and ground water	soil	biota
Suspended particles (aerosols)	+				
Sulphur dioxide	+				
Ozone[1]	+				
Carbon monoxide	+				
Nitrogen oxides	+				
Hydrocarbons	+				
Benzo(a)pyrene	+	+	+	+	+
Organochlorines (DDT etc.)	+	+	+	+	+
Heavy metals (lead, mercury, cadmium, arsenic)	+	+	+[2]	+	+
Carbon dioxide	+				
Freons	+				
Biogenic elements (nitrogen, phosphorus)		+	+	+	+
Anions and cations[3]		+			
Radionuclides		+			

[1] Integral amount and concentration in the surface air layer
[2] Including methylmercury
[3] According to WMO program

a possible maximum disturbance of the natural geochemical equilibrium. Evidently such a disturbance, even if it is not causing serious damage at present, can have serious consequences in the future, i.e., it presents a certain potential danger. When some anthropogenic pollutants enter the natural environment, they can greatly alter the existing natural background level of these pollutants, thereby causing present or future (possibly even remote future) damage (here only pollutants which are already present in the natural environment are referred to).

In order to asses the change in the natural cycle of the substances produced by anthropogenic activities, the concept of the 'technophilic nature' is introduced into geochemistry. The latter is determined by the ratio of the annual production of the given chemical element (in tons) to its percent abundance in the litosphere. However, coefficient K_{li}, characterizing the ratio of the total amount of the ith substance emitted into the considered part of the environment (in tons) to its percent abundance in the lithosphere, is

Table 5.3. Quantities characterizing the technophylic nature of various elements.

Element	K_{li}	Emissions into atmosphere, tons/year	Mean content in lithosphere, %
Hg	4.5×10^9	1.5×10^4	3.3×10^{-6}
Cd	1.6×10^9	2.5×10^4	1.6×10^{-5}
Sb	1.0×10^9	2.0×10^4	2.0×10^{-5}
Br	9.0×10^8	2.0×10^5	2.2×10^{-4}
Pb	6.4×10^8	1.0×10^6	1.6×10^{-3}
Se	3.6×10^8	5.0×10^3	1.4×10^{-5}
Zn	1.7×10^8	9.0×10^5	5.1×10^{-3}
Ag	1.6×10^8	8.0×10^2	4.8×10^{-6}
As	1.1×10^8	2.0×10^4	1.9×10^{-4}
Fe	8.3×10^5	3.0×10^6	3.6
Al	1.75×10^5	1.4×10^6	8.0

evidently more significant [33].

Table 5.3 gives such data, in the order of decreasing coefficient K_{li} (for the atmosphere). It is clear from the table that, despite the relatively small emissions into the atmosphere of such elements as mercury, cadmium, antimony, and lead, the geochemical equilibrium can be serously disturbed by the anthropogenic activity, because of the low contents of these elements in the Earth's crust. On the other hand, in spite of the considerable emissions of iron and aluminum, these do not play a substantial part in disturbing the geochemical equilirium.

It would be of great interest to assess the priority of the 'anthropogenicity' of the various elements transported with aerosols. This characteristic takes into account both the amount of anthropogenic pollutants emitted into the atmosphere and their ability to spread over for long distances (but without taking into account passage into other parts of the environment). Such a criterion, allowing for the necessity to identify the ecosystem responses to anthropogenic impacts, is of particular importance.

In order to determine such a priority, the following enrichment factor, or fractionation factor (relative to some reference element), was suggested in [33]:

$$K_i = \frac{C_i}{P_i} : \frac{C_0}{P_0}, \tag{5.3}$$

defined as the ratio of the concentration of the element being investigated to the concentration of the reference element in the sample, normalized to their percent abundance (some time later, a similar approach was described in [4]).

As a reference element we selected scattered elements with low K_{li} (such as thorium, cesium, and scandium).

Table 5.4 shows the fractionation factors K_i for aerosols collected in sev-

Table 5.4. Fractionation factors of various elements (relative to Th) in areas of Central Asia.

Element	$K_{(a+l)_i}$ (industrial area)	K_{li} (background area)	K_{ai}
Sb	7.0×10^3	1.4×10^3	5.0
As	6.6×10^2	1.1×10^2	6.0
Hg	4.2×10^3	2.1×10^2	20.0
Cr	4.1×10^1	1.2×10^1	3.4
Br	6.4×10^3	8.9×10^3	0.72
Fe	6.0	2.7	2.2
Co	3.4×10^1	2.9×10^1	1.2
Zn	1.6×10^3	5.5×10^2	2.9
Cs	1.9×10^1	2.6×10^1	0.71
Cd	1.1×10^4	3.5×10^3	3.1
Mo	2.7×10^2	5.4×10^1	5.0
W	5.5×10^1	3.7×10^1	1.5
Se	2.5	3.8	0.66
Th	1.0	1.0	1.0

eral industrial and 'background' areas of Central Asia ($K_{(a+l)_i}$ and K_{li} respectively). A mountain glacier (Abramov glacier in the Pamirs) was taken as a background area. The values of the fractionation factors in background areas can exceed unity due to natural processes, and they can characterize the degree of volatility of a chemical element. In order to determine the priority, allowing for this effect, the values of $K_{ai} = K_{(a+l)_i}/K_{li}$ are given in the Table as well.

As seen from Table 5.4, the elements in the aerosol pollution can be ordered according to their degree of 'anthropogenicity':

$$\begin{array}{ccccc} \text{Hg} > & \text{As, Sb, Mo} > & \text{Cr, Cd, Zn} > & \text{Fe, W} > & \text{Co, Th, Sc, Cs, Br.} \\ 20.0 & 5\text{–}6 & 2.9\text{–}3.4 & 1.5\text{–}2.2 & 0.6\text{–}1.2 \end{array}$$

The data in Table 5.2 show that some elements possessing the highest priority, according to the above criterion (mercury, arsenic, cadmium), were already included in the program of ecological monitoring on the basis of other (sometimes very similar) characteristics.

Monitoring the biotic component of the biosphere (biotic monitoring) consists in recording the biota responses and evaluating the impact on the biota of pollutant levels close to the background level. This is a fundamentally new problem; some approaches to the solution of this problem are discussed in [20, 54], and a program for this part of the ecological background monitoring in biosphere reserves is presented in [54].

As noted in Section 5.2, ecological background monitoring consists of a program of field observations, an experimental program, and a program of mathematical modeling. Obviously, it is practically impossible to evaluate the background pollution effects on the biota only from field observations, because data collection over a very long time is needed to detect these low level impacts with sufficient reliability.

In this connection, it is suggested to include in the experimental program experiments showing how the impacts (identified during the field experiments), produced in special chambers (ecostats), held under controlled (constant) conditions, affect the ecosystem elements of the biosphere reserves.

The reproduction factor is selected as an integral indicator to plot the 'dose-effect' curves, in accordance with considerations given above (see Section 5.2). The choice of experimental objects and the interpolation of the results can be carried out in the coordinates of the phylogenetic tree, for minimization of the interpolation errors. These recommendations are based on mathematical modeling. The experimental investigations and the modeling can lead to a prediction of qualitative and quantitative changes of the ecosystem in time.

Observations of species, especially in order to predict their loss or reduced abundance, are recommended as a program of field observations of the biotic section of ecological monitoring, aimed at evaluating and monitoring the current state of the ecosystem.

Recently significant progress has been made in the scientific establishment and realization of programs of ecological monitoring. Our program of ecological monitoring of anthropogenic pollution [16, 20, 37, 54] has been considered and debated at national and international (including WMO and UNEP) levels [16, 22, 61, 66]. It has already been implemented in a network of biosphere reserves. The specific content of this program, which is a result of the above scientific premises and the principles developed in [23, 24], will be set forth below.

The main attention in the given program is concentrated on monitoring pollutants and their ecological consequences. The aim of the program is not only to establish the factors observed but also to predict them (that is why the program includes the development and use of prognostic ecological-mathematical models [41]). Abiotic monitoring and biotic monitoring are included as constituent parts of the program.

Quite detailed information is available about the selection of the chemicals to be determined at background stations in natural environments, as well as the criteria for their selection. A list of the chemical substances is given (see Table 5.2). This part of the ecological monitoring is already being realized, and observations are being carried out in Soviet biosphere reserves (see, e.g. [25]).

The program of abiotic monitoring also includes hydrometeorological and other geophysical observations.

The hydrometeorological and geophysical characteristics which are to be measured at background monitoring stations should include the quantities required for interpreting data on the pollutant concentrations in individual parts of the environment and for studying biogeochemical cycles and cycles of chemical substances, as well as for implementing climate monitoring. These quantities are:

(1) wind velocity and direction, atmospheric pressure and air temperature, humidity, and amount of atmospheric precipitation;
(2) intensity of solar radiation (direct, scattered, and total) including ultraviolet radiation;
(3) water level and flow rate, water temperature, flow rate of suspended matter;
(4) humidity and thermal balance of soil.

The frequency of atmospheric observations varies from 1 to 5 days, and atmospheric precipitation is collected once every 10 days or once a month. An integral sample of the snow cover is taken once a year, before the spring snow melts. Surface waters and suspended solids are sampled six times a year during characteristic hydrological periods. Bottom sediments are sampled once a year during the summer low-water period. The soil is investigated once or twice a year and the biota twice a year (content of substances).

The monitoring of biota responses to effects of anthropogenic pollutants at levels close to the background constitutes a new and complicated problem.

As already noted, the species reproduction factor, which does not depend on the density, is selected in the program as a control indicator, in order to assess the state of a biological species. A change of this indicator produced by a change in the pollution level of the environment serves as a good integral characteristic of the biota response to the pollution impact. Furthermore, the above indicator can be used directly to predict the exogenous succession.

In accordance with the objectives of ecological monitoring, the program of biotic monitoring includes:

(1) assessment of the current state of the biota (periodical measurements of the reproduction factors of some species);
(2) prediction of biota respones (establishing the relationship of the biota sensitivity to the anthropogenic pollutants);
(3) monitoring the state of the biota with the help of field methods.

The implementations of these three subprograms are interrelated.

The first subprogram reduces to a continuous recording of changes in the reproduction factors of the populations of test species. The testing takes place in vegetation chamber ecostats, where the conditions of the abiotic environment (temperature, humidity, illumination) are regulated.

In the first stage of realization of the program it is recommended to use only species from certain groups, selected because they are representative, widespread, and sensitive to the pollutant impact.

The relative increment in the abundance of a thinned culture will serve as a specific indicator for protozoa, bacteria, and algae. The relative increase of biomass will serve as a specific indicator for mosses, lichens, and higher plants.

The periodicity of the measurements for each specific species is individual and depends on the duration of generation. The pollution levels of the experimental parts of the environment (air, water, soil) correspond to the current background pollution in this area.

The above measurements are carried out at background monitoring stations in each biosphere reserve. For each of the above groups one specific test species is selected. When selecting this species, we should take into account the abundance of the species in the regional ecosystem, its response to the pollutant impact, and its importance for the regional ecosystem.

In the case of protozoa, bacteria, and algae (one species of each) the following frequency of observations (number of tests per year) is recommended: twelve tests each of the relative increment in abundance per 24 hours; one test each of the relative increment of biomass per year for mosses and lichens; and four tests each for higher plants.

A reliable variation (increase or decrease) of the control parameter, the reproduction factor, and the existence of a corresponding persistent trend, are indicative of a change in the state of the population of the control species in the region of the background station, produced by the change in the state of the pollution.

Monitoring the biota response to the pollutant impact involves a complex of experiments to determine the species reproduction factor for different pollutions, according to individual polluting substances. All the other abiotic conditions are fixed. The experiments with each of the species selected are carried out in vegetation chamber ecostats.

The reproduction factor of each species is determined individually for each priority pollutant on the list.

The principle of selection of the system of species (test objects) is minimization of the error in interpolating the entire biota response to the pollution, based on data on the sensitivity of the test objects.

Report [61] presents some sets of classes, from which one test species each should be selected at the corresponding stages of development of this subprogram.

The obtained information on the field of the sensitivity of the biota to a change in the pollution of the environment (dose-response relation) can be used for a very simple tentative prediction.

The need to monitor the state of the biota with the help of field methods is explained by the errors and impreciseness which creep in during the assessment and prediction of the state of biocenoses within the framework of the previous subprograms. Those biological species which are most sensitive to the pollutant impact are precisely the ones which should be monitored.

Academician M.S. Gilyarov [12, 76] believes that anthropogenic activities affect the composition and the ratio of elements of the soil fauna. This feature, as well as individual bioindicators, namely the concentrators of various elements (worms concentrate cadmium, ground beetles concentrate lead; wood lice concentrate copper), can be used in the monitoring.

According to [12], the animal population in the soil should be one of the major objects of ecological monitoring. It is noted in [12] that soil invertebrates (which are characterized by a high abundance, biomass, and species diversity: 200 to 300 species per m^2 and a total number of up to 2 million individuals) can be widely used for these purposes. Preference is given to large soil invertebrates, many of which live under a wide range of ecological conditions. Worms, snapping beetles and their larvae, large predatory ground beetles, some species of wood lice and diplopods, and, finally, cockroaches and darkling beetles in arid ecosystems turn out to be most suitable for monitoring purposes. The abundance and activity of the soil fauna should be determined three times in a season. A two-step monitoring system is convenient: monitoring of some species of large soil invertebrates at the population level and periodical soil and zoological surveys of typical habitats.

Gerasimov [10, 11] suggests how to study the functional characteristics of 'reference' (natural) background ecosystems in biosphere reserves, by a comparison with their anthropogenic modifications in intermediate (*vis à vis* the anthropogenic impact) zones and with natural and technological systems.

Observations are proposed both of climax ecosystems and of succession series of ecosystems, which develop both under natural conditions (e.g., post-fire recovery) and as a result of anthropogenic short-term impacts (postcutting recovery) or continuous impacts (under conditions of anthropogenic pollution). Comparative observations are recommended of spatial combinations of natural ecosystems formed under various conditions, e.g., studies of watershed, slope, and lowland ecosystems [8].

When monitoring the state of land ecosystems, it is recommended [56] to study thoroughly the energy balance and substance reserve of the ecosystem; this presupposes a determination of the amount of organic matter produced in the ecosystem, of how much of it gets into the soil, of the amount of decomposed substance, and of the absorption of living plant material by other organisms. Major pollutants in the viable parts of the ecosystems should also be recorded (e.g., sulfur and heavy metals are studied in the needles, lichens, forest litter, upper soil layer, and earthworms of a pine-forest ecosystem). The soil cover and plant cover should be described in detail (so as to relate the ecosystem under study to larger biogeophysical formations).

Annual photography of constant ecological profiles is proposed in [43], in order to monitor the succession dynamics of biocenoses (so as to identify anthropogenic effects). The photos help us to study the separate stages of the succession, and also to detect variations in the forest erosion base, eutrophication in overwatered soils, deterioration of normal vitality in cenoses, etc. It is proposed to observe the dynamics of the animal population (in constant test areas) and the soil mesofauna by means of layer-by-layer soil removal, with analysis of the litter and humus. This program provides a real basis for studying and monitoring the state of biocenoses.

The proposals concerning the program for ecological monitoring in biosphere reserves and at regional and baseline background stations [62] are of considerable interest.

It is suggested that the amounts of chlorinated pesticides (DDT and others), PCBs, and heavy metals (lead, mercury, cadmium, iron), as well as calcium, magnesium, sodium, and aluminum, present in the atmosphere be carefully monitored. In addition, the amounts of SO_4^{2-}, NO_3^-, NH_4^+, K^+, Ca^{2+}, Mg^{2+}, PO_4^{3-} and radionuclides, as well as the pH, are to be measured in atmospheric precipitation.

Tropical forests, forests of the temperate zone (detailed studies are possible), meadows, arid zones, marine, mountain, and island ecosystems, and urbanized systems are all suggested as biomes to be investigated. The components of an ecosystem are: the major predator, located at the top of the food pyramid (birds, mammals), the dominant herbivorous animal, the dominant plant, earth worms, the top (1 cm) soil layer and the soil profile, different groups of vegetation (tree, bush grass, moss), soil organisms, and deposits.

This describes, on the whole, the background ecological monitoring on land.

5.4. Ecological Monitoring of the Ocean

The planning and organization of integrated ecological monitoring of the ocean constitutes an independent problem [31]. Its solution calls for a system of observations, analysis, evaluation, and prediction of the state of the ocean. It includes physical, geochemical, and biological components.

Ecological monitoring of the ocean (physical component) aims at a systematic analysis, observation, and prediction of thermodynamic processes and processes of transport of anthropogenic pollutants, which determine the ecological situation in the ocean. The physical component of ecological monitoring is closely connected with climate monitoring, so that investigations in this field constitute an element of the World Climate Program. Within the framework of this program, a major task is to investigate the effects of the ocean characterstics and dynamics on the thermal exchange and gas exchange between ocean and atmosphere, on the global cycle of heat, moisture, and various chemical compounds, particularly carbon dioxide, in the climate system. Its objectives also include a determination of the effect of different anthropogenic impacts on major geophysical phenomena in the ocean, on the processes of interaction between ocean and atmosphere, and on the state of the climate system, including the cryosphere.

The ecological monitoring of the ocean (geochemical component) incorporates a systematic monitoring, evaluation, and prediction of the pollution levels of marine ecosystems, including the rate of arrival of pollutants in the World Ocean, the pollution concentration in seawater, the accumulation of pollutants in suspended matter, bottom sediments, and the biota, and the rate of removal of pollutants from seawater via biogenous sedimentation and microbial metabolism.

Ecological monitoring of the ocean (biological component) consists of systematic observations, evaluation, and prediction of the biological consequences of anthropogenic pollution and other adverse impacts, as well as the identification of 'critical' impact factors and the most vulnerable links in the biotic component of marine ecosystems.

The main objectives of ecological monitoring of the ocean are: establishment of a system of observations of sources and of factors of anthropogenic impacts and biological effects in marine ecosystems, and also determination of the permissible loads on ecosystems (based on an evaluation, analysis, and prediction of the state of the ocean).

The scientific aspects of organizing ecological monitoring of the ocean are given by us in detail in [21, 65].

The modifications in communities and populations produced by anthropogenic pollution can be sudden (as in the case of oil spills from ships, when a substantial amount of the pollutant is removed by rivers, etc.), or chronic, when slight, albeit long-term, pollution effects are observed. In the latter case the biological consequences can be very inertial in character; then, the selection of suitable biological objects for observation and monitoring is of great importance.

The optimum category of hydrobionts for such purposes comprises [21]:

(1) marine microflora with a high reproduction rate and with various types of physiological activity;
(2) young of the neuston community inhabiting the near-surface microbioton of the ocean, where pollutants are concentrated;
(3) macrofauna and meiofauna of the benthos, many forms of which integrate the impact effects over a long time period;
(4) macroalgae in tidal zones and shelf regions.

The biological indication of injury to ecosystems is of great importance in the ecological monitoring system. The advantage of bioindication when analyzing the ecological situation in the ocean is that biological processes reflect the integral effect of the environmental changes; biological processes make it possible to determine the rate and direction of changes in the environment; the marine biota indicates both the paths of migration and the areas in ecosystems where chemical substances concentrate.

Biological indication serves as a basis for biotesting, a biological method of assessing the character and extent of the marine environmental pollution.

Various types of phytoplankton, zooplankton, mollusks, some representatives of ichthyofauna, and marine microorganisms are widely used as bioassays.

As already mentioned, biological monitoring is implemented at the individual level and the population-biocenotic level.

The biological indicators at the individual level should include morphological, ethological, biochemical, physiological, and genetic characteristics typical of the given species.

The size composition of hydrobionts, the state of the gill membrane of fish, and the shapes of microorganism cells can be used as morphological indicators. The way hydrobionts are oriented in space and the frequency of gill movements can be used as ethological indicators. The rate of a given enzymatic process and the characteristics of individual enzymes can be used as biochemical indicators. The oxygen consumption, the reproduction rate, the growth rate, the osmotic regulation and the ratio of the oxygen consumption to the metabolically released nitrogen can be used as physiological indicators. The mutation rate, etc., can be used as genetic indicators.

Monitoring at the level of communities determines the state and stability level of an ecosystem. It is precisely this aspect of biological monitoring which is of particular importance, since monitoring at the individual level can reveal some alarming signs of changes, but it does not characterize the modifications of the ecosystem as a whole.

Monitoring at the level of populations and communities includes a study of the structural and functional characteristics. The determinations of the biomass, the number and kind of species, and the species distribution in time and space are usually taken as the structural indicators. The production and respiration, characterizing the total amount of energy consumed by the given community, are used as the functional indicators.

When analyzing the state of the ecosystem, particular attention is focused on the following characteristics, revealed during the observations:

(1) the ratio of the major taxonomic groups in the community, the diversity index;
(2) the change in the mean biomass of the dominant groups (substantial changes, or modifications, are indicative of a 'stress' state);
(3) ratio of the processes of production and destruction of organic sub-

stances in the ecosystem;
(4) the ratio of the producer and consumer organisms, characterizing the trophic structure of the community and determining its degree of stability;
(5) microbiological indicators (numbers, biomass, microorganisms, indicator forms); a change in the ratio of the various physiological groups of microorganisms indicates the presence of a modification in the communities;
(6) the state of the indicator and critical forms of hydrobionts.

The system of biological indicators for ocean monitoring should cover all the major marine communities: neuston, plankton, periphyton, and benthos.

When carrying out efficient ecological monitoring, we have to possess a good knowledge of the hydrometeorological regime of the water area (and of the other physical characteristics), to ensure representativeness of all parts of the biotic component, and to be familiar with the 'behavioral' features of the chemical substances under investigation.

It is this information which determines the optimum system of observation and sampling (frequency, representativeness, seasonal nature, etc.), that is, the monitoring regime.

An important role in the biological monitoring system is played by model experiments, applied to the monitored ecosystems, which recreate the pattern of the ecosystem response to the studied impacts.

A block diagram of the geochemical and biological monitoring of a marine environment (based on data from [18, 21]) is shown in Figure 5.2.

The Soviet Union has already proposed organizing a number of stations (sites) for long-term observations of the background levels of changes in the marine environment, within the framework of the 'Biosphere Reserves' program. One such Soviet station has already been organized in the Atlantic.

In order to organize stations (sites) for the background monitoring of a marine environment, it was proposed [18] to select regions with ecosystems having different trophic levels, regions located far away from pollution sources, and regions with vulnerable ecosystems.

The authors of [18] also proposed the following regions as possible sites for background monitoring stations in the Pacific: the Bering Sea, The Gulf of Alaska, the area near Japan, the vicinity of the Oregon upwelling, the area near the Marianas, the Equatorial Current, the Peru upwelling, the Great Barrier Reef, and the open oligotrophic area near the southern Line Islands (Figure 5.3).

At the symposium on 'Integrated Global Monitoring of the World Ocean', held in Tallin (USSR) in October 1983, the urgent need to develop integrated global monitoring of the ocean was noted. In a number of papers presented by both Soviet and non-Soviet scientists various approaches and plans for

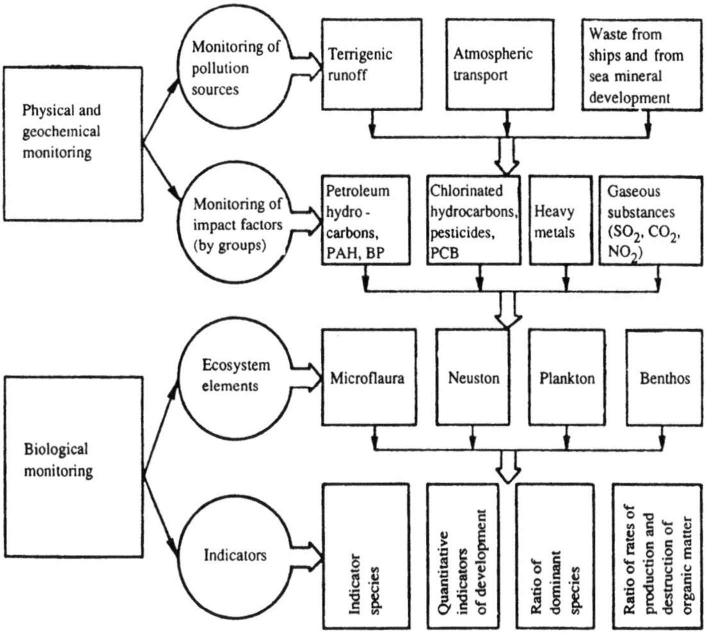

Figure 5.2. Block diagram of geochemical and biological monitoring of marine environment.

developing integrated monitoring were formulated (including physical and ecological aspects) [31].

The recommendations adopted at the symposium called upon GESAMP to develop aspects of the program for global monitoring of the ocean relating to the ecological consequences of pollution and marine environmental protection. It was noted that the climatic aspects of the program of integrated global monitoring could be developed within the framework of the 'World Climate Research Program'.

The symposium recommended that the eutrophication of the ocean be monitored by measuring the microbiological and biochemical parameters directly related to the rate of the cycle of biological elements. The use of mollusks (*Mytilus* sp.) as indicators was advised for monitoring the pollution level of the marine environment.

This symposium also emphasized the need for a further study and development of the techniques of integrated global monitoring of the ocean, including the use of biological indicators of the environment and the assimilative capacity of marine ecosystems. Accordingly, the symposium recommended that GIPME (Global Investigation of Pollution in the Marine Environment) be given the task of developing a program of fundamental investigations of the spreading of pollutants in the open ocean.

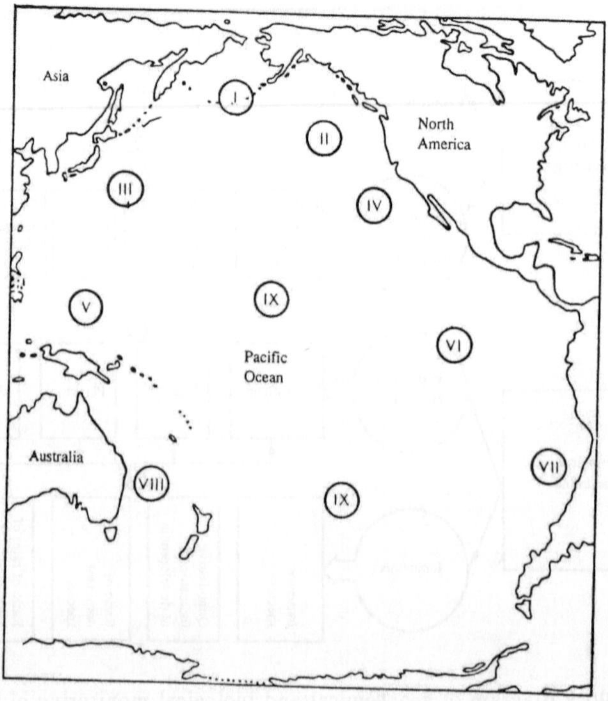

Figure 5.3. Zones in Pacific of proposed organization of monitoring of ocean: I. Bering Sea; II. Gulf of Alaska; III. Area near Japan; IV. Oregon Upwelling; V. Mariana Islands region; VI. Equatorial Current region; VII. Peru Upwelling; VIII. Great Barrier Reef; IX. open sea areas.

5.5. The Use of Satellites in Ecological Monitoring

It is obvious that integral indicators and the characteristics of ecosystems averaged over large scales (in space and time) become of great importance when implementing ecological monitoring; remote techniques ensuring a direct supply of data on large-scale changes should play a major role, when organizing a monitoring system.

One such important method is based on the use of space devices (satellites), which have already found their place in the study of natural resources; this field is known as 'space geography.'

Special satellites and satellite systems are being created for these purposes: *Meteor* and *Meteor-Priroda* in the USSR; *ERTS* and *Landsat* in the USA. In the early 1970s pictures from the *Meteor* satellite indicated with sufficient accuracy for practical purposes the state of the pasture vegetation in the Central Asian republics.

An enormous amount of information on various natural resources and processes at the Earth's surface has been obtained from other space systems and individual satellites.

Particular attention should be given to the manned orbital *Salyut* space stations, where, in addition to the information mentioned above, search and detection of the given phenomenon are carried out by the crew, with the help of various remote techniques.

The integrated use of information from space systems, aircraft, and ground-based systems can provide a most effective evaluation of anthropogenic impacts.

The satellite information used to organize ecological monitoring includes information on the state of forests, agricultural lands, land vegetation, marine phytoplankton, and the earth's surface (soil cover, disturbances of the surface produced by human activity, erosion processes, urbanized areas), as well as on the redistribution of water resources, and the pollution of the atmosphere, sea, and land. Concise data indicative of the practical capabilities of satellite systems for solving certain problems of ecological monitoring will be given below.

Besides the main task of space geography, namely photographing the Earth's surface with a subsequent decoding of the images, new directions are being developed aimed at discovering new properties of the terrestrial objects under study. Multizonal surveys used for this purpose enable us to view one and the same element (region) in various spectral ranges and to identify (due to the difference in the spectral brightness of the image) considerably more features of the given element [5, 26, 67, 73]. Such surveys make it possible to determine the abyssal characteristics of observed water objects, due to the penetration to great depths of radiation in a certain wavelength range (e.g. in the range from 0.5–0.7 μm to 20 m).

For multizonal spectral surveys we use radiation in the wavelength range corresponding to the 'transmission windows' of the atmosphere. Ultraviolet radiation (with wavelengths shorter than 300 nm) is not used for such surveys, since it is practically completely absorbed by the atmosphere. Visible radiation passes readily through the Earth's atmosphere (except for some narrow absorption bands), and therefore it is used successfully in spectral surveys, as is radiation in the near infrared, which is mostly used to study natural resources and the state of the environment, and also to distinguish anthropogenic effects. The infrared range from 8 to 15 μm is suitable for multipurpose sounding, since there is practically no atmospheric absorption at these wavelengths. Thermal radiation and radio-frequency radiation (up to the decameter range) are used as well. On the whole, the value of any information obtained about the underlying surface and the atmosphere depends on the correct selection of one or several wavelength ranges.

The described methods make possible an integrated investigation of geosystems, defined as aggregations of natural and natural-technological formations recurring regularly over the earth's surface (or natural and cultivated landscapes), i.e., they constitute a new means of developing modern geography [38]. Aerospace techniques are also used for the investigation of nature reserves [6]. Satellite information has already revealed new morpho- and geostructures, and it is also being used to determine patterns of land use and to evaluate the state of the land. A multispectral survey can be used to identify genetic types and the mechanical composition of the soil.

The soil cover can be studied remotely from satellites via soil mapping or soil zoning according to certain signs [38], as well as by studying the main characteristics of the soil cover: humidity, soil type, salinity, humus content, mechanical composition of soil, its disturbances (produced by agricultural activities and soil erosion), and the presence of vegetation [5].

Soils can be distinguished by direct measurements of the characteristics of the surface, or indirectly, according to geological, geobotanical, and geomorphological indicators. Such direct characteristics as the temperature, humidity, and humus content are very important when determining the type and state of the soil.

Soil temperature measurements are based on measurements of the radiation surface temperatures in the infrared range (in the transmission windows of the atmosphere), and also of the radio-brightness temperatures in the microwave range with an accuracy up to 2.0°C (1974) and 0.5 – 1.5°C (during the International Meteorological Experiment [52]).

The soil humus content is determined with the help of an optical method (in open areas), there being a correlation between the spectral brightness (in the visible range) and the humus content.

The soil salinity is also determined from the brightness, using large-scale photos. A multispectral survey makes the solution of these problems much easier.

The brightness on soil photos decreases abruptly with an increase in humidity (approximately 0.02 brightness for a 1% humidity increase up to soil moisture contents of 18-24% [5]).

Measurements in the microwave range provide information about the soil humidity in a layer and they can be used to obtain humidity profiles.

A space (satellite) survey has great potential for geobotanical zoning, for studies of the vegetation (natural and agricultural), and for the determination of biomass reserves and productivity. As already mentioned, the productivity of pasture vegetation in the Central Asian republics was satisfactorily determined, even using photos with a low resolution (*Meteor* Satellite in the early 1970s).

The relationship between the coefficient of spectral brightness and the biomass is very pronounced in the wavelength range from 0.59 to 0.68 μm. Multispectral surveys are very promising for productivity determinations.

Some results of observations of plant development are given in [5] (with recording at certain time intervals). Forest tracts, bogs, paludal forests, floodlands, and agricultural lands were identified from TV pictures taken in winter by the *Meteor* satellite. The purpose of this was to develop techniques for the zoning of bogs and paludal forests in the northern taiga region.

For ecological monitoring, a determination of the state of the vegetation (e.g., phenology, diseases) is of particular importance. Such investigations can be carried out most effectively with the help of a multispectral survey (in the range from 0.28 to 14 μm) in narrow spectral bands. Some applications of multispectral-survey data to determine the state of agricultural and forest vegetation have already been developed [57, 63]. Infrared surveys (including the far range of the spectrum from 8 to 14 μm) and radiothermal surveys are of interest for these purposes. A multispectral survey with a relatively high resolution (of the order of 200 m) is the most promising for ecological monitoring. It is precisely such surveys which make it possible to judge the state of the vegetation in considerable detail and, what is most important, with a sufficient level of probability.

The identification of signs indicating anthropogenic changes in the state of the soil and vegetation is of particular importance. This is why observations of the most diverse landscapes, particularly manmade ones, should be organized.

Observations of the sizes of settlements and populated areas give us an idea of the human population and its increase [46] (the settlement area is an exponential function of the number of people). In [74] it was proposed to determine the approximate total energy consumed by man on the basis of the nighttime brightness of lights detected from outer space. If the energy released by the settlement is measured in the IR and radio ranges, then it will apparently be possible to evaluate the energy consumption by man, including the industrial energy consumption.

For example, 3 to 4°C anomalies of the radiation temperature were observed in the New York area with a satellite [71].

Observations of water reservoirs and natural water bodies enable us to evaluate very accurately (according to the location of the banks and the water level) changes in the state of these objects, including anthropogenic changes.

For instance, we considered [67] the dynamics of the Kara-Bogaz-Gol after its separation from the Caspian Sea (early in 1980). Figure 5.4 shows the change in the boundaries of the drying gulf according to the space data. Not only the boundaries, but also the depth (and consequently the volume), of the gulf were determined with the help of these photos (At the beginning of 1983 the depth varied from 10 to 65 cm).

Photos taken in February 1984 in the wavelength range from 0.7 to 1.1 μm reveal an almost completely dry gulf and also the boundary of the the last stages of its drying [77].

The separation of the gulf from the Caspian Sea resulted in a decrease of

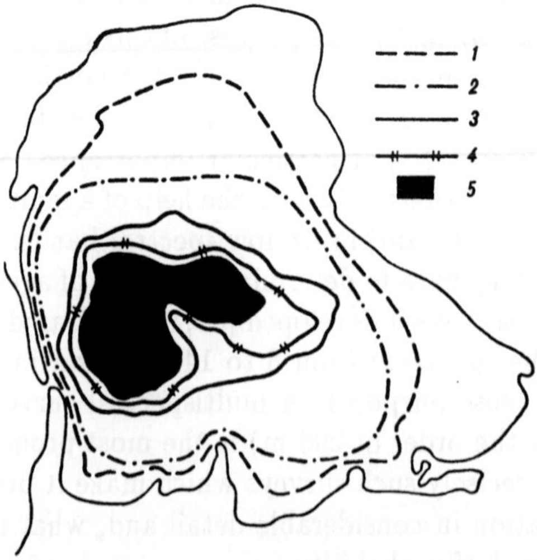

Figure 5.4. Boundaries of Kara-Bogaz-Gol at different times: (1) line of brine before separation, (2) line of brine in September 1980, (3) in October 1981, (4) in September 1982, (5) region under water at beginning of 1983.

more than 30 km^3 in the evaporation from the total water area of the sea in 1980–1984.

Since it is thought that the ecological conditions of the gulf should be maintained in their former state (equilibrium state of buried brine, decreased ability of wind transport of salts from surface of dried gulf bottom, etc.), 11 pipes of large diameter with a regulating device were installed in the stream crossover region (as provided for in the dam project), allowing up to 5 km^3 of water to flow from the Caspian into the gulf (if necessary). At first it was planned to discharge into the gulf up to 2 km^3 of water each year (*Pravda*, 16 Sept. 1984).

Satellite techniques are useful for indentifying many anthropogenic changes in the environment, present changes as well as those in the past, e.g., forest fires (plumes indicate present fires, and ashes indicate fires in the past), pollution of the atmosphere and the underlying surface (various changes in multispectral photos: dust emissions clearly identified in the visible range and thermal emissions in the IR range indicate present changes, while typical changes in the vegetation indicate past changes), soil erosion, etc.

Dust layers, and particularly dust clouds during dust storms, are clearly identified.

Observations of smoke release make it possible to determine (according to the degree of transparency) the particle density in these plumes [5, 70]. The ingredients of such plumes can be determined from the radiation absorption in the respective zones of absorption by various gases.

In Chapter 4, during our discussion of climate monitoring and anthropogenic impacts on the climate, the possibility of measuring the contents of ozone, aerosols, and minor gas constituents (such as halocarbons, nitrous oxides, etc.) from satellites was considered as well. In this case measurements of the thermal or scattered radiation of the atmosphere near the Earth's disk are the most promising, as well as measurements of the attenuation of sunlight [30].

Satellite photos can provide observations of some phenomena just as they start, when the high concentrations of matter or energy make measurements much easier.

Smoke damage of coniferous vegetation was observed with the *ERTS-I* satellite, according to the change in spectral brightness in the 0.5-0.7 and 0.8-0.9 μm ranges; similar observations made it possible to estimate the chlorophyll concentration in water (according to the ratio between the coefficients of spectral brightness in the 0.52-0.57 and 0.41-0.48 μm ranges), and the eutrophication development in water bodies; satellite measurements of water pollution by suspended particles, petroleum, and other surface-active films are described in [55, 59]. In the latter case the spectral-angular distributions of the reflected radiation were used.

The centers of origin of forest fires are found both from photos in the visible range (smoke plumes) and from nighttime IR photography (burning is evident from the color, which even indicates the 'age' of the fire).

The results of our investigations using multispectral photography with a resolution of 250 m, intended to identify and observe pollutants and other anthropogenic effects on the biosphere, i.e., the results of using satellite photography in ecological monitoring, are given in [67]. The main results of these and more recent investigations [77] will be given below.

Satellites of the *Meteor-Priroda* system were used. Such satellites (e.g. the one launched in August 1981) are equipped with multispectral scanners, spectrometers, and microwave radiometers, which can detect plumes of anthropogenic aerosols in urban and industrial areas, surface-water pollution, etc. The use of active radar, realized as early as 1983 aboard the Soviet *Cosmos-1500* satellite, provides some interesting possibilities. Regular additional information provided by the manned *Salyut* space station makes possible not only a qualified interpretation of the information obtained but also monitoring the field of anthropogenic pollution and other impacts. Such work is already being carried out over European territory.

These studies show that anthropogenic trails of aerosols can spread over hundreds, and in some cases thousands, of kilometers. The photos reveal that under certain meteorological conditions practically the whole territory of Europe is covered by an anthropogenic 'haze' (the trails of haze, or smog, can be compared with those from forest fires, although in the case of industrial pollution the aerosol concentrations are naturally lower).

Anthropogenic changes on land are determined not only by pollution but also by increased urbanization, the extension of agricultural areas, communication development, cutting of forests, redistribution of water resources, strip mining, etc. These changes lead to a change in albedo, and, consequently, to a change in spectral brightness. The total 'anthropogenic effect' is revealed effectively by satellite photos. Its separate components (e.g. pollution layers) can be singled out against the background impact.

Anthropogenic changes in the surface waters of seas and oceans can be identified quite readily.

The interactive processing of satellite videoinformation and the relationships between the spectral brightness and the extent of 'anthropogenicity', obtained as a result of calibrations and field experiments, were used for the analysis. The set of obtained optical densities of a picture being analyzed is divided into classes, according to different density levels (256 gradations were used); as many as 1024 gradations can be used.

The results of processing satellite photos (with a resolution of 250 m) clearly show zones of increased anthropogenic effect ('anthropogenicity').

Photos taken in the 0.5–0.7 μm range reveal the atmospheric pollution by aerosols more clearly.

Trace recordings of the optical-density variations along any horizontal or vertical sections of the picture were used to analyze the spatial and temporal dynamics of the anthropogenic impacts. Trace recording of the 'anthropogenicity' along a horizontal section through the picture shows zones subjected to a strong anthropogenic impact and the corresponding maximum values on the trace are evident. The remaining areas correspond to a different degree of anthropogenic impact, which can be estimated quantitatively from the trace recording.

The examples considered give an idea of the new direction (particularly effective in combination with ground-based observations) which has been taken in the development of ecological monitoring.

It is noteworthy that so far a great deal of experience has been gained on the use of satellite information (primarily multispectral data) to study ecological anthropogenic changes and to identify the causes of these changes.

CHAPTER 6

Climate Monitoring

6.1. Main Objectives

A system of global and regional climate monitoring, as well as services for the acquisition of climate data, are necessary for the practical use of climate information to manage and develop the economy, and to study the climate and its possible changes and fluctuations, so as to optimize the relationship between man and nature.

Many national meteorological services have for a long time been collecting such data and providing it to interested organizations and individuals. The World Meteorological Organization (WMO) provides an international data exchange and facilitates the practical application of climate data.

The WMO Executive Committee in 1977 and 1978 (at its 29th and 30th sessions), while considering the World Climate Program, cited climate monitoring and the provision of climate data as the main objective of the program [16].

A system of environmental monitoring has been defined as an information system making it possible to detect changes in the state of the biosphere due to human activity (cf. Chapter 4). Thus a 'climate monitoring system' will be understood to be an information system which makes it possible to single out anthropogenic changes and fluctuations of the climate.

Clearly, observations, assessments, and predictions of such anthropogenic changes will be needed in order to detect the anthropogenic components of the climatic changes and fluctuations.

If the climate is considered as an aggregate of states of the atmosphere, as a recurrence of weather conditions, then it can be described by a set of statistical characteristics of the possible states of the atmosphere. The state of the atmosphere is determined by a set of meteorological quantities, characterizing the aggregate of atmospheric phenomena and processes. Therefore, observations and measurements of the characteristics of the atmospheric state, averaged over a certain time interval, will directly correspond to *climate monitoring*.

The state of the atmosphere, and the nature of the processes developing in it, depend on the physical characteristics and composition of the atmosphere, as well as on the factors influencing it, and they can change significantly when the atmosphere interacts with all the elements of the biosphere, first and foremost with the underlying surface.

This is precisely why it is necessary, if we are to understand climatic changes and fluctuations, to have data on the atmosphere-ocean-land system (with rivers and lakes), the cryosphere, and the biota, as well as on the interactions between the various elements of the climate system over a long period, i.e., to carry out *climate monitoring.*

The natural variability of the climate obviously has to be thoroughly studied, too, in order to distinguish the anthropogenic climatic changes and fluctuations.

The acquisition of climate data from the past can be considered to be climate monitoring as well. This requires a system of acquisition and study of (fossil and other) data on the possible climatic variations and fluctuations during the last centuries, millennia, and even longer time intervals (analysis of tree rings, including those of long dead trees, bottom sediments, moraine deposits, ice cores, pollen, etc.). All this will make it possible to study the influence of variations in the climate system on the climate of the past.

In order to investigate anthropogenic climatic changes, we have to study the effect of anthropogenic changes in the characteristics of the underlying surface (large-scale hydraulic structures, changes in the area of forests, urbanization), and to study anthropogenic changes in the state and optical properties of the atmosphere (because of the emission of aerosol particles and various gaseous substances), and also the possible effect of intensive thermal emissions. Assessment of the global atmospheric pollution and its effect on climate is recognized by the UNEP as being one of the chief goals of the Global Environmental Monitoring System (GEMS) [15].

Natural and anthropogenic climatic changes can in turn affect the state of the biosphere, producing various ecological consequences, and they can influence the normal functioning of individual plant and animal populations, as well as human activity and ultimately man's health and well-being. This can have both economic and social consequences [17].

This type of monitoring forms a part of ecological monitoring. A special system of observations should be organized for its implementation: a study of the impact on ecosystems in different regions will evidently require joint observations in zones unaffected by local human activity, for example, in biosphere reserves.

Climate monitoring and services for the acquisition of climate data can be used to solve various practical and scientific problems. For instance, a vast amount of climate information is needed to solve the applied practical problems in agriculture, water management, power engineering, construction, marine applications, and other types of human activity. A service for climate-data acquisition is clearly needed for such purposes, and it is evident that such a service goes beyond the framework of the monitoring of anthropogenic climatic variations.

Voluminous data on the individual characteristics of the elements of the biosphere and on the processes determining climatic variability, are required to study climatic changes and fluctuations, to comprehend these changes, and to distinguish the anthropogenic components. First and foremost, this pertains to the study of the space-time climatic variability on different scales.

The prediction of seasonal and interannual climatic fluctuations requires a special global observation system, which is nonuniform in both space and time.

Obviously, observations designed for the study of climatic variability should also take into account the inertial nature of the climate system.

The analysis and assessment of the present climate, and the prediction of its possible changes and fluctuations, require a large amount of data and call for a comprehensive analysis of the state of the environment [5] and climate model [29].

Therefore, the most important tasks of climate monitoring are: data acquisition; analysis and assessment of natural and anthropogenic climate changes and fluctuations (including a comparison of past and present climates) and of changes in the state of the climate system; identification of observable anthropogenic effects in these climate changes; and identification of natural and anthropogenic factors promoting climatic changes and of critical elements of the biosphere which, when affected, can most readily lead to climatic changes.

In order to solve these problems, in addition to the climate monitoring system, it is necessary to implement a broad-based research program, with modeling of climate changes and fluctuations.

As already noted, climate data are widely used in man's practical activity, since the ever-developing economy is intimately related to the climatic conditions. In this connection, along with climate monitoring, this section will also discuss those aspects of the services for acquisition of climate data which provide information outside the sphere of the above definiton of monitoring.

Nevertheless, it is noteworthy that all these information systems are closely interconnected.

Taking all of the above into account, we can group a wide class of problems of climate monitoring and subjects pertaining to possible climate changes and fluctuations as follows:

1. Measurement of main meteorological quantities; study and analysis of atmospheric phenomena and processes characterizing the corresponding weather conditions (including climate data for use in organizing and managing the economy).
2. Monitoring the state of the climate system; acquisition of data char-

acterizing the response of the climate system and its elements to any natural and anthropogenic impacts.

3. Monitoring external and internal factors (particularly monitoring of anthropogenic factors) affecting the climate and the state of the climate systems; monitoring the sources of these effects.
4. Monitoring possible physical and ecological changes in the environment induced by climatic changes and fluctuations (monitoring the majority of the indirect indicators of climatic variability is included here).

6.2. Climate Data and Information Required for Analysis of Climate Variability

The first part deals with measurements of air temperature (including the daily extreme values), atmospheric pressure, air humidity, wind direction and velocity, precipitation rate, and cloud cover (including the lower cloud boundary). Measurements of the main meteorological quantities (sometimes together with additional measurements) make it possible to monitor atmospheric phenomena and processes (including the general circulation of the atmosphere) and to observe the hydrologic cycle. These data are obtained by the national meteorological services at their stations with the help of various engineering techniques.

This part should include the acquisition of hydrological data, data on the snow cover, soil moisture, depth of frozen soil, and some others. All these data are obtained both at meteorological (climatological) stations and at hydrological stations and posts.

According to approximate estimates, there are at present 40.000 climatological stations and 140.000 raingauge stations operating in the world [19]. Unfortunately, however, these stations are distributed very irregularly over the world: major gaps exist on some continents.

The main objective of the World Weather Watch (WWW) of the World Meteorological Organization (WMO) is to provide an international exchange of principal weather data. The World Weather Watch includes a global observation system, a global telecommunication system, and a global data-processing system. The Soviet observation system contributes significantly to these systems. The system is designed to record, store, and provide data. The global observation system consists of ground-based and satellite subsystems.

The ground-based subsystem is based on a reference synoptic network (synoptic and aerological stations). This subsystem also includes information from research vessels and aircraft, meteorological radar, and various atmosphere-sounding systems.

The reference synoptic network of the WMO observation stations, which provides international data exchange on a regular basis, consists of about

2500 ground-based stations (the total number is more than 9400 [18]) and 967 aerological stations (data for 1978) [6, 7]. The quality of the work of these stations can be judged from the following data: the number of ground-based observations performed at the reference times amounts to 91% of the required number of observations. This number at the aerological stations is 79% (at 0 and 12 hr) [21]. 2600 maritime weather reports (from more than 7000 vessels) and 3000 aircraft weather reports (1978) arrive daily from regions of the World Ocean for the global exchange.

Monthly climatological reports with ground-based data arrive from 1572 stations, and with aerological data from 509 stations (1977) [6, 7]. In 1983 their numbers were, respectively, 1663 and 499 [18]. For comparison we note that in 1974 there were 1399 and 391 such stations, respectively. At the request of the WMO, climate data from approximately 1250 ground-based synoptic stations and 430 aerological stations are published in the USA (Asheville) on a regular basis.

The satellite subsystem consists of two systems: satellites in circumpolar orbits (Soviet *Meteor*, American *NOAA*, etc.) and geostationary meteorological satellites (there are five at present) [18]. A total of 230 stations (according to data for 1982 [21]) receive satellite information, which includes data on the vertical profiles of temperature and humidity, on the temperature of the sea surface, the land surface, and the upper cloud layer, as well as data on the wind field (obtained from cloud-movement observations), the cloud amount, and the upper cloud limit, and on the snow and ice cover and the radiation balance.

On an experimental basis, information on the soil moisture, the sea state, pollutants, the particle distribution in the atmosphere, and other data will be included (for more detailed information on satellite climate monitoring, see Section 6.4 below).

Data from rocket sounding of the atmosphere (up to a height of 60–80 km) are also supplied on a regular basis.

Let us note again that these global meteorological data are used primarily for weather forecasting, although they are undoubtedly used to prepare climate information as well.

The program for observation of the quantities characterizing the state of the atmosphere and its interaction with the ocean was expanded considerably during the First Global Experiment (GARP). Such observations are also carried out during the regional experiments within the framework of GARP.

The ground-based observation subsystem also includes stations for measuring the solar radiation and the background atmospheric pollution; solar-radiation monitoring will be included in the next part of climate monitoring, while observations of the atmospheric composition can apparently be included in the first part. Measurements of atmospheric components which can change so much that they can substantially affect the climate should also be included in this part. These components include, first of all, carbon dioxide (CO_2), ozone (O_3) (in the stratosphere), and various gaseous pollutants. Aerosol particles of natural and anthropogenic origin, electromagnetic radiation, and thermal pollution can be considered as factors influencing the climate or climate system (they will be considered in more detail below).

There are not many baseline stations were atmospheric carbon dioxide changes are observed, only four [25]. The observations are carried out at sites rather far away from the local (natural and anthropogenic) sources of carbon dioxide (at baseline monitoring stations). This is a very important requirement, particularly if we take into account that these scattered stations are considered to be representative of the whole atmosphere of the Earth.

Ozone monitoring is much more extensive: data (world ozone data are published in Canada) on the total ozone amount are exchanged among approximately 80 stations, while data on the vertical ozone distribution from ozone-sounding observations are provided by 10 stations [25].

The second part of climate monitoring, namely monitoring the state of the climate system, is very important and extensive. It is similar to the corresponding part of ecological monitoring. This system should include essentially the whole biosphere, but it must be so designed as to make it possible to distinguish precisely those effects which are directly related to the anthropogenic climatic change (and to study its natural variability).

This pertains to monitoring of the so-called climatogenic factors, as well as monitoring of the quantities characterizing the response of the climate system and its elements (which can affect climatic changes) to different impacts, mainly anthropogenic.

It is necessary to obtain data on the state of the underlying surface, which determines the surface albedo, and to monitor the energy exchange and mass exchange between the atmosphere and the underlying surface. For this purpose we require measurements of the area of sea, river, and lake ice, as well as of glaciers; of the area and volume of snow cover on the plains and in the mountains; of the surface of land water bodies; of the surface and biomass of the plant cover; and of the area of desertification. In addition, measurements are needed of the soil and plant water content, as well as observation and study of the ocean circulation, the optical characteristics and mass of the atmosphere, and the state of the ozonosphere. Changes in land-use practices should also be monitored, since they can affect the carbon dioxide and nitrogen oxide contents of the atmosphere. In addition, they determine the albedo of the underlying surface, and the humidity of the soil and air (because of the difference in evapotranspiration), the soil temperature, etc.

The study of the water balance on a large scale and its effect on the climate variability (using a large volume of hydrological data) also pertains to this part of the monitoring.

All these factors can be considered to be climatogenic, and a change in the quantities mentioned corresponds to the response of the climate-system elements to the effect.

Many of the above factors can be measured with the help of remote techniques, using satellites, while some (area of snow cover and ice cover, changes in vegetation area) are measured with satellites more accurately (integrally) and, what is most important, in the field. The use of satellites in these measurements is very promising (such observations make up a field known as space geography).

Observations of the state of the atmosphere, and its composition, were covered above.

Monitoring the state of the ocean involves measurements of the temperature of the surface and of the surface layer of the ocean, the salinity and chemical makeup of the water, the waviness, and the currents at various depths. These measurements are performed within the framework of the WWW and IGOS (Integrated Global Oceanic System), WMO/IOC, from about 7400 vessels (according to data [18] for 1983, from 7517 vessels), from 6 fixed oceanic stations, and 40–50 buoys [18, 19].

Koshlyakov and Monin [23] consider that, when monitoring the ocean state, it is necessary to measure the temperature of the ocean surface and the enthalpy of the mixing layer, and to study the water circulation in the ocean, the atmospheric fields (temperature, humidity, pressure, wind speed, cloudiness), in order to determine the intensity of the energy exchange between atmosphere and ocean, as well as the fluxes of short-wave and long-wave radiation to the ocean surface.

In order to obtain data on the interaction between the atmosphere and the ocean (and to study this interaction), regular marine climatological measurements of the air and sea temperature, dew points, visibility, wind direction and force, atmospheric pressure, cloud amount, and waviness are made. The temperature of the surface layer of the ocean is measured with a bathythermograph.

During the first GARP experiment a program was discussed and approved which was aimed at increasing the number of the above measurements, with the aid of more vessels, oceanic stations, and buoys.

The cryosphere affects substantially the various impacts and climatic changes and is in turn very sensitive to them. In addition to monitoring the snow cover, glaciers, sea, river, and lake ice, monitoring the permafrost area should be included in the monitoring of the state of the cryosphere.

It is very important to monitor the factors affecting the state of the climate system and the climate (the third part of climate monitoring), as well as the sources of the impact factors (the causes of the appearance or change of the factors). The above factors can be divided into external factors (relative to the Earth's climate system) and internal factors, while the sources of the internal factors can be divided into natural and anthropogenic.

The factors controlled by the Sun and cosmic radiation are considered to be external.

The factors produced by the sun include electromagnetic solar radiation over a wide spectral range (including ultraviolet radiation, and hard electromagnetic radiation), corpuscular fluxes of various energies, and the magnetic field.

The intensity of the external impact factors depends on the solar activity, the parameters of the Earth's orbit, and Earth's rotation velocity.

The effects of the impact are determined by the intensity of the impact factors, by the properties and composition of the Earth's atmosphere, and by the characteristics of the Earth's surface (its albedo).

Although the variations of the total solar radiation over a wide wavelength range (solar constant) are quite insignificant (less than 1% of its value), solar-activity changes often turn out to be related to different processes in the atmosphere and to climatic changes. For example, changes in the magnetic field related to changes in the solar-wind boundaries apparently affect the tropospheric circulation; changes in solar activity on a scale of several years or tens of years can influence the intensity of droughts in different parts of the world; long-term (on a scale of centuries) variations in solar activity can to a significant extent determine such changes as a 'little ice age'.

All these things place very strict demands on measurements of the solar radiation (solar constant), measurements in different spectral ranges, and measurements of corpuscular radiations. Apparently, the accuracy of solar-constant measurements should be of the order of 0.1%, while in different parts of the solar-radiation spectrum (particularly in the ultraviolet), it can be up to 1%.

Solar radiation is measured by the World Actinometric Network (more than 900 stations [6]). The direct, scattered, and reflected radiation is measured; the integral transparency of the atmosphere and the albedo of the underlying surface are determined; the total radiation and the balance of shortwave and longwave radiation are calculated. The data are presented to the World Data Center in Leningrad, where they are published yearly.

Existing actinometric stations are distributed irregularly throughout the world: in Europe and North America such stations are spaced 100–400 km apart, but in Africa, South America, and Asia this spacing is more than 1000 km. WMO experts believe that the average separation between such stations should be approximately 500 km, which means that the global network of stations should be enlarged by around 20%.

The ultraviolet radiation is measured by ozonometric stations; the X-ray and hard electromagnetic radiation, and the fluxes of solar corpuscular radiation, are measured with satellites; cosmic rays are measured at special stations and with satellites.

The internal factors affecting the climate and the climate system (as well as the climate variability) can be assumed to include thermal emissions and the discharge into the biosphere of various substances of both natural origin (volcanic eruptions, weathering of rocks) and anthropogenic origin (heat release from human energy consumption, biosphere pollution, emission from various anthropogenic sources), or the redistribution of these substances among the different parts of the environment. The above-mentioned impact factors lead to a change in the properties of the climate system, that is, in the albedo of the underlying surface and the atmosphere, the hydrologic cycle, and the heat exchange and gas exchange between the underlying surface and the atmosphere.

Thermal emissions lead to warming of the atmosphere. The total global emission of anthropogenic heat is around 10^{-4} of the solar heat, and the global anthropogenic warming can apparently be neglected for the time being. However, the local effects of atmospheric warming can be very significant (for instance, in large cities).

For instance, if the average temperature over five years (1971–1975) in the vicinity of Moscow (Serpukhov, Kolomna, Mozhaisk) is compared with that of Moscow itself, then the temperature in Moscow will be much higher. In the former case the average temperature is +4.8°C, and in the latter it is +5.6°C. The substantial difference (0.8°C) is evidently due to the effect of the large city (mainly due to thermal emissions and to a change in albedo).

When measuring changes in the atmospheric composition and studying the possible effects of these changes on the climate, we have to observe carefully the content and change in the CO_2 concentrations in the atmosphere, as well as the the CO_2 exchange between the ocean and land biota.

An increased CO_2 concentration in the atmosphere leads to an increased absorption of the departing long-wave radiation, and, consequently, to warming of the atmosphere (greenhouse effect), i.e., to a possible climatic change.

Precise measurements of the carbon dioxide content in the atmosphere (for instance, at Mauna-Loa station, in Hawaii) indicate that it is continually increasing. This increase is related to human activity (the burning of fossil fuel).

Many scientists [1, 2, 12] attribute the warming in the Northern Hemisphere to the anthropogenic increase of atmospheric CO_2, and they predict significant warming during the next decades: of the order of 1°C (some predict even greater warming). Such warming could affect the biosphere and human activity significantly (see Section 8.1).

The increased content of CO_2 and a number of other gaseous pollutants is now considered to be the most important anthropogenic factor which can affect the climate in the near future [1, 11]. Therefore, special attention should be given to thorough measurements of the atmospheric concentration of CO_2 (as well as of nitrous oxide, freons, and tropospheric ozone), to studies of the carbon balance in the biosphere, the exchange of CO_2 with the abyssal ocean layers, and the effect of a petroleum film on the gas exchange between ocean and atmosphere.

An increased content of stratospheric aerosols has, as a rule, the opposite effect: a possible cooling, since part of the incoming solar radiation is reflected. Furthermore, aerosol particles can affect the climate indirectly, promoting cloud formation.

Therefore, monitoring the aerosol component in the troposphere and stratosphere has to be included in this part of the climate monitoring.

The content of aerosol particles in the atmosphere can change, either due to natural causes (volcanic eruptions, sandstorms) or due to human activity (emissions from industrial sources, massive land plowing with subsequent weathering).

Of the other (anthropogenic) pollutants emitted into the atmosphere, substances which can affect the ozonosphere (by disturbing the natural nitrogen, hydrogen, and chlorine reaction cycles in the stratosphere) deserve special attention. This is because of, first of all, the penetration into the lower atmosphere of chlorofluoromethanes (halocarbons, freons), affecting the chlorine cycle, and nitrous oxide, affecting the nitrogen cycle. These substances are inert in the troposphere (and water insoluble): however, they diffuse and reach the stratosphere. In the stratosphere the freons photodissociate, and the released chlorine is incorporated into the chlorine cycle, connected with ozone destruction (see Section 8.2).

A characteristic feature of the above reaction cycles is their catalytic nature. When atoms of nitrogen, chlorine, and hydrogen in the form of different radicals interact repeatedly with ozone, the natural equilibrium between the processes of ozone formation and ozone destruction is disturbed.

Refrigerators and air conditioners are sources of chlorofluorocarbons. The latter are also used in aerosol sprays (total world consumption of around one million tons per year). A considerable amount of nitrous oxide is produced by the use of nitrogen fertilizers.

High-altitude aviation releases nitrogen oxides and water vapors directly into the stratosphere. As a result of photolysis, water molecules also produce radicals which participate in the reactions with ozone.

The disturbance of the equilibrium in the ozonosphere and the decrease of the ozone content change the distribution of the ultraviolet radiation fluxes in the stratosphere at different altitudes. This can result in a temperature change at these altitudes and can affect the nature of the circulation in the upper atmosphere, as well as the thermal balance of the whole atmosphere, i.e., it can lead to certain changes in both weather and climate.

Direct monitoring of chlorofluorocarbons or nitrous oxides in the stratosphere is rather difficult, because of their low concentrations and their chemical inertia. It would be desirable to have a wide network of ozone observations on a regular basis (total amount of ozone in air column and height distribution).

The changes of the main atmospheric constituents and pollutants are measured at the national level in many countries. This type of measurement is the most important element in the global monitoring system (it is discussed in other sections of this book) and in the global observation system of the WWW.

The background atmospheric pollution is measured by the network of background stations of the World Meteorological Organization, consisting of 176 stations (soon to be 197 [18]). Seventy-five of these stations transmit data on the chemical composition of clouds, and sixty stations provide data on the turbidity of the atmosphere [6].

Special measurements of the aerosol constituent of the atmosphere are carried out, and indirect data on aerosols are obtained by comparing the direct and scattered radiations during actinometric observations.

Thermal emissions are estimated from the temperature change (in the air and water) in the emission zone. Volcanic activity is carefully recorded, an inventory being made of all the major volcanic eruptions throughout the world.

Changes in the characteristics of the elements of the climate system constitute climatogenic factors, and they can be considered to be factors affecting the climate and the climate system. These include changes in the nature of the underlying surface due to anthropogenic causes (construction of irrigation systems, water-mass redistribution, soil plowing, changes in the area of forests and agricultural land, urbanization) and natural causes (change in area of snow cover or sea ice), as well as changes in the atmospheric composition and properties (transparency), due to the discharge of aerosol particles and various other substances.

It is noteworthy that in many cases some of the above factors can be considered external relative to the part of the system where the impact factors and changes are determined. For instance, in [29] a situation is considered in which the internal and external systems also differ. In addition to the solar flux, the external system in [29] includes the characteristics of the land surface and albedo of the vegetation cover.

As already noted, many of the above factors influencing climate can be measured (or observed) successfully using satellites.

In the fourth part of climate monitoring, the consequences of climatic changes and fluctuations (in the biosphere) are monitored.

Changes and fluctuations of the climate can significantly affect the state of the biosphere and, consequently, human economic activity. Therefore, it is of practical importance to study changes in the biosphere produced by climate variability, including changes of ecological nature.

Obviously, some effects in the biosphere (changes in the hydrologic cycle, in the total biomass of ecosystems, etc.) produced by climatic changes can either enhance or weaken the effect of other factors (due to the presence of feedback).

Monitoring changes in the biosphere connected with climatic changes and fluctuations is becoming an important aspect of environmental monitoring.

Changes in the elements of the climate system and the ecological consequences of climatic changes are, as a rule, sensitive indicators of the mere fact of a change (or fluctuation) in climate.

The elements which are most susceptible to climatic changes are elements of the biosphere located in polar latitutes (polar ice, tundra ecosystems, ecosystems in shelf polar zones), in arid zones (lakes in arid regions), and ecosystems in desert areas, as well as high-mountain ecosystems (Alpine ecosystems).

Such characteristics of the changes in the biosphere are often called indirect indicators of climatic change. Indirect indicators include many factors which are relatively simple to measure, such as: the levels of seas, lakes, and rivers; shifting of shorelines and of boundaries of various natural zones; changes in annual layers of lake-bottom sediments and in annual layers of glaciers; changes in the snowline and the regions of permanent snow accumulation in mountains or in the boundaries of spreading of glaciers.

A number of ecological signs can be considered here as well: a change in the nature of the vegetation or in the yield of different crops; a change in the marine microflora and microfauna; a change in insect populations or in the way that animal and plant diseases spread, first of all in zones most sensitive to climatic changes. When organizing observations of such signs, it is important to exclude local anthropogenic impacts (for example, the level of a river or lake can suddenly drop, because part of the flow is withdrawn for irrigation).

The data described above are needed for a comprehensive analysis of the state of the environment and for climate modeling. It should be emphasized, that a comprehensive analysis of the state of the environment and climate modeling together enable us to distinguish the critical impact factors and the most sensitive elements of the biosphere (from the viewpoint of the subsequent effect on the climate) which facilitates optimization of the climate-monitoring system.

6.3. Priority and Accuracy of Measurements

The priority in selecting the parameters and factors when organizing climate monitoring, as well as the measurement accuracy, will depend on the specific objectives for which the obtained information is intended.

As already mentioned, the information obtained can be used to solve applied problems in various fields of human activity (in agriculture, construction, power production, municipal activity, etc.), or it can be used to model climate, with a view toward determining its sensitivity to changes in various parameters; to predict possible climate variability, to detect impending climatic changes, and to distinguish the anthropogenic component in these changes and determine the causes of these changes.

The selection of the quantities, needed first of all to solve different applied problems, and the required accuracy of their measurement should be determined for each kind of human activity, taking into account its specifics, technical level, and local peculiarities. This work is usually done by national meteorological services. It will be advisable in the future to move toward eliminating the gaps in the existing network of climatological stations and toward organizing climate data banks.

It is important to select the quantities and to give priority to the most important objective: climate modeling. This issue was thoroughly studied in connection with the second task of the Global Atmosphere Research Program (GARP) and the preparation for the First Global Experiment of GARP.

Summary Table 6.1 (taken from [29]) represents the opinions of experts (from different countries) on the order and accuracy of the measurements needed during and after the First Global Experiment of GARP for climate modeling (the required and desirable ranges of measurement accuracy are given). The given requirements were formulated in addition to those available for data acquisition on the basis of the World Weather Watch (WWW).

Besides the quantities in Table 6.1, it is also necessary to measure the surface-current characteristics, the deep circulation of the ocean, and the thickness and drift of sea ice, and also to observe the state of the polar ice sheets (the change in their boundaries and thickness).

A wider range of requirements was formulated in the Report of a Meeting of a Group of Government Experts [27], although these requirements are practically the same as those in Table 6.1.

These developments are essentially a formulation of the requirements for the measurement accuracy of a future system of observations.

Table 6.1 also contains the required and desirable measurement resolution in both space and time (shown as ranges), which should be relatively uniform all over the world.

However, it appears that, when selecting the space and time resolution, their priorities have to be known. For instance, Academician G.I. Marchuk, who developed new approaches to hydrodynamic long-term weather forecasts based on the simultaneous solution of the equations of hydrothermodynamics for the atmosphere and ocean and on the use of adjoint equations, showed that when forecasting temperature anomalies (one or two seasons

Table 6.1. Quantities (according to their priority) needed for climate modeling, and their required measurement accuracy [29].

Quantity measured	Accuracy	Resolution	
		in time, days	in space, km
Radiation balance of Earth-atmosphere system	2–15 w/m^2	5	100–500
Cloudiness	5% according to amount, 1°C according to temperature of upper part	5–15	100–500
Surface temperature and enthalpy of mixing layer of ocean	0.5–1.5°C 1–3 kcal/cm^2	5–10	100-500, 200
Extent of snow cover and sea ice		5–15	50–100
Albedo of Earth's surface	0.01–0.03	5–15	100–500
Precipitation over land and sea	1–3 mm/day	5–15	100–500
Soil moisture and runoff from main river basins	10%	5, 15–30	100–500
Surface temperature of soil and ice	1–3°C	5	100–500
Gaseous atmospheric constituents and particles			
water vapor	1-5%	1–5	500
ozone (total amount, height distribution)	± 0.5 ppm, 2 km along vertical	1	100–500
carbon dioxide, tropospheric and stratospheric aerosols	± 0.1 ppm, Volume density 5%	15	Sparse network
opacity of atmosphere	1%	7	Sparse network
Tangential wind stress	0.1–0.4 dyne/cm^2	5–10	200
Sea level	2–10 cm	5–10	200

ahead) in given areas of the world, using apriori information on the climatic regime of the ocean and atmosphere and actual data, the influence of regions of the world (of the World Ocean) on these anomalies varies [14, 24]. It is shown that the weather in a number of regions of the Northern Hemisphere is determined to a large extent by the state of the ocean and its interaction with the atmosphere in the equatorial Atlantic and in some areas of the Pacific.

This results in stricter requirements for monitoring the state of the atmosphere and ocean, and also for observing cloud fields and radiation fields in these areas.

The zones of the World Ocean which affect the climate of some region especially markedly should evidently be covered by a denser (in time and space) observation network. The accuracy of the measurement of different parameters in these zones should also be higher. The measurement accuracy of the principal parameter, the sea-surface temperature, should in these zones be at least 0.2°C. For instance, the collection of experimental data in ten such zones (Atlantic and Pacific) has already been started from Soviet weather vessels, within the framework of an extended program of studying the interaction between ocean and atmosphere, for the purposes of long-term weather forecasting and climate theory [8].

The Thirtieth Session of the WMO Executive Committee [16] supported the Soviet proposal to organize an international research program based on the above ideas (*Sections Program*).

In order to determine the possible climatic changes, we have to select both direct and indirect indicators of variations in those elements of the biosphere which are most sensitive to climatic changes (on global and local scales).

Some works, for example, [3, 12], describe in detail the possible consequences and transformations of the most sensitive elements of the biosphere, connected with possible climatic changes. The climatic-change indicators include: the average air temperature, especially for high latitudes; the sea-ice boundaries in polar regions and their extent; the boundaries of glaciers in the high and middle latitudes; the level of inland seas and lakes; the precipitation; and the moisture content of the soil.

The most complicated problem seems to be distinguishing the anthropogenic component of the possible climatic changes, as well as finding the causes of these changes and the sources which affect the variation of the factors most strongly.

The most serious anthropogenic causes of a possible climatic change are: an increased amount of atmospheric carbon dioxide and other gaseous pollutants absorbing long-wave radiations and affecting the ozonosphere; anthropogenic heat release; emission into the atmosphere of particles which form stratospheric and tropospheric aerosol layers. Anthropogenic aerosols change the reflectivity of the atmosphere, and, acting as condensation nuclei, they affect precipitation-formation processes. Various types of land use and changes in the vegetation alter the Earth's albedo. However, it is not enough just to study these processes as a whole; what is needed is regular monitoring of these effects in space and time, as well as monitoring of the effects they produce.

In order to distinguish anthropogenic climatic changes, we have to determine the elements most susceptible to the anthropogenic impact: some components of the radiation balance, the atmospheric transparency, the contents of various gaseous pollutants and aerosols in the atmosphere and so forth.

Obviously, local anthropogenic climatic changes are easier to identify than global changes. The changes at polar latitudes are larger in amplitude than at middle or low latitudes, so that anthropogenic effects are easier to identify at high latitudes.

High-accurcay observations are needed in order to discern the anthropogenic effects. The accuracy requirements can be determined with the help of modeling.

The formulated requirements raise the problem of observing and measuring an enormous number of quantities; if such a system begins operation, it will provide a great volume of data. Consequently, considerable discrimination of the data obtained should be carried out; at an earlier stage the requirements placed on the observation system and measurement accuracy (the priorities) should be specified [25, 29].

Climate monitoring and the acquisition of data related to climatic changes and fluctuations require the application and development of the newest observation techniques, particularly remote sensing systems (including radar and laser technology) and automatic data-processing systems. Satellites play an exceptional role in developing such systems and in providing the corresponding data. It is advisable to combine ground-based observations with observations from space.

6.4. Satellite Climate Monitoring

In this section we consider the possibility (and advisability) of using satellite systems to obtain information about the Earth's climate and about the state of the climate system.

At present most meteorological quantities and major characteristics of the climate system can be measured using satellites. However, sometimes these measurements are difficult to perform and their accuracy is not high enough. On the other hand, some satellite observations are more successful than those using ground-based measurement techniques.

Allowing for the capabilities of existing satellite systems and those being developed, as well as the need to organize measurements providing more accurate information on the Earth's climate and the state of the climate system, let us list the following functions of a satellite system:

1. Measuring meteorological quantities and obtaining other data essential for understanding climatic changes and fluctuations in places where ground-based observation techniques are available;
2. Measuring these same quantities in areas almost inaccessible for ground-based observations:
   (a) in continental areas,
   (b) in oceanic areas;

3. Measuring quantities and factors which are difficult to determine or cannot be determined directly from the Earth's surface:
   (a) integral characteristics of underlying surface (albedo, characteristics of energy exchange and mass exchange between underlying surface and atmosphere);
   (b) some components of radiation balance of Earth-atmosphere system (reflected solar radiation and long-wave outgoing radiation from terrestrial objects);
   (c) corpuscular and hard electromagenetic solar and cosmic radiation;
4. Use of satellites for operational data transmission from almost inaccessible areas of the world.

The following priorities are proposed when using the above functions (1–3) of satellite systems to obtain information on climate:

Function	1	2(a)	2(b)	3(a)	3(b)	3(c)
Priority	III	II	I	II	I	I

The priorities were determined allowing for already existing ground-based observation systems.

At present satellites observe (or can observe) such important quantities as: fields of cloud and wind; temperature and humidity of air at different altitudes, ocean-surface temperature, extent (boundaries) of sea ice and seasonal snow cover of land; zones covered with vegetation (and characteristics of their state) on land and with plankton at sea; soil humidity; precipitation zones and precipitation intensity; main components of radiation balance (see, for instance, 4, 22, 26]. The works cited also include data on the future development of satellite systems. Cloud fields have been identified for more than ten years. The spatial distribution and structure are determined by photography in the visible and IR ranges. Successful measurements of the phases of clouds are being carried out (by comparing measurents of the radio-brightness temperature of clouds in the absorption band of liquid water in the 0.8 cm range and the radiation temperature in the transparency window at 10–12 μm) [22], as well as of the temperature and height of the upper cloud boundaries (with the help of photometric and radiometric methods).

The accuracy of the cloud-height measurements should be increased, and this should be accomplished by the International Project on the Satellite Climatology of Clouds [28].

The temperature profiles in the atmosphere are determined from an analysis of data on infrared or microwave radiation in the absorption bands for gases with known concentrations (carbon dioxide, oxygen). The accuracy of the reconstructed profiles reaches 2 or 3°C up to altitudes of 30 to 35 km.

Satellite information on the wind field, reconstructed with the help of the drift of clouds in the lower and upper levels, is of considerable interest. The wind velocity at sea can be determined from the increase in the radio brightness temperature, connected with the increased amount of foam (foam formation begins at a wind velocity of 7 m/sec). This effect makes it possible to reconstruct the wind velocity with an accuracy of ± 2 m/sec in the interval from 5 to 30 m/sec [13]. The use of IR radar sounding is promising, where the accuracy can be enhanced. To reconstruct the atmospheric pressure, the absorption of SHF radiation is used (in the 30–70 GHz range). The accuracy here can be as high as 1.5–2.0 mb.

The accuracy of satellite determinations of precipitation is not high, since they are based on the relationship between cloud brightness and precipitation intensity; here the method of the future is undoubtedly active radar.

The temperature of the surface layer of the ocean is determined by measuring the radiation in transmission windows in the IR and SHF ranges; the measurement accuracy is ± 1 to 1.5°C, and it is limited (particularly in the IR range) by cloud absorption, dust layers, and other atmospheric interference. The importance of determining the temperature of the surface layer should be emphasized, since its value can be used to determine approximately the enthalpy of the mixing layer of the ocean. The latter value can in turn be used for quantitative discussions about the turbulent exchange of heat and moisture between the atmosphere and the ocean. Therefore, the temperature of the surface layer of the ocean should be measured with an accuracy of several tenths of a degree.

Methods for vertical sounding of the temperature and density of the subsurface ocean layer (down to the boundary of the mixing layer) are being developed. The accuracy of measurement of the land surface temperature is somewhat poorer.

The boundaries and extent of the ice cover and snow cover can be determined quite accurately in the visible range. By making simultaneous observations in the visible and IR ranges, we can determine various types of polar ice, while measurements in the microwave range (at wavelengths around 1.5 cm) make it possible to distinguish with a high accuracy areas covered with ice or water, and ice of various ages and thicknesses.

Exceptionally interesting data are obtained with side-scan radars (active radar), which have been in operation aboard Soviet meteorological satellites since 1983.

Important properties of the surface (vegetation, amount of plankton in ocean) are determined from multispectral pictures.

The amount and distribution of plankton in the ocean can be determined from chlorophyll-measurement data, on the basis of the absorption band of the reflected solar radiation in the 0.43 μm range and the intensity of phytoplankton luminiscence.

Soil humidity and precipitation can be determined with the aid of microwave radiometers.

There are a variety of possibilities for measuring the characteristics of the climate-system elements and other parameters subject to anthropogenic impacts. Anthropogenic impacts can significantly affect the atmospheric turbidity, and can promote an increase in atmospheric carbon dioxide, as well as the appearance of pollutants affecting the ozonosphere (halogenated hydrocarbons, nitrous oxide). Dust layers (atmospheric turbidity) are identified from satellite pictures of the horizon region and from the angular distribution and polarization of scattered solar radiation.

The optical thickness of the atmosphere is determined by measuring the angular distribution of the reflected solar radiation in the visible and near IR ranges of the spectrum (0.55; 0.74; 1.0; 2.2 μm).

The distribution of aerosol particles is determined from the vertical distribution of the aerosol attenuation factor in the absorption bands of atmospheric gases with a constant concentration.

The total amounts of some gaseous components of the atmosphere, such as water vapor, carbon dioxide, and ozone, can be determined using the spectrometry of the absorption bands of the emitted thermal radiation (in the IR range for ozone 9.6 μm) and reflected (and scattered) solar radiation in the UV range. The vertical ozone distribution can be determined as well. The water vapor can be determined in each layer from measurements of the absorption bands at 6.3 and 20–25 μm in the IR range and at 1.35 cm in the SHF range.

Pollution by petroleum products, and the formation of films on the ocean surface, which disturb the mass exchange between the ocean and atmosphere, can influence greatly both the interaction between the ocean and atmosphere and the albedo of the ocean surface.

Data on pollution of the sea surface are obtained by measuring the backward solar radiation and the self-radiation. Systems for remote detection of petroleum and other pollutants make use of radiation in the UV, visible, IR, and SHF ranges. The seawater salinity can be estimated from SHF radiometric measurements.

As noted in the previous chapter, the role of satellite data in obtaining information on anthropogenic changes in the land surface is of great importance. These data are needed in order to understand the causes of possible climatic changes. Satellite data can be used to estimate changes in the vegetation cover caused by forest cutting, desertification, and changes in the nature of agricultural crops. This gives us an idea of the reasons for the change in the Earth's surface albedo. Urbanization affects the albedo as well. Major irrigation systems and the redistribution of water resources affect the nature of the hydrologic cycle and the surface albedo; these changes, as well as snow-cover changes in urban and industrial regions, can be easily discerned from satellite pictures taken in the visible and IR ranges.

The components of the radiation balance corresponding to reflected solar radiation (in the 0.3–3 μm spectral range) and to thermal radiation from the Earth's surface (in the 3–100 μm range) depend markedly on anthropogenic factors (the anthropogenic variation of the albedo of the Earth's surface and the intensity of the thermal radiation of terrestrial objects).

All the components of the Earth–atmosphere radiation balance are determined with satellites, and some components can even be determined more accurately than from the ground.

Anthropogenic changes in the outer space around the Earth are also determined with the aid of satellite observations. For instance, the artificial terrestrial radiation belts were determined radiometrically by satellite-borne equipment.

In this chapter we have summarized and generalized the current viewpoints on climate monitoring and climate-data acquisition expressed by specialists from different countries and international organizations [9, 10, 19, 20, 27–29].

In conclusion, it should be emphasized once again that, of the great number of existing environmental characteristics and impact factors, we have to single out the most important ones, those which are of high priority with regard to providing the basic climate data needed to solve applied problems, with regard to determining the characteristics of the state of the climate system and the impact factors affecting this state, in order to study and understand climatic changes and fluctuations on different scales, and with regard to singling out anthropogenic impacts and effects of climate changes.

The most urgent task at present is to establish a monitoring system which would make it possible (naturally, in combination with climate modeling and other techniques) to distinguish the anthropogenic and other factors which have the greatest effect on the climate and its variation.

CHAPTER 7

Implementation of Monitoring Systems

7.1. Monitoring Operations within the Framework of the Global Environmental Monitoring System

This chapter will discuss the Global Environmental Monitoring System (GEMS) and the way it has been implemented recently under the aegis of UNEP and other international organizations. When this problem was raised at the UNEP level, it was questioned whether such a system could handle the monitoring of any changes (and trends) in the state of the environment (both natural and anthropogenic), as well as a wide range of observations of the state of the environment, including those which have been performed globally for many decades (for example, meteorological observations). Some scientists [10, 37, 54] felt some of these observations to be beyond the framework of GEMS. For example, the organization of an early warning system for disasters, although extremely autonomous in nature, can hardly agree with the general logics of the global monitoring system, as a system for assessing mainly anthropogenic changes in the environment. The author's point of view with respect to montoring was discussed in detail in Chapter 4. However, the procedure and terminology of global monitoring, discussed at international meetings and adopted by the UNEP Secretariat (see, for instance [53]), will be fully retained in this chapter. It should be interesting for the reader to become acquainted with the state of art in this field.

In 1972 at the UN Conference on the Environment held in Stockholm it was proposed to organize the Earthwatch, which is now one of the elements of the United Nations Environment Program (UNEP). The Global Environmental Monitoring System (GEMS), and the assessment, study, and exchange of information were proposed as components of Earthwatch [61]. The main objective of the GEMS was defined as the early warning of anticipated natural* and anthropogenic changes in the state of the environment, which can cause direct or indirect damage to human health or well-being.

A monitoring system was defined as 'a system of continuous observations,

---

* As already mentioned, observations and assessments of natural changes in the environment (e.g., weather changes) lie mostly within the sphere of existing geophysical services. Therefore, it is the author's opinion that they should not be included in the GEMS (although these data will be needed to interpret anthropogenic changes).

measurements, and assessments for certain purposes' [54] (see also Section 4.1 above, for the definition of a monitoring system).

During subsequent years the idea of the GEMS was discussed in detail. First, the discussion focused mainly on the problem of monitoring pollutants (this was considered to be the most important, first phase of monitoring). Individual ecological problems (including the effects of pollutants on the biosphere), and problems connected with the monitoring of renewable natural resources, were then not discussed at length.

The report prepared by Munn [50], in which the accent is on monitoring the anthropogenic pollution of the environment, is very interesting and constructive.

The Intergovernmental Meeting on Monitoring held at Nairobi in February 1974 represented an important step in the development of the GEMS. An attempt was made there to equalize the GEMS objectives (i.e., to give equal attention to anthropogenic pollution, to ecological problems connected with the impact of pollutants and other impacts, to the monitoring of renewable natural resources, and to warning of disasters and other phenomena dangerous to human health and well-being). Seven basic objectives of the GEMS program were formulated [60]:

1. organization of an extended system for warning of danger to human health;
2. assessment of the global pollution of the atmosphere and its impact on climate;
3. assessment of amount and distribution of pollutants in biological systems, particularly in food chains;
4. assessment of critical problems arising as a result of agricultural activity and land use;
5. assessment of response of terrestrial ecosystems to the effect of the environment;
6. assessment of ocean pollution and the impact of pollution on marine ecosystems;
7. establishment of an improved system for warning of disaster on an international scale.

The formulated objectives of the GEMS program provide for operation in different directions; in some directions successful research and observations have been carried out for a long time, while in others the work is just beginning. As already mentioned, although some directions can be considered to be observations of the state of the environment, still they are hardly suitable for inclusion in the GEMS system. For instance, the World Weather Watch of the World Meteorological Organization, which is already in operation, is only partially connected with the GEMS; its various components provide many countries with important hydrometeorological information, but not all of this information is significant for the identification of anthropogenic changes in the environment.

To avoid duplication of the operation of the GEMS with already existing systems, the following important principles were adopted at the Nairobi Meeting [60]:

1. Intergovernmental cooperation in the field of monitoring should as far as possible be based on existing national and international systems.
2. It is necessary 'as much as possible' to use special UN agencies as starting points for coordinating and implementing the monitoring program.
3. Priority should be given to global and regional (multinational) monitoring.
4. The monitoring system should be so developed as to satisfy clearly formulated requirements, while the data assessments should be a joint effort of the project of this system.

Thus, for many objectives there has been no need to organize new services; rather, it is a question of maximum use of already existing systems, of maintaining and developing these, and of efficiently using information and coordinating efforts within the framework of the structures being developed.

An Interagency Working Group on Monitoring for coordinating the activities of different agencies in this field has been established. Within the framework of the UNEP, there was established a Program Center of Activities, which should not have to deal with the internal coordination of individual monitoring, for which the agency is responsible; therefore, this center should coordinate the different fields of monitoring and its role should not be operational [54]. It has been noted [46] that the GEMS is not a system of different, closely interconnected components. Rather, it is an ensemble of various kinds of activities, each of which is directed by the Center with a view toward obtaining the necessary answer.

In connection with the very wide circle of problems based on the GEMS major objectives, the UNEP Governing Council asked the UNEP director to discuss how the seven objectives of the system could be combined, and to outline the short-term and long-term plans of implementation for each given area.

Actually, these directions are closely interconnected. During their implementation in the UNEP the above objectives were divided into two main groups: one group included all issues related to pollution monitoring, and the other included issues related to the monitoring of renewable natural resources. In [54] the latter was called 'ecological monitoring',* but it is

* In the author's opinion, 'ecological monitoring' has a broader meaning (see Chapter 5). For instance, the monitoring of pollution and of biota responses to pollutants is an essential element of ecological monitoring.

actually the monitoring of renewable resources. The problem of disaster warning was left as an independent direction.

In UNEP paper [5] (1978), the five basic objectives of the system were defined:

- monitoring climatic conditions;
- monitoring large-scale transport and deposition of pollutants;
- monitoring for health purposes;
- monitoring renewable natural resources;
- monitoring the ocean

A later work [46] describing UNEP activity (1982) uses this same approach. Evidently, to solve problems in each group of the monitoring system, different approaches should be applied.

The ultimate objectives of the global and regional monitoring of pollution within the GEMS are:

1. to determine the levels of selected critical pollutants in a certain part of the environment, and their distribution in space and variation in time;
2. to get an idea of the size and velocity of the fluxes of selected pollutants and harmful products of their transformations;
3. to compare sampling and analysis techniques of different countries, including developing countries, so as to provide comparable results and an exchange of experience in the organization of monitoring systems;
4. to provide on a global or regional scale the information needed to adopt management decisions in the battle against pollution.

It was more difficult to determine how to sum up the output data from the monitoring of renewable resources. Within the framework of the GEMS a special methodology was developed for this, and a regional study of the natural resources on a small scale was carried out (for example, of the state of the soil or of forests). It was thought that this would lead to future applications of the experience gained to different countries and to the organization of a broad network of such monitoring. In any case, it is noteworthy that recently the center of gravity within the GEMS has moved from pollution monitoring toward a more balanced approach, with regard to both the monitoring of renewable resources and the solution of some additional ecological problems.

During the period after the GEMS concept was formulated, much was done internationally to organize observations and studies of changes in the state of the environment. Some operations were carried out within the GEMS, and some were autonomous (most were supported by the UNEP). A great deal has been accomplished by such international organizations as UNESCO, WMO, WHO, and FAO. The Council of Mutual Economic Assistance (CMEA) contributed significantly to the realization of the GEMS.

CMEA member countries began establishing background stations (within the framework of the GEMS) in their territories. A Coordination Center for the Global Environmental Monitoring System was established in the Soviet Union, and a plan for participation in the GEMS was worked out.

On the whole, a great volume of material has been collected on changes in the environment, and a review of it is interesting from the viewpoint of realizing the GEMS goals and objectives. These data are discussed in [5, 46, 53], which also consider the state of the art with regard to international efforts to achieve global environmental monitoring.

It is noted in [5, 37] that quality control of the data used is important for the success of the system. In this connection, all similar measurements should be reliably intercalibrated (up to 25% of the funds allocated for monitoring can be spent for this). The efforts of specially selected laboratories should be directed toward accomplishing this goal. The program of sampling must be valid and reasonable, with regard to statistics and the representativeness of the samples.

Data storage and retrieval will be one of the main stages in the implementation of the GEMS. Sample data from different parts of the environment should be preserved and cataloged, and made accessible to various laboratories if checks and verifications are needed. Sample data should be stored for a long time. Specially selected research institutes can be used for the storage.

The basic results within the framework of the GEMS and other international activities connected with environmental study and monitoring (in accordance with the data of [5, 46, 53]), as well as some proposals concerning the development of the GEMS [37, 5, 46, 53], will be considered below.

The first group incorporates material about the monitoring of renewable natural resources, in this work called 'ecological monitoring' (the fourth and fifth original objectives of the GEMS). It includes the results of monitoring the soil and vegetation cover, monitoring the water resources and marine resources, as well as some aspects of monitoring the background state of the biosphere.

In this part of the monitoring, when considering the actions recommended, we stressed the need to monitor the change in the area of tropical and deciduous forests. These data can serve as an indicator of the state of the global biological system and the anthropogenic impacts on it.

Data obtained by mapping zones subjected to a heavy anthropogenic impact (agricultural zones, urban regions, mining regions) can be used to estimate changes in the state of ecological systems on a large scale, taking into account critical problems of land-use practice. For instance, droughts limit agriculture on the African continent to 44%, soil degradation in North America reduces productivity by 22%, and floods in Southeast Asia cause a 19% reduction [46].

Let us note some results of the above studies.

With regard to the *global assessment of soil degradation*, FAO and UNESCO, jointly with UNEP, started collecting existing data (using remote sounding data); maps of soil degradation are being compiled, in particular, for the parts of Africa north of the equator. Soil degradation and the danger of such degradation (erosion, salinization, etc.) are assessed from these maps. It was demonstrated that 50% of Algeria and 75% of Nigeria are at risk with regard to wind erosion. Almost all of Sierra Leone and 50% of Liberia are endangered by water erosion. The compiled maps can be used to determine which areas are at risk, and recommendations for land-use planning are made accordingly.

The suitability of the land for cultivation in various agroecological zones and the state of the pastures in arid and semiarid zones are evaluated by FAO. This work is closely connected with the project on desertification research, since pastureland is very vulnerable to the desertification process. A large-scale project to study the state of pastureland was launched in 1979 in Senegal.

Monitoring the tropical forest cover is being implemented jointly by FAO and UNEP in some African countries; the purpose of this is to obtain data on the areas of existing forest cover, its state, and qualitative and quantitative variations (as a result of anthropogenic impacts, such as wood cutting, land cultivation, and disturbance of the hydrological regime); zoning and classification of the vegetation with respect to that of other areas of the world is being carried out. This work is supposed to be extended to cover the whole tropical belt (using satellite data).

The territory of three countries (Togo, Benin, and Cameroon) which is about 0.5 million km^2, has been studied (including the use of satellite data) and shown on the map. The boundaries of 15 ecological zones have been determined on the basis of satellite pictures and information.

In order to continue studying the manmade damage to tropical forests, a Project for Assessing the Global Resources of Tropical Forests was established within the GEMS; the project covers 26 countries: information is gathered on natural forests, forests which have been cut out, reduced forests, and replanted forests. Global reports on the state of tropical forests in Asia, Latin America, and Africa were published in 1982.

It was found that about 80 000 km^2 of tropical forests are being cut annually, which will result in a 12% reduction by the end of the century; however, this cutting is irregular: nine countries can in 25 years turn out to be completely without forests, while eleven others can lose half their forests. This can have serious consequences at both the national and global levels [46].

Monitoring of water resources is implemented within the framework of the International Hydrological Decade (IHD) and is a part of the UNESCO International Hydrological Program (IHP). The water balance was studied, as well as hydrological regions. This activity is being carried out jointly with the WMO (within the framework of the Operative Hydrology Programs). UNESCO, jointly with UNEP, are making a global inventory of glaciers.

Monitoring the background state of the biosphere is being implemented under the aegis of the UNESCO Program 'Man and the Biosphere' (MAB). Most activities are carried out along the line of MAB projects which study both man's impact on the resources of lakes, rivers, bogs, deltas, and riversides and the impact of pollutants on the biosphere (in particular, MAB Project 14). Investigations according to various types of such monitoring are being carried out in 35 countries. Some of the important subjects from the monitoring viewpoint are: water body eutrophication and the related nutrient flux, the cycle of nutrients in forest and meadow ecosystems, the effect of land use on water quality, and the impact of atmospheric pollution on land and water ecosystems.

The establishment of biosphere reserves is of great importance. By 1977 there were 118 such reserves in the world. Most biosphere reserves are established on territory where the monitoring of both natural ecosystems and background pollution is possible. An attempt is made to monitor the changing state of various endangered species.

For the *monitoring of living marine resources* the International Council on Exploration of the Sea collects data on fish catches and estimates the fish stocks in the northeastern Atlantic, and in the adjoining waters of the Arctic Basin and the Baltic Sea. These data include catches and stocks of herring, cod, haddock, hake (merluccius), halibut, salmon, deep-sea shrimp and some mollusks.

The second large group [53] includes material on pollution monitoring (including monitoring related to human health and climate, and the monitoring of ocean and land ecosystems). This group covers objectives 1–3 and 6 of the GEMS.

When discussing the recommended measures, much attention was paid to *monitoring possible climatic changes* (second objective) and their causes (see Chapter 6). The Seventh Congress (1979) of the World Meteorological Organization adopted a resolution on implementing the World Climate Program; climate monitoring and gathering climate data are the primary objectives of this program.

Much attention should be paid to the monitoring of various factors affecting the climate and the climate system, and to monitoring the processes needed to understand possible climatic changes. Also important are monitoring the carbon dioxide and the atmosphere turbidity (in order to assess the Earth's radiation balance), studying the state of the ozonosphere, and measuring various trace pollutants in the atmosphere, such as chlorofluoromethanes and nitrous oxide, which can affect the atmospheric ozone layer. A specially organized network of WMO background pollution monitoring stations is used [64]. The glaciers of the world are also taken into account (at 750 glacier stations in 21 countries) [46].

During the discussion, considerable attention was given to the organization of a system for *monitoring the state of terrestrial ecosystems*, and for assessing the mass flow of anthropogenic compounds affecting various biological systems. It was proposed to organize the monitoring of various components of the atmosphere, oceans, and terrestrial ecosystems, as well as the monitoring of food chains and the geographical movement of food products, with a view toward assessing the mass flow of these substances. Also proposed was the monitoring of biological systems, with a proper selection of representative carnivorous animals at the top of the trophic pyramid, and cosmopolitan birds, which may be the best indicators of the paths of anthropogenic pollutants in the food chains.

Monitoring of human health also serves the objectives of this group, provided special attention is given to obtaining data on diseases produced by changes in the natural environment (such diseases are sensitive indicators of the environmental impact on human health).

For instance, it was found that cancer of some organs, as well as some birth defects, can be brought about by environmental factors [48, 68]. Cardiovascular and pulmonary diseases are sensitive indicators of the content of chemicals in the air we breathe [38]. Recently a significant amount of data related to the monitoring of atmospheric pollution has been obtained all over the world.

Systems for monitoring air pollution are being established in cities of more than 50 countries (WHO jointly with WMO and UNEP), and for monitoring water quality at 300 stations in 70 countries [46]. The WHO publishes data on 'Air quality in selected urban areas', and a project designed to measure the pollutant content in the human organism has been launched.

The WMO is implementing a broad program using the Background Air Pollution Monitoring Network [64], which is a part of the World Weather Watch. This network consists of 176 stations, including 75 stations which transmit data on precipitation chemistry [6]; UNEP supports this program. The following types of stations are being prepared for incorporation into the network:

(1) baseline stations for global monitoring of very low background concentrations of the most significant components of the atmosphere;
(2) regional stations for monitoring long-term changes in the composition of atmospheric air, produced by the regional impact of human activity (land use, etc.);
(3) regional stations with an extended program.

The minimal monitoring program at baseline stations includes measurements of the atmospheric turbidity, the conductivity of the air, the carbon dioxide in the air, and the precipitation chemistry (including an analysis of SO_4^{2-}, Cl^-, NH^-, NO_3^{2-}, Ca^{2+}, Mg^{2+}, Na^+, K^+, and heavy metals), as well as pH measurement. The minimal program at regional stations includes observations of the atmosphere turbidity and the precipitation chemistry. The extended programs of the baseline and regional stations provide additional measurements of SO_2, H_2S (or total sulphur), NO and NO_2, NH_3, particle composition, CO, CH_4, N_2O (total ozone and its concentration near the ground), heavy metals in the air, and Aitken nuclei.

At present about 10 baseline stations are in operation, and several more are being prepared. The pollution of the atmosphere and precipitation, as well as of other parts of the environment, will be analyzed at these stations (see Figure 4.6 in Section 4.4).

During the measurements at all these stations much stress is placed on the application of unique techniques of analysis and intercalibration. The measurement data from the stations are published on a regular basis.

The UN Economic Commission for Europe (ECE) has contributed much to the organization of measurements and assessments of the long-range transboundary transport of sulphur dioxide in Europe (see Chapter 9). A Coordinating Chemical Center has been established in Norway, as well as two Meteorological Synthesizing Centers (one in Western Europe, Oslo and one in Eastern Europe, in Moscow). This project was set up because polluted air masses moving over long distances can impact vast areas. The long-range transport of SO_2 is dangerous both to human health and to land and water ecosystems (because of acid rain and acidification of the surface water and soil). SO_2 is monitored at 70 stations. Plans for monitoring residual pesticides in biological systems are in the works (FAO and UNEP).

WHO, jointly with UNESCO, WMO, and UNEP, has organized a network for water-quality monitoring, in order to identify the most dangerous pollutants, to track pollutant transport and to monitor water-body eutrophication. This aspect of the monitoring is related to the GEMS objectives (fourth objective).

The monitoring of food products and animal fodder, organized by FAO, WHO, and UNEP, is aimed at determining the flows of various pollutants along this chain (including the transport of food products). The first project, for ten countries, provides monitoring of chlorinated hydrocarbons in certain dairy products and monitoring of lead in some vegetables and fish products.

The environmental impact on human health is kept track of by monitoring the pollutants in the fluids and tissues of the human body, in breast milk, and in human hair; the monitoring of pollutants in 'targets' is performed by WHO, UNEP, and IAEA (International Atomic Energy Agency). This monitoring includes measurements of lead, cadmium, mercury, and organochlorine compounds (as well as arsenic and pesticides) in the targets.

A widespread identification of the factors affecting human health is performed by WHO via epidemiological investigations. Since 1955, the UN Scientific Committee on the Effects of Atomic Radiation has been coordinating measurements of the environmental radioactivity levels and radiation doses in biological objects from all existing sources (natural radioactivity, radioactive fallout, industry, medical practice, etc.), as well as an assessment of the danger of somatic and genetic effects from such radiation. This committee regularly publishes reports on *Ionizing Radiation: Levels and Effects.*

Since 1961, IAEA (jointly with WMO) has been gathering data on the content and distribution of tritium, deuterium, and ^{18}O in precipitation. For many years more than 100 meteorological stations in 65 countries have been collecting monthly precipitation samples, which are sent for analysis to the Hydrological Laboratory of the International Atomic Energy Agency in Vienna. IAEA publishes the measurement results on a regular basis (5 volumes have already appeared).

A vast amount of material relating to possible climatic changes and fluctuations has been gathered. A great deal is being done by the World Weather Watch of the WMO. Within the framework of the WWW, an extensive international exchange of data from the network of climatological stations is taking place; observational data on solar radiation and atmospheric ozone are collected at common centers. A special observation program was established in connection with the second objective – the study and modeling of climate – within the framework of the Global Atmosphere Research Program (GARP). This program was implemented especially actively during the Global Weather Experiment (1979). Observations of the atmospheric composition, of a number of pollutants in the atmosphere, and of the aerosol component are also incorporated into the program, in an attempt to understand possible climatic changes. On the whole, the gathering of climate data is organized by WMO, with the support of UNEP. Many of these data are of great importance, being used for applied purposes at the optimum level of organization and management of the economy, construction, and natural-resource development.

Because of the significance of monitoring climate data, this topic will be dealt with in Chapter 6.

It is noteworthy that, although in many cases the collection of climate data is now supported by UNEP, the work carried out before the GEMS concept was formulated was successfully implemented by the WMO. Much progress in this field will certainly be attained when the World Climate Program is implemented. Meteorological satellites can be (and are) of great importance for gathering climate data.

Monitoring of ocean pollution is very important, and international activities and GEMS have already contributed much to it. Of particular value is international cooperation, aimed at obtaining information on the levels and trends in sea and ocean pollution on regional and global scales. This pollution, which is potentially dangerous to human health, has an adverse effect on marine ecosystems and has an effect on the energy and mass exchange between the atmosphere and ocean.

The selective monitoring of several particularly important and representative substances in the open ocean and in coastal regions was supposed to be sufficient to answer many questions concerning the state of the oceans and marine ecosystems. It was assumed [37] that 8 to 12 stations would be needed for this, to measure the water pollution by petroleum, the concentrations of plutonium, americium, nitrates, phosphates, and fluorochloromethanes, the primary productivity, and the carbonate equilibrium.

A program of observations of some marine organisms, such as bivalved mollusks (providing an 'integration' of the pollution in time), was recommended. Such observations were also to include measurements of the content of heavy metals. Pollution measurements at marine hydrological stations are to be referred to this same type of monitoring.

Monitoring of the open water of the ocean is implemented on the basis of the Integrated Global Ocean Services System (IGOSS), which has been in operation for many years.

The goal of IGOSS, which was established jointly by the International Oceanographic Commission (IOC) and WMO, is to obtain information about the state of the oceans, various processes in the ocean, and its interaction with the atmosphere. Observations of different physical parameters of the atmosphere and the upper ocean layer are performed within the framework of the global system of oceanic stations. Data are transmitted throughout the world on a real-time scale. A total of 7500 merchant vessels and about 10 weather vessels, which form the basis of the system, perform observations as a part of the WMO World Weather Watch (in certain areas of the World Ocean).

Recently IGOSS, jointly with WMO, started implementing the monitoring of marine pollution by petroleum products (MARPOLMON) in different points of the World Ocean (forty countries are participating in the study).

A Group of Experts on the Scientific Aspects of Marine Pollution (GESAMP), which is supported by UNEP and other international organizations, prepares reviews on the health of the oceans [63].

The WMO Executive Committee in 1978 approved a Soviet initiative to organize integrated monitoring aboard one of the weather ships (in the North Atlantic). In addition to pollution monitoring, this would include observations of the state of the marine ecosystem, i.e., monitoring according to a program similar to the biosphere-reserves program [30].

The preparation of a global inventory of rivers discharging pollutants and suspended solids into the ocean, which was initiated in 1974 by UNESCO and UNEP, is also of interest with regard to ocean monitoring. Information is being collected from 270 large rivers.

The plan of action developed and implemented by UNEP for regional marine areas and regional seas (the Mediterranean, the Red Sea, and the Caribbean) is very important for monitoring the seas and oceans [46].

UNEP coordinates monitoring operations in the Mediterranean Sea implemented by 16 countries. The program of activities includes the monitoring of petroleum and petroleum hydrocarbons in seawater, the monitoring of heavy metals (particularly mercury and cadmium), DDT, PCBs, and other chlorinated hydrocarbons in marine organisms, the study of pollutant impacts on marine organisms and ecosystems, the study of the coastal transport of pollutants, and the regular checking of coastal waters. The Mediterranean countries take part in these activities, and the investigations of pollution of the sea from continental sources provides countries located on the shores of the Mediterranean with information on the type and quantity of pollutants from land sources and rivers, on the characteristics of the discharge, and on the practical monitoring of water pollution. An inventory of pollution sources was made during the first phase, and now actual data on discharges into the sea are being collected, after which all of the collected information will be assessed.

Extensive investigations of pollution have been carried out in the North Sea for about ten years (by the IOC, jointly with other organizations). These investigations include a study of the physical, chemical, and biological transformations of pollutants, including petroleum, chlorinated hydrocarbons, and metals (particularly mercury, copper and zinc). Pollution sources associated with industrial wastes, municipal wastes (including garbage), and atmospheric precipitation are being considered. Pollution of marine organisms (fish, mollusks) are being studied as well. The result of joint investigations carried out by 10 laboratories in eight countries are being published in a special report.

Investigations in the Baltic Sea are similar to those in the North Sea; seven Baltic countries are participating in this operation.

Extended investigations are included in a large-scale interdisciplinary experiment in the open sea, BOSEX-77, in which scientists from all the Baltic countries are participating. The experiment will provide sufficient knowledge for the monitoring and for the control measurements which are to be made within the framework of the Helsinki Convention on Monitoring the Pollution of the Baltic Sea.

Quite extensive investigations are being carried out in the northeastern and northwestern Atlantic. These are concerned with the pollution levels in the living resources of the sea and with the pollutant input into the ocean. All the countries along the North Atlantic are participating in this operation.

Some interesting recommendations were made at the First International Symposium on the Integrated Global Monitoring of the World Ocean

(MONOC), held in Tallin in October 1983 [51]. A special working group was established in March 1984 at a session of GESAMP, for the development and realization of these recommendations.

Warning of disasters (this operation is referred in [53] to the third group), which is one of the objectives of GEMS (the seventh objective), has been provided for a long time, being up to now the job of specialized organizations (and national services). In order to establish a warning system, it is necessary to set up a system for the forecasting and warning of natural disasters, with data collection and processing on a real-time scale. The purpose is to modify and develop existing systems in cases where the monitoring system does not meet the specified objectives (or where there is no system).

Presently there are operational systems for tropical cyclones (WMO) and tsunamis. Earthquakes, floods, volcanic eruptions, and snow avalanches are also recorded.

Many international organizations and countries are investigating the establishment and development of systems for monitoring the background state of the environment. Project 14 of the UNESCO Program 'Man and the Biosphere' is completely devoted to this subject. The project is called 'A Study of Environmental Pollution and Its Effects on the Biosphere'. The Soviet Union initiated this project, and the first meeting of the working group was held in Moscow in April 1974.

Project 14 provides for an analysis of the paths and processes of pollutant transformation, a study of the effects of pollutants on the biota, and the identification of indicators of the state of the environment and the critical elements of the biosphere, which are most sensitive to the impact [16]. This approach, together with a search for the most dangerous pollutants (from the viewpoint of their impact on the ecosystem), forms a part of the scientific basis of ecological monitoring. The observation stations for the above investigations are located either in the biosphere reserves or at baseline stations.

Under the aegis of ICSU/SCOPE (where ICSU is the International Council of Scientific Unions), at Chelsea College in London in 1974 there was established a Center for the Study of Various Aspects of Environmental Monitoring. This work, which is supported by UNEP, has four principal goals:

1. to study the requirements for monitoring and assessment;
2. to study the significance and possibilities of regional monitoring;
3. to carry out monitoring via the dynamics of processes in the environment (for instance, using the concept of the 'expected' dose);
4. to investigate possible future changes in the environment.

The third task is the most significant one, with regard to the GEMS objectives. The interrelationship between pollutants in different parts of the environment and the paths of their transfer from one part to another are studied with the aid of modeling. Since many pollutants, such as heavy metals, are encountered in natural geochemical cyles, and are also produced by man's activity, it is important to establish the natural variability of the background pollution levels produced by these substances (independently of human activity). For this purpose, data collected over many years, from before the start of any significant anthropogenic changes, are analyzed.

Later the Center began to study problems connected with the measurement of different variables in different parts of the environment (for instance, in air, water, biota) at these same monitoring stations (multipurpose monitoring with an integrated approach), and in addition cooperation was begun with the WMO, in order to study the possibility of organizing such monitoring at the stations of the WMO Background Air Pollution Monitoring Network [64].

Such an integrated approach to monitoring was introduced at the WMO background stations on the initiative of the Soviet Union [30]. Here the concept of a comprehensive analysis of the environment was applied [15]. This approach was adopted for monitoring as a whole at the International Symposium on the Integrated Global Monitoring of Environmental Pollution (Riga, December 1978).

7.2. The Present Situation in the USSR and Elsewhere

In order to organize and develop in an optimum manner a system for monitoring the pollution of the environment, we must first analyze the present state of environmental pollution in the Soviet Union and in other countries. This analysis should be based on a knowledge of the hydrometeorological and physical-geographical peculiarities of the region under consideration. The characteristics of a given ingredient (its mobility, susceptibility to transformations, and penetration into other parts of the environment) must be studied. The toxicity of the ingredients and the population size, the amount of living organisms exposed to radiation or to any other impact of the given source, the susceptibility of various structures to the impact, etc., should be taken into account. Thus, the organization of a scientifcally justified monitoring system (including a selection of priority pollutants and of the places where control points have to be set up) should be based on a comprehensive analysis of the environment and should be directed toward regulating the quality of the environment, but allowing for the maximum possible development of industry and of the national economy as a whole.

Such an analysis has been conducted in the Soviet Union, and the system for monitoring the pollution level in the USSR is being improved on the basis of the results of this analysis.

When organizing a network of observations of urban air pollution (and particularly a network for monitoring pollution sources), it should be kept

Table 7.1. Average annual concentration.

Country	mg/m^3
USA	0.12–0.29
Japan	0.08–0.22
Germany	0.04–0.16
USSR	0.1–0.2

in mind that some low-lying sources (automobile transport) and even small-sized sources (chimneys of houses) can affect the local pollution level to a greater extent than high-rising sources (emissions from high stacks) [31]. Here not only the total emission and concentration of pollutants in the surface air layer must be taken into account, but also the size and the number of biological objects subject to the impact. Such an approach is discussed in [9].

The analysis indicated that sulfur dioxide, dust, carbon monoxide, and nitrogen oxides account for the main part of the pollution (up to 85%) in Soviet cities. Specific substances, usually produced by certain industrial enterprises and therefore observed in a comparatively small number of places where these industries are located, account for the rest. These are hydrogen sulfide, chlorine, carbon disulfide, ammonia, fluorine compounds and hydrocarbons.

The urban air pollution by sulfur dioxide is in many countries (on the average) close to the maximum permissible concentrations, and during some periods even exceeds these values. For instance, according to [39, 40], in 101 regions, out of a total of 247 being monitored in the USA, SO_2 concentrations either exceeding or close to the established standards are observed. Table 7.1 gives the maximum concentrations of SO_2 in cities of the USSR, USA, Japan, and West Germany [2, 39, 44, 52, 58] in 1970–1972 (the present levels are somewhat lower). Significant levels of pollution by nitrogen oxides, dust, and carbon oxides are recorded in the urban air of many countries.

The nitrogen oxide content in industrial Soviet cities is 0.02–0.04 mg/m^3 (MPC = 0.085 mg/m^3), and the CO content in large cities is 1–3 mg/m^3 (MPC = 3 mg/m^3). At the peak hours, because of autotransport, it can reach 5–9 mg/m^3 in some places [23].

Air pollution by benzo(a)pyrene (from vehicle exhausts) in large cities is a serious problem. In some cities of Western Europe the maximum permissible concentrations have reached: 0.4–0.7 $\mu g/m^3$ [33]. The concentrations in Soviet cities are much lower. For example, the benzo(a)pyrene concentrations in Leningrad are on the average only 1/7 to 1/10 of those in the large industrial cities of Great Britain and the USA [23].

Heavy metals are of particular interest, since they are emitted by both metallurgical factories and fossil-fuel power plants.

In many cities the atmospheric lead concentrations produced by the exhausts of vehicles using ethylated fuel are quite high. For instance, the average annual atmospheric lead concentration in cities of the US was, according to [49, 57], as high as 4.5 $\mu g/m^3$ (in rural regions it was only 0.1 – 0.5 $\mu g/m^3$).

When organizing the Soviet network for monitoring the air-pollution level, scientists took into account the spreading of pollutant emissions, their total amount (contribution to the total emission), and also the toxicity of the emitted ingredients. An analysis demonstrated that sulfur dioxide, dust, carbon monoxide, and nitrogen oxides are the major pollutants in Soviet cities (up to 86%) [19]. In accordance with this, sulfur dioxide and nitrogen dioxide account for 23% of the measurements, carbon monoxide for 18%, dust for 11%, and phenol for 7% [23].

Industrial and municipal wastewater and agricultural runoff (from fields and livestock farms) are the principal sources of river and lake pollution.

Petroleum products, phenols, iron, copper, zinc, heavy metals, and many other harmful substances are observed in rivers, lakes, and water reservoirs. The incidence of organic matter into river and lake water reduces the amount of dissolved oxygen.

The pattern of water-reservoir pollution depends markedly on the distribution and amount of discharged polluted wastewater, as well as on the hydrological and ice conditions, which change from year to year and from area to area.

Just as in the case of air-sample analysis, the analysis of water samples in previous years was carried out in accordance with the abundance and significance (toxicity) of the given ingredient. For instance, most chemical analyses were performed to determine the concentrations of petroleum products, phenols, ammonium, nitrates, and iron in the water bodies of the country (15 000 to 20 000 analyses per year); the same number of analyses was made in order to determine the amount of dissolved oxygen and BOD_5 [23].

As mentioned above, heavy metals are found in river water and bottom sediments. Of particular danger is mercury, most of which is present in a very toxic methylated form. The mercury accumulation factor in hydrobionts is 10^5 or higher. The mercury content in some water bodies of the USA reaches the maximum permissible level [36], while in some Swedisch lakes it exceeds the permissible values [67].

Another serious problem is water-body pollution by organochlorine pesticides (DDT, etc). These substances can accumulate in the final links of the food chains (at higher trophic levels). The accumulation factor in the food chain for pesticides can be as high as 5×10^5 (for adsorption of pesticides from the water by detritus, and concentration in the tissues of detritophages and small fish, and then in fish-eating birds). Table 7.2, based on data from

Table 7.2. Average content of pesticides in water and hydrobionts.

Water, hydrobionts	Content of DDT and other pesticide, ppm
Water	$0.5–1.0\times10^{-4}$
Plankton	0.01–0.04
Aquatic invertebrates	0.1–0.9
Freshwater fish	1–2
Birds eating freshwater fish	22–26

[55, 69], gives the mean contents of DDT and other organochlorine pesticides in water and hydrobionts characteristic of the present time.

Still higher are the accumulation factors for polychlorobiphenyls (PCBs), which are actively absorbed by sediments and are included in biological cycles. The accumulation factor for PCBs in aquatic invertebrates and fish can reach 7×10^4, and in birds of prey it can be $10^8 - 10^9$. The corresponding accumulation factors for polybrominated biphenyls (PBBs) are around 10^4 [62].

The problem of eutrophication of water bodies has become general. It is caused to a large extent by the great amounts of nutrient substances discharged with municipal and other wastewater (great amounts of fertilizers are washed out of agricultural fields). While in the soil, fertilizers are (as a result of bacterial activity) rapidly transformed into nitrates, which when discharged into water bodies present a great danger to the population and speed up eutrophication (eutrophication begins when the nitrogen nitrate concentration is only 0.3 mg/l). If we take into account that in the USA the permissible standard for nitrates in drinking water, 10 mg/l, is often exceeded, then the urgency of this problem becomes evident [34] (the average concentration of nitrates in the surface waters of the USA is around 2 mg/l [35]).

The pollution of a river can be observed throughout almost its whole length (in industrialized areas) [8]. For example, in the USA 24% of all rivers are polluted over more than half their length. The extent of the pollution in 48% of the rivers is 20–50% [50]. However, river mouths and estuaries are the most vulnerable. This can be seen from the detailed description of the pollution of the Potomac [34]. A similar situation is observed in Great Britain, where the water quality in more than 70% of the estuaries is poor or doubtful (this information pertains to 1974–1977).

Such areas have to be carefully monitored, according to both chemical and hydrobiological indicators.

Along with large-scale monitoring of the pollution in rivers, lakes, and other water bodies, there should also be observations of the 'background' pollution of the surface waters in areas far away from local pollution sources. With regard to this, much attention should be given to monitoring the pollution produced by the air transport from neighboring regions. It is this pollution path which appears to be the main one for water bodies.

The problem of surface-water acidification, caused by the washing out by precipitation of sulfur dioxide and its conversion products, has become an urgent one for many European water bodies. This effect is often observed at quite considerable distances from the SO_2 emission sources (see Chapter 9). The problem of pollution of the World Ocean will be dealt with in Chapter 11.

Recently much attention has been given to the development of a system of observations of the background pollution of the atmosphere and other parts of the environment. The first line of background monitoring was established in the USSR. The World Meteorological Organization, which is implementing a special global system of background air pollution monitoring (BAPMON), has contributed much to this activity [47, 64] (see Figure 4.6).

The national background pollution monitoring system in the USSR is designed to study and observe the composition, cyle, and migration of pollutants, as well as the biota response to the background pollution, and also to assess the observed state of ecosystems and to forecast changes in this state.

The stations carrying out the integrated background monitoring in biosphere reserves and in other reserves are placed at optimum locations. At present this network includes stations in the Berezina, Caucasus, Prioka Oka-Terrace, Central Chernozem, Sari-Chelek, Repetek, and Sikhote-Alin biosphere reserves, as well as in the territories of the Borovoe, Abramov Glacier and Barguzin reserves.

The standard program of observations for stations carrying out integrated background monitoring covers all parts of the environment (atmosphere, soil, water) and their biological components.

Ozone, dust, nitrogen and sulfur compounds, lead, mercury, arsenic, cadmium, polycyclic aromatic hydrocarbons, and organochlorine hydrocarbons are classified as priority pollutants. These substances are measured systematically. The results of studies of the background state of the environment in the USSR and elsewhere are given below [17, 27, 28].

The background average hourly values of the ozone concentrations vary from 4 to 160 $\mu g/m^3$, except for some areas of the US, where values up to 240 $\mu g/m^3$ have been recorded; the average daily ozone concentrations for various background regions in the world vary from 84 to 115 $\mu g/m^3$, and in the USSR from 30 to 80 $\mu g/m^3$.

The average concentration of suspended particles over the European USSR is 30 $\mu g/m^3$, and the regional background in Western Europe is 50 $\mu g/m^3$. In remote areas, such as in high-mountain regions, and over seas and oceans the dust content is 6–12 $\mu g/m^3$.

The background content of nitrogen dioxide (mean daily concentrations) has typical values of 0.1–12 $\mu g/m^3$. In most cases the concentrations over the ocean surface are lower than over continental areas. This fact may be due both to the high intensity of microbiological processes at the ground in continental regions and to a possible effect of urbanized regions.

The sulfur dioxide content in nonindustrial regions of the world changes over a wide range: from 0.05 to 14 $\mu g/m^3$. The lowest sulfur dioxide concentrations (0.04–4 $\mu g/m^3$) are recorded over the ocean, in Antarctica, and in the USSR (in biosphere reserves and mountainous regions of Georgia, in the backcountry of Lithuania). Background concentrations of sulfur dioxide above 4 $\mu g/m^3$ are characteristic of Western Europe and America.

The range of change of the sulfate concentrations in the air over backcountry regions is, as a rule, from 0.5 to 10 $\mu g/m^3$. An increased sulfate content in the air is characteristic of the European region (from 5 to 15 $\mu g/m^3$).

The highest background lead concentrations (above 100 ng/m^3) are characteristic of the plains regions of Western Europe. The lead concentration in the air over the Soviet backcountry is 2–40 ng/m^3. In mountainous regions this value is much lower, and at an altitude of 3000 m above sea level it is 2–6 ng/m^3 in Western Europe and 4.6–21 ng/m^3 in North America.

The gaseous mercury concentration in background regions varies from several ng/m^3 to 50 ng/m^3.

The cadmium content in the air over background regions is low and approximately uniform: 0.02–1.6 ng/m^3. The highest cadmium concentrations are recorded in Western Europe. The arsenic concentrations in various background regions of the world vary from 0.2 to 3 ng/m^3. Arsenic, like cadmium, is quite uniformly distributed.

The concentrations of 3,4-benzopyrene in the air over nonurban areas are 0.1–0.5 ng/m^3 or lower. The concentrations of 2,4-benzopyrene in US national parks and nature reserves are within 0.01–1.9 ng/m^3, and the lowest concentrations (on average 0.01 ng/m^3) are observed over the ocean. The content of 2,4-benzopyrene in the air over various background regions of the USSR is from 0.06 to 0.8 ng/m^3.

The global levels of organochlorine pesticides in the air basin vary over wide limits: from 0.001 to several ng/m^3. The lowest background concentrations of DDT and HCCH are characteristic of the atmosphere over the ocean; over the continents the global level is 0.5–1 ng/m^3. In the USSR the level of background concentrations is 0.05–1 ng/m^3.

A statistical analysis of the results of observations at background stations (in particular, at Borovoe station in the Northern Caucasus) clearly revealed a bimodal nature of the pollutant distribution on histograms, i.e., there were two lognormal distribution laws, attributable to two groups of factors. An analysis of the effect of weather conditions during the observation period revealed a relationship between relatively high pollutant concentrations and negative energy of instability of the atmosphere; low pollution levels were grouped with days having a positive energy of atmospheric instability. The latter data ought to be considered the most representative when assessing the regional background level of atmospheric pollution [18].

The lead content in the soils of background regions of various countries varies from 1.1 to 71 mg/kg. The mean present-day global level is estimated as 17 mg/kg, which is markedly higher than the Clark content (10 mg/kg). The background concentrations of cadmium in the major soil types vary from 0.01 to 1.4 mg/kg, and for mercury they are 0.01 to 0.8 mg/kg. The concentrations of 3,4-benzopyrene in soils of background regions seldom exceed 20 μg/kg. In the USSR the content of 3,4-benzopyrene in soils of background regions is 0.1–4.8 μg/kg.

Most pesticides (90–95%) are concentrated in the topsoil (10-cm) layer. The DDT concentration level outside the agricultural zones of the USA varies from 0.01 to 0.6 mg per kg of dry mass, in Canada from 0.09 to 0.5 mg, and in the USSR from 0.004 to 0.4 mg. The HCCH content in the soil is approximately 1/5 to 1/10 as high.

The concentration of 3,4-benzopyrene in water bodies far away from pollution sources emitting carcinogenic polycyclic aromatic hydrocarbons is, as a rule, not higher than 0.1 ng/l. On the whole, the concentrations in Europe and the USA are higher (0.3–100 ng/l), in the USSR they are 0.1–4 ng/l.

The global background content of DDT and HCCH in the surface waters varies from 0.01 to 0.1 μg/l. The highest concentration level (up to 0.6–1.2 μg/l) is characteristic of countries in Western Europe. In the USSR the background level is 0.02–0.2 μg/l.

Herbaceous plants are characterized by a low and rather uniform content of metals. The lead content is 1–5 mg/kg. The leaves of woody plants have a somewhat higher content of the elements: lead 0.5–10 mg/kg; cadmium 0.05–0.3 mg/kg; mercury 0.01–0.12 mg/kg. The background content of 3,4-benzopyrene in plants varies from 0.1 to 50 ng per g of dry plant mass, depending on the plant species.

On the whole, the background pollution of such large regions as North America, Western Europe, and Central Europe is, according to practically all the indicators, markedly higher than for the eastern Eurasian continent (the eastern USSR), and much higher than the background pollution of the polar zones and oceans.

7.3. Radioactive Contamination of the Atmosphere and the Earth's Surface (Including Water) from the Chernobyl Nuclear Power Plant Accident

Below, we give the results of measurements of the natural environment contaminated with radioactive products from the Chernobyl power plant accident, obtained from available airborne and ground-based facilities within the USSR system of monitoring of the natural environment.

As a result of the accident a significant amount of radioactive material stored in the reactor during its operation was released outside the site.

At the moment of the accident a cloud was formed, and then a plume, which produced radioactive fallout at the Earth's surface both westward and northward in response to the prevailing meteorological conditions typical of mass transport through the air. For a long period of time (about two weeks) a plume of gaseous, volatile and aerosol products continued to be released. The most powerful plume was observed during the first 2–3 days after the accident, northward from the site. The height of the plume on 27 April was above 1200 m. On subsequent days the height of the plume was 200–400 m.

Radioactive products were then released from the NPP zone and transported mainly southward up to 7–8 May, producing radioactive fallout in this direction.

The gamma-ray survey of the radiation situation was performed from aircraft and helicopters equipped with roentgenometers and gamma-ray spectrometers, as well as from ground-based facilities of the USSR Goskomhydromet [3].

As a result of the gamma-ray spectrometric analysis of the air samples, fission products accumulated in the reactor were detected, as well as induced activity isotopes ^{239}Np and ^{134}Cs. The characteristic peculiarity of the radioactive products released into the atmosphere was their enrichment in iodine and cesium.

As a result of the distribution of radioactive products in the atmosphere and their deposition on the underlying surface in the period from 26 to 30 April, a radioactive fallout zone was formed close to the site. The map in Figure 7.1 shows the distribution of radiation zones at the Earth's surface on 29 May.

Based on the gamma-ray dose rate at the Earth's surface at various times the total amount of radioactive products deposited in the near-site trace of the radioactive fallout were estimated.

Table 7.3 gives the results of the integrated areas limited by various dose rate isotopes $(R/h)m^2$, as well as the total amount of radioactive products in the near-site trace (outside the industrial site of the power plant up to a range of 80 km), expressed in energy rate units (MeV/S) and in activity units (Ci).

Figure 7.2 shows the contour of the radioactive contamination over the USSR along the 0.05 mR/h isopleth (and higher) obtained by the aircraft gamma-ray survey during the first ten days of May 1986. The amount of radioactive products deposited over the USSR European territory amounted to approximately 1.2×10^8 $(R/h)m^2$ or 5.0×10^{17} MeV/S beyond the near-site trace. Thus, the total deposition in the near-site and remote zones was

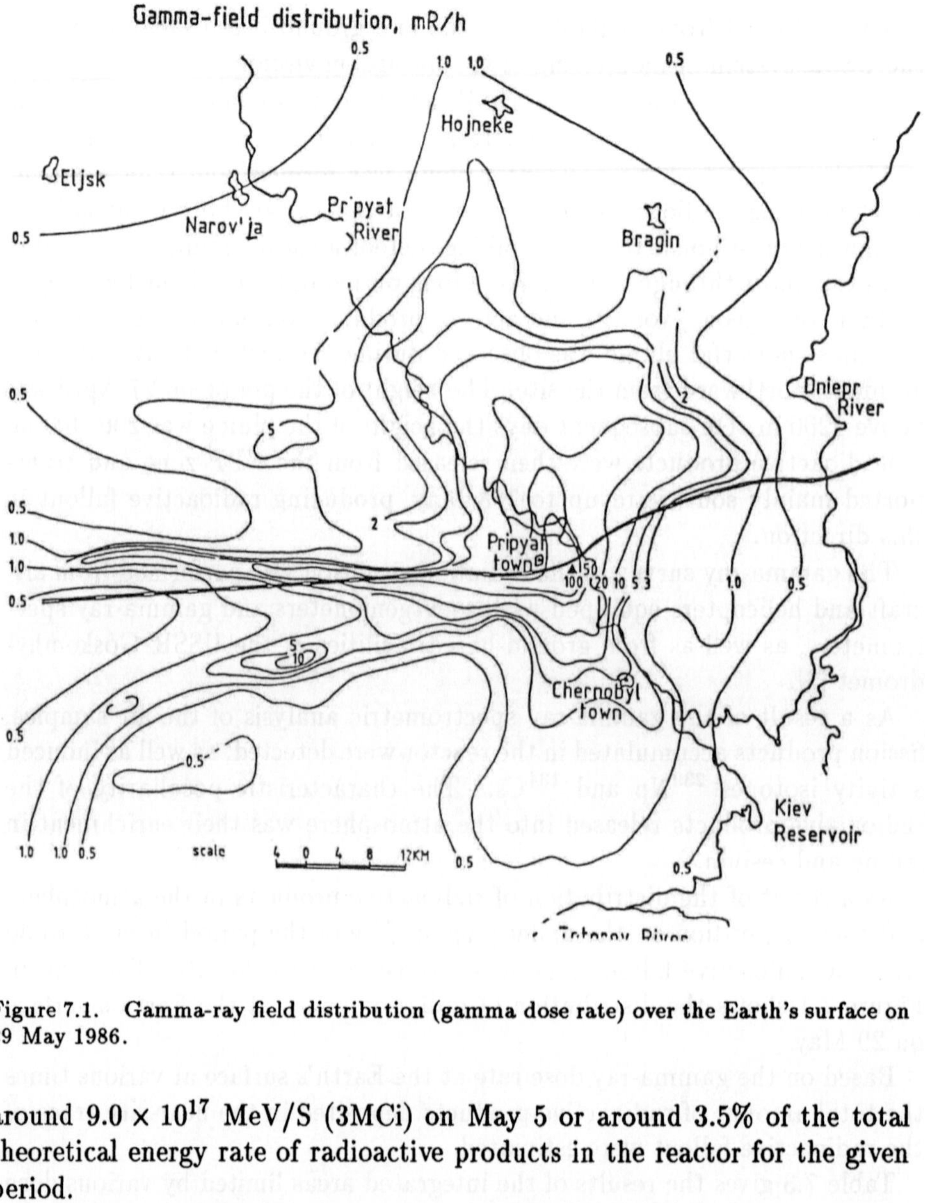

Figure 7.1. **Gamma-ray field distribution (gamma dose rate) over the Earth's surface on 29 May 1986.**

around 9.0×10^{17} MeV/S (3MCi) on May 5 or around 3.5% of the total theoretical energy rate of radioactive products in the reactor for the given period.

According to [1] the total amount of volatile products released into the atmosphere was: ^{137}Cs = 13%, ^{132}Te = 15%, refractory products = 3% or less.

It is worth noting for comparison that during the Three Mile Island nuclear power plant accident 25% iodine, 53% cesium and 0.5% ruthenium isotopes were released outside the reactor containment (rather than into the environment, which received a much lower amount of radionuclides).

The radionuclide composition of the atmospheric aerosol was determined by air sampling on various filters, followed by laboratory analysis with high-resolution gamma-ray spectrometry and radiochemistry. Individual respi-

Table 7.3.

Date of measurement in trace, 1986	Integral (R/hour)m^2	Amount of radioactive products in trace MeV/S	Ci	Fraction of summary emission of radioactive products in the reactor zone for the moment, %
11 May (D 15)	7.9×10^7	3.3×10^{17}	1.2×10^7	1.6

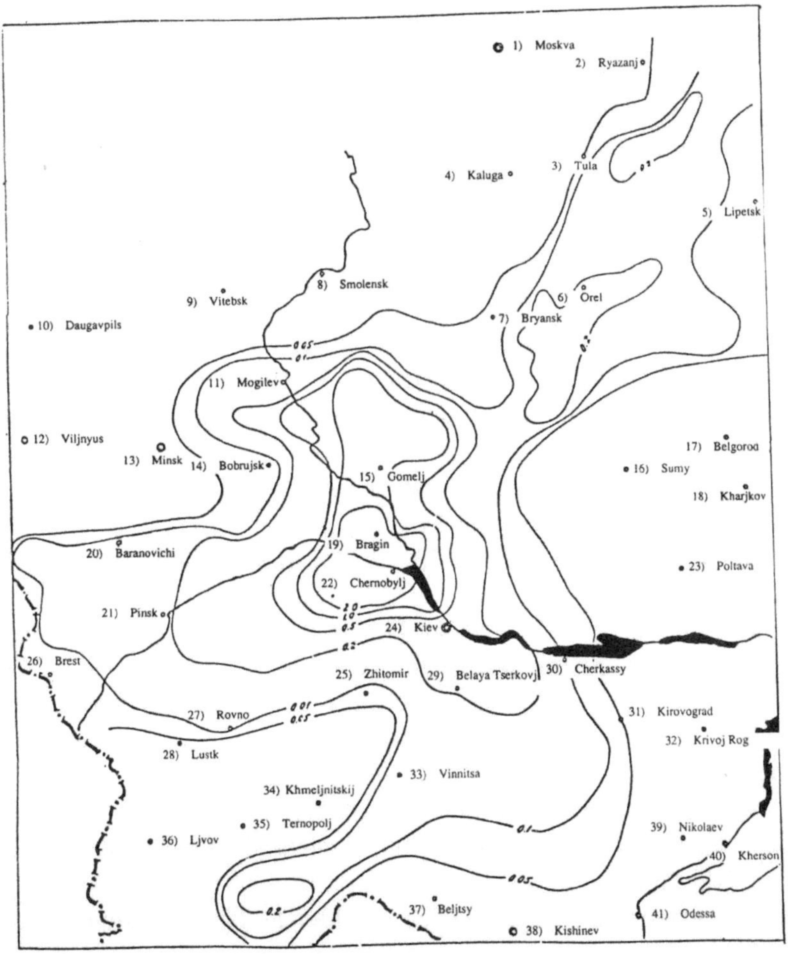

Figure 7.2. Contours of the gamma-ray field (in μR/h) over USSR territory in the first ten days of May 1986.
1 – Moskva, 2 – Ryazanj, 3 – Tula, 4 – Kaluga, 5 – Lipetsk, 6 – Orel, 7 – Bryansk, 8 – Smolensk, 9 – Vitebsk, 10 – Daugavpils, 11 – Mogilev, 12 – Viljnyus, 13 – Minsk, 14 – Bobrujsk, 15 – Gomelj, 16 – Sumy, 17 – Belgorod, 18 – Kharjkov, 19 – Bragin, 20 – Baranovichi, 21 – Pinsk, 22 – Chernobylj, 23 – Poltava, 24 – Kiev, 25 – Zhitomir, 26 – Brest, 27 – Rovno, 28 – Lustk, 29 – Belaya Tserkovj, 30 – Cherkassy, 31 – Kirovograd, 32 – Krivoj Rog, 33 – Vinnitsa, 34 – Khmeljnitskij, 35 – Ternopolj, 36 – Ljvov, 37 – Beljtsy, 38 – Kishinev, 39 – Nikolaev, 40 – Kherson, 41 – Odessa

Table 7.4. Radioisotope activity ratio to the ^{95}Zr activity in the plume over the Chernobyl nuclear plant.

Sampling date	^{144}Ce	^{141}Ce	^{140}Ba	^{131}I	^{103}Ru	^{106}Ru	^{137}Cs
28 IV	0.66	0.78	0.98	1.11	0.68	0.20	
29 IV	0.38	0.31	0.66	1.02	0.48	0.19	
30 IV	1.59	1.30		9.48	4.92	0.78	2.35
1 V	3.80	2.21		9.08	3.56	0.57	0.69
2 V	0.65	0.82	0.86	0.61	0.44	0.12	0.15
4 V	0.73	0.92	1.82	1.17	0.98	0.20	0.19
6 V	0.98	0.94	0.46	0.37	0.58	0.34	0.09
7 V	1.10	1.10	0.55	2.43	4.04	1.22	0.39
8 V	1.30	1.20	1.57	6.50	8.37	2.30	0.88
9 V	0.87	0.75	0.44	2.14	1.41	4.77	0.24
11 V	0.93	0.50		0.50	0.59	0.52	
Average	1.18	0.98	0.92	3.13	2.37	0.97	0.62
Theoretical ration	0.96	0.95	1.05	0.60	1.07	0.57	0.11
Fractionation factor	1.23	1.03	0.88	5.22	2.21	1.70	5.64

rators, filter cloth from ground-based filter ventilation facilities and filters from cars and samplers aboard aircrafts were used while sampling. All the samples were taken in the near-site zone.

The radionuclide composition was represented by a wide range of fission products with a half-life varying from 67 hours (^{99}Mo) to 30 years (^{137}Cs).

Besides fission products, ^{239}Np and ^{134}Cs were reliably determined to be present in gamma-spectra.

To check the entry of the radioactive products from the reactor zone into the atmosphere during the periods following the accident on 5 June, samples were taken from a specially equipped aircraft using a multilayer filter with a controlled air flow. The approximate air volume sampled from the plume was about 10 m^3.

The content and concentration of separate radionuclides in the air sample and the amount of radioactive products released into the atmosphere are given in Table 7.5.

Data on surface air contamination outside the near-site zone (more than 100 km) were also obtained. Depending on meteorological conditions the radioactive products entered the surface atmosphere at different times for various products.

A large amount of 'hot particles' were found in samples and fallout. Alpha-activity was detected in some of them.

Plutonium contaminated samples were studied with a specially equipped vehicle for taking air, soil and grass samples along the circle around the Chernobyl NPP, outside the 30 km zone (May 1986).

The radiochemical analysis of plutonium was performed under stationary

Table 7.5. Radionuclide content in the radioactive plume from the damaged block of the Chernobyl nuclear plant on June 5.

Radionuclide	Concentration Ci/m^3	Ratio to ^{95}Zr	Radionuclide activity released into the atmosphere, Ci/day
^{95}Zr	5.1×10^{-11}	1.0	30
^{95}Nb	8.6×10^{-11}	1.7	50
^{103}Ru	2.6×10^{-11}	0.51	16
^{106}Ru	1.0×10^{-11}	0.2	6
^{131}I	2.4×10^{-12}	0.046	1.5
^{134}Cs	2.9×10^{-12}	0.056	–
^{136}Cs	6.5×10^{-13}	0.013	–
^{137}Cs	6.2×10^{-12}	–	–
^{140}La	7.3×10^{-12}	0.14	4.4
^{141}Ce	2.2×10^{-11}	0.42	13
^{144}Ce	3.4×10^{-11}	0.67	20
Total			135

laboratory conditions. At all sampling points plutonium concentrations were below permissible levels (MPC for ^{235}Pu is 3×10^{-17} Ci/l).

The average value of the coefficient of the secondary rise (in the air) for the measured samples $K = 1.3 \times 10^{-16}$ I/m.

The atmospheric fallout was monitored within the GOSKOMGYDROMET network using trays exposed for 24 hours. The maximum occurred at different times for various points: Kiev – 30 April – 2 May (beginning 29 April), Gomel – 28–29 April, Baryshevka – 30 April – 1 May etc.

In the Berezina reserve (Belorussia) at the integrated background monitoring station, 120 km northeast of Minsk, in the period from 26 April to 21 May the aerosols were sampled in the surface air layer (at a height 1.5 m above the underlying surface) to determine uranium fission products in the air. The samples were taken daily on filters by forced air pumping for 24 hours. The volume of the air pumped through the filter was 1200 m^3. The following radionuclides were found: ^{131}I, ^{103}Ru, ^{106}Ru, ^{95}Zr, ^{140}La, ^{140}Ba (cf. Figure 7.5, where ^{131}I and ^{95}Zr are shown).

The aim of the studies at the initial stage consisted of obtaining data on the distribution and density of radioactive deposition from the accident at the Earth's surface. The studies were performed by taking soil samples at open sites. At each point 4–10 soil cover samples were taken from an open site (bare of vegetation); the area of each sample was 400 cm^2, the sample cover was 2–3 cm.

The isotopic composition of gamma-ray emitting radionuclides in soil samples was determined using a high-resolution semiconductor gamma-ray spectrometer. In the spectra measured during 10–30 days after the accident gamma-ray lines of the following radionuclides were identified: ^{239}Np, ^{99}Mo (^{99}Te), ^{132}Te, ^{132}I, ^{131}I, ^{140}Ba, ^{140}La, ^{141}Ce, ^{103}Ru, ^{95}Zr, ^{95}Nb, ^{144}Ce,

Table 7.6. Radionuclide composition in soils of contaminated area in sector 'South'.

	Zone I $P_\gamma \geq 10$ mR/hr			Zone II $2 \leq P_\gamma \leq 10$ mR/hr			Zone III $P_\gamma < 2$ mR/hr		
Radionuclide	A_i/A_{95}	$f_{i,95}$	σ_i/P_γ	A_i/A_{95}	$f_{i,95}$	σ_i/P_γ	A_i/A_{95}	$f_{i,95}$	σ_i/P_γ
^{239}Np	8.0	–	2.5	–	–	–	1.4	–	1.4
^{99}Mo	0.45	0.5	0.07	–	–	–	2.5	2.8	0.9
^{132}Te	1.4	1.9	1.1	0.9	1.2	0.8	1.3	1.8	0.8
^{131}I	0.6	1.0	4.0	0.6	1.0	5.0	2.0	3.3	8.0
^{140}Ba	0.8	0.8	9.0	0.9	1.0	13	1.0	1.0	8.0
^{141}Ce	1.1	1.2	18	1.1	1.2	25	1.0	1.0	13
^{103}Ru	0.7	0.7	13	0.7	0.7	17	0.8	0.7	10
^{80}Sr	0.8	0.4	5.0	1.6	3.0	6.0	1.2	2.5	16
^{91}Y	0.3	0.45	5.0	–	–	11	0.8	1.2	12
^{95}Zr	1.0	1.0	20	1.0	1.0	27	1.0	1.0	15
^{144}Ce	0.7	0.7	16	0.6	0.6	20	0.6	0.6	10
^{106}Ru	0.3	0.5	7.0	0.2	0.35	7.0	0.3	0.5	3.8
^{134}Cs	0.01	0.07	0.2	0.015	0.1	0.5	0.03	0.2	0.5
^{90}Sr	0.02	0.3	0.35	0.16	2.0	0.8	0.15	2.0	2.2
^{137}Cs	0.03	0.3	0.5	0.04	0.4	0.9	0.1	0.9	1.2

Notes: A_i/A_{95} = ith radionuclide activity ratio to the zirconium-95 activity at the moment of the accident; $f_{i,95}$ = fractionation factor; σ_i/P_γ = contamination density ratio to the dose rate at a height of 1 m on 10.05.

^{106}Ru, ^{134}Cs, ^{137}Cs.

The radiochemical determination of strontium radionuclides was performed according to standard techniques.

When generalizing isotopic analytical data all soil samples were grouped by the three main directions of radioactive product distribution in the period from 26 April – 9 May, exactly, in sectors 'South', 'West', and 'North'.

Making allowance for the radionuclide composition of radioactive products falling out over the Earth's surface (depending on contamination density and distance from the source) three zones were singled out within each sector, based on the gamma dose rate criterion (on D + 15): I ($P_\gamma \geq 10$ mRh), II ($2 \leq P_\gamma \leq 10$ mRh) and III ($P_\gamma < 2$ mRh).

Tables 7.6–7.8 present the results of isotopic analysis of soil samples in the form of averages for each group (sector, zone) activity $(A_i/A_{95})_0$ ratios (on 26 April) and average (σ_i/P_γ) values (on 10 May). Here one can also see averages for the group fractionation factors, characterizing the radionuclide composition variation in individual groups of samples as compared with the calculated theoretical composition.

Average values of parameters $(A_i/A_{95})_0$ and (σ_i/P_γ) over groups of samples demonstrate a certain dependence on both the direction from the contamination source and the level of the Earth's surface contamination within each sector. The most significant characteristic of a radionuclide's behavior is its fractionation factor. The deviation of this factor from unity is indicative of a degree of radionuclide volatility as compared to the refractory ^{95}Zr.

Table 7.7. Radionuclide composition in soils of contaminated area in sector 'West'.

	Zone I $P_\gamma \geq 10$ mR/hr			Zone II $2 \leq P_\gamma \leq 10$ mR/hr			Zone III $P_\gamma < 2$ mR/hr		
Radionuclide	A_i/A_{95}	$f_{i,95}$	σ_i/P_γ	A_i/A_{95}	$f_{i,95}$	σ_i/P_γ	A_i/A_{95}	$f_{i,95}$	σ_i/P_γ
^{239}Np	6.0	–	1.7	9.0	–	2.0	5.0	–	0.9
^{99}Mo	1.4	1.6	1.0	1.4	1.6	0.3	1.5	1.7	0.3
^{132}Te	1.4	1.9	1.0	3.8	5.3	1.8	3.6	5.0	1.8
^{131}I	0.9	1.5	4.0	3.0	5.0	10	3.0	5.0	7.0
^{140}Ba	1.1	1.0	10	1.6	1.5	10	1.4	1.3	14
^{141}Ce	0.9	1.0	13	1.1	1.2	17	0.9	1.0	7.5
^{103}Ru	0.8	0.7	13	0.7	0.7	11	0.7	0.7	14
^{80}Sr	0.14	0.3	2.5	0.6	1.3	6.0	0.7	1.5	8.5
^{91}Y	0.2	0.3	4.0	0.6	0.9	9.0	0.9	1.4	11
^{95}Zr	1.0	1.0	20	1.0	1.0	19	1.0	1.0	20
^{144}Ce	0.6	0.6	11	0.7	0.7	15	0.6	0.6	7.0
^{106}Ru	0.25	0.4	5.0	0.3	0.5	6.2	0.2	0.35	2.6
^{134}Cs	0.02	0.13	0.4	0.14	0.9	1.5	0.13	0.9	1.7
^{90}Sr	0.02	0.3	0.3	0.03	0.4	0.4	0.1	1.3	1.6
^{137}Cs	0.05	0.5	1.0	0.2	1.8	2.1	0.2	1.8	2.7

Notes: A_i/A_{95} = *i*th radionuclide activity ratio to the zirconium-95 activity at the moment of the accident; $f_{i,95}$ = fractionation factor; σ_i/P_γ = contamination density ratio to the dose rate at a height of 1 m on 10.05.

Table 7.8. Radionuclide composition in soils of contaminated area in sector 'North'.

	Zone I $P_\gamma \geq 10$ mR/hr			Zone II $2 \leq P_\gamma \leq 10$ mR/hr			Zone III $P_\gamma < 2$ mR/hr		
Radionuclide	A_i/A_{95}	$f_{i,95}$	σ_i/P_γ	A_i/A_{95}	$f_{i,95}$	σ_i/P_γ	A_i/A_{95}	$f_{i,95}$	σ_i/P_γ
^{239}Np	16	–	5.0	–	–	–	2.0	–	3.0
^{99}Mo	–	–	–	–	–	–	–	–	–
^{132}Te	1.7	2.4	3.8	10	14	11	5.0	7.0	5.0
^{131}I	3.0	5.0	13	6.0	10	35	5.0	8.0	40
^{140}Ba	1.0	1.0	7.0	1.1	1.0	17	1.0	1.0	13
^{141}Ce	1.1	1.2	16	1.1	1.2	27	1.1	1.2	27
^{103}Ru	1.0	0.9	14	1.6	1.5	37	1.4	1.3	31
^{80}Sr	0.2	0.4	3.0	0.3	0.6	7.0	0.2	0.4	3.0
^{91}Y	0.3	0.45	4.5	0.5	0.8	9.0	0.1	0.15	2.0
^{95}Zr	1.0	1.0	14	1.0	1.0	30	1.0	1.0	30
^{144}Ce	0.65	0.7	12	0.6	0.6	21	0.7	0.7	22
^{106}Ru	0.3	0.5	6.0	0.4	0.7	11	0.3	0.5	9.0
^{134}Cs	0.4	0.7	1.5	0.18	1.2	5.0	0.13	0.9	4.0
^{90}Sr	0.02	0.3	0.4	0.03	0.4	0.9	0.02	0.3	0.6
^{137}Cs	0.2	1.8	3.5	0.3	2.7	9.4	0.3	2.7	7.8

Notes: A_i/A_{95} = *i*th radionuclide activity ratio to the zirconium-95 activity at the moment of the accident; $f_{i,95}$ = fractionation factor; σ_i/P_γ = contamination density ratio to the dose rate at a height of 1 m on 10.05.

Table 7.9. Radionuclide activity in the local section of the trace (up to 40 km) on D + 15.

Radionuclide	Radionuclide activity in the local section of the trace, Ci	Radionuclide part in the local section of the trace compared with the total activity in the reactor on D+15, %
^{132}Te	2.5×10^6	5.0
^{131}I	1.3×10^6	5.1
^{140}Ba	9.1×10^5	1.4
^{141}Ce	1.7×10^6	1.7
^{103}Ru	1.5×10^6	1.4
^{89}Sr	6.2×10^5	1.2
^{91}Y	6.4×10^5	0.85
^{95}Zr	1.8×10^6	1.5
^{144}Ce	1.3×10^6	1.0
^{106}Ru	5.7×10^5	0.8
^{134}Cs	1.3×10^5	0.6
^{137}Cs	2.8×10^5	1.9
^{90}Sr	8.5×10^4	0.85
Total	1.1×10^7	Average 1.6
Total energy rate, MeV/S	3.3×10^{17}	1.6

According to this characteristic, tellurium, iodine and cesium are the most volatile radionuclides. As it follows from the tables, cesium fractionation is very pronounced to the north.

Based on data on the Earth's surface contamination density in various zones, the activity of some radionuclides in the area nearest to the trace was estimated in comparison with the dose rate on D + 15 (Table 7.9).

To assess the change in the gamma-ray dose rate in various areas of the radioactive trace the gamma-ray activity of the total radionuclides deposited on the Earth's surface was calculated with time. At each of nine predetermined segments of the trace an average for the segment A_i/A_{95} ratio was assumed as a generalized characteristic of the radioisotope composition.

The results of calculations of external dose loadings are given in Table 7.10 (with no allowance being made for screening inside the building).

The Earth's surface contamination with isotopes ^{134}Cs and ^{137}Cs (particularly ^{137}Cs, which spread over large distances from the place of the accident: cesium is a highly volatile isotope) is worth noting.

A 'spotted' form of Earth surface contamination was produced by rain falling in the zone of the radioactive cloud and the plume.

The most contaminated zone, besides those close to the accident area (Figure 7.3), is the one located at the 'crossing point' of three regions: Gomel, Mogilev (Belorussia) and Brjansk (RSFSR) (Figure 7.4) and in Tula, Orel and Kaluga' regions. Table 7.11 presents areas with different density of

Table 7.10. The external gamma-radiation dose from the polluted Earth's surface related to the dose rate I mR/h on D + 15 for different areas of the radioactive trace, R.

Sector	Zone	Dose from the total isotope composition		Dose from ^{137}Cs	
		from D + 1 to 1 g	from 1 g to 50 l	from D + 1 to 1 g	from 1 g to 50 l
	I	2.5	6.9	0.21	6.05
North	II	2.5	8.3	0.25	7.2
	III	2.5	9.1	0.28	8.2
	I	2.4	1.6	0.038	1.1
South	II	2.1	1.9	0.049	1.4
	III	2.4	3.7	0.11	3.11
	I	2.2	2.0	0.054	1.6
West	II	2.3	5.7	0.16	4.8
	III	2.3	6.0	0.18	5.2

Table 7.11. Areas contaminated by ^{137}Cs, including the zone size (km^2).

Republic	Ranges of contamination density (Ci/km^2)		
	5 – 15	15 – 40	> 40
Belorussia	10160	4210	2150
Ukraine	1960	820	640
RSFSR	5760	2060	310
Total	17800	7090	3160

contamination by ^{137}Cs.

The total area of contamination by ^{137}Cs above 5 Ci/km^2 was around 28 000 km^2 and that above 1 Ci/km^2 lay around 100 000 km^2. Radioactive contamination of the atmosphere and Earth surface was also observed outside the Soviet Union [5].

River water radioactive contamination occurred primarily as a result of aerosol fallout from the plume, spreading from the nuclear plant zone onto the water body surface mostly in the first days following the accident (first period), then due to rain washoff from the contaminated watersheds.

Observations of the dynamics of variations in radionuclide concentrations in the Pripyat' river were indicative of abrupt concentration reduction when the aerosol contamination period was over. The total β-activity of the Pripyat' river declined from 10^{-7} Ci/l at the beginning to 10^{-9} to 10^{-10} Ci/l by July – August.

According to the forecast for the river and water reservoir contamination made in May 1986 the contamination was not expected to exceed that of the first period (due to the atmospheric aerosol deposition), and this was

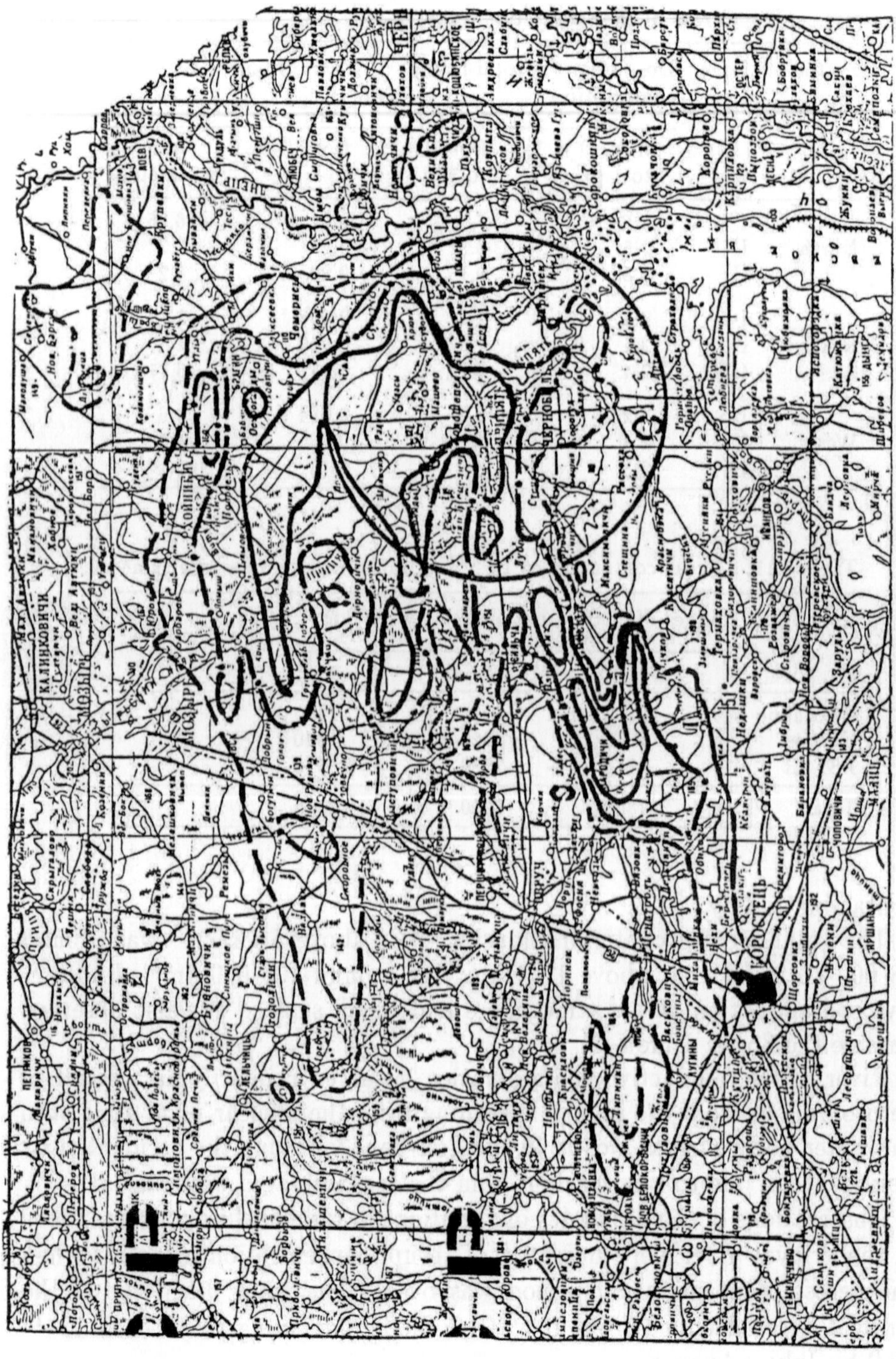

Figure 7.3. Earth's surface contamination density by ^{137}Cs in the emergency region. The 30-km zone is shown by a circle. /legend ——— isoline 40 Ci/km^2. –·–·– 15 Ci/km^2. – – – – – – 5 Ci/km^2.

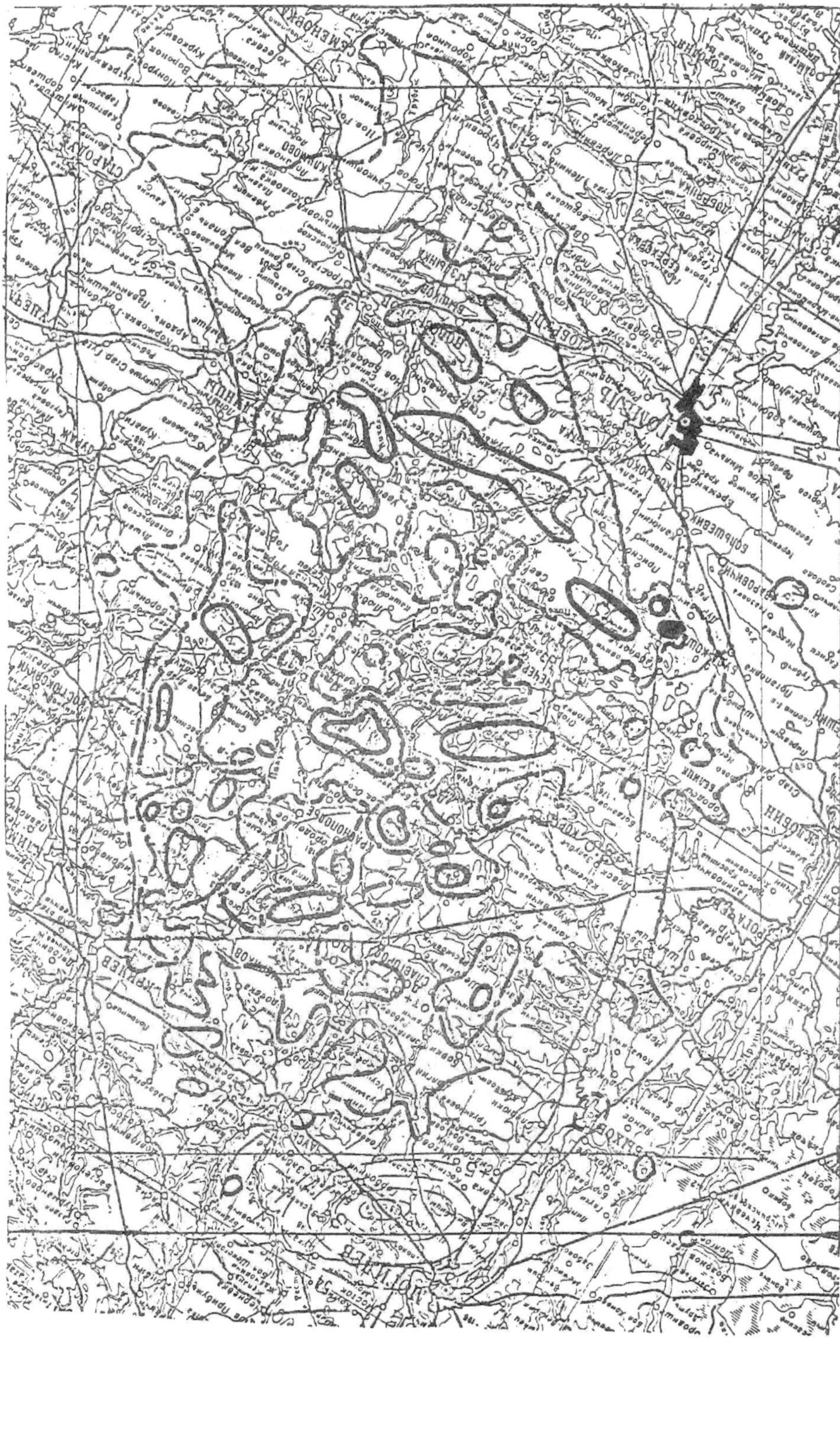

Figure 7.4. Contamination density at the 'crossing point' of three regions: Gomel, Mogilev (Belorussia) and Brjans (RSFSR). The legend is the same as in Figure 7.3.

Table 7.12. Radioisotope content in rivers in the region of the accident, 10^{-12} Ci/l.

Rivers	^{131}I	^{137}Cs	^{134}Cs	^{103}Ru
Dnestr, Termetsy settlement	240 ± 25	120 ± 20		140 ± 15
Desna, mouth		410 ± 310		160 ± 35
Irpen', Kozarovichi settlement	220 ± 40	240 ± 75	170 ± 30	100 ± 20
Teterev	420 ± 150	220 ± 110		80 ± 15
Pripyat'	430 ± 110	490 ± 70	230 ± 50	240 ± 80
Uzh	630 ± 80	350 ± 150	220 ± 100	190 ± 10

confirmed later on.

Variations in the total β-activity in the Pripyat' river after the aerosol contamination periods were caused by variations in radionuclide washoff from the watershed area.

Table 7.12 shows average data on gamma-ray emitter composition in the principal rivers running into the Kiev water reservoir, and in the Dnieper river from 3 to 10 July.

During the precipitation-free period from 6 to 14 June the radioactive concentration varied within $(1-2) \times 10^{-9}$ Ci/l in total β-activity. On 15 June, a brief shower (5.2 mm), observed at the Chernobyl weather station from 17.50 to 18.25, caused a tripling of the total β-activity for 5 hours. The second concentration increase, due to the 'late arrival' of the contaminated water from higher watershed areas, was observed from 17 to 19 June (up to 5.5×10^{-9} Ci/l). Long-term precipitation over practically the whole area of the Pripyat' watershed (35–75 mm) during the period from 20 to 24 June resulted in another increase in the total β-activity of the Pripyat' water.

The Kiev water reservoir did not display such an abrupt variation in the total β-activity during the precipitation as did the Pripyat' water. For instance, in June, the total β-activity of water varied within the range of $(1-6) \times 10^{-9}$ Ci/l. During precipitation fallout from 20 to 25 June the concentration did not exceed 4×10^{-9} Ci/l.

During the precipitation event of 28 to 30 July, samples were taken from the Pripyat' river and the ^{137}Cs content was determined. In suspended matter and in the filtrate the ^{137}Cs concentration amounted to 1.1×10^{-9} Ci/l (before that, it amounted to 2.0×10^{-10} Ci/l), and the activity in suspended matter made up 70% of the total activity. No increase occurred in the ^{90}Sr content. Its concentration was about 3.0×10^{-11} Ci/l, the same as previously.

The radionuclide concentration (total β-activity) in November in the rivers Dnieper, Desna, Uzh, Teterev and Irpen was lower than 5×10^{-10} Ci/l. The relationship between ^{103}Ru, ^{137}Cs, ^{144}Ce and ^{95}Zr in suspended matter of the Pripyat' river and in the soils of the Chernobyl area in different periods was indicative of a rather good (within 50%) coincidence of the ratios A_i/A_{95} in soils and suspended matter in July.

The most dangerous radionuclide, ^{90}Sr, has been measured in various waters since 1 May. Maximum radioactive levels were observed in the first days of May in the Pripyat' river. At that time, the ^{90}Sr concentration in the Pripyat' amounted to MPC (4×10^{-10} Ci/l). The drastic reduction of the radioactive products released from the source into the atmosphere resulted in a decreased radionuclide concentration in the Pripyat' – lower than MPC – and from the end of May to July the ^{90}Sr content in the Pripyat' remained at $(3-5) \times 10^{-11}$ Ci/l. Similar levels are also observed in the Kiev water reservoir. One may conclude that the ^{90}Sr concentration in the Kiev water reservoir was related to its content in the Pripyat' river, since in other rivers running into the Kiev water reservoir the ^{90}Sr content was considerably lower than in the Pripyat'.

The ^{90}Sr content in the Dnieper water reservoirs was studied during 14–20 May. In May the ^{90}Sr concentration distribution in the Kiev water reservoir was rather homogeneous. Down the cascade, ^{90}Sr concentration decreased and at the end of the Kremenchug water reservoir it amounted to approximately 5×10^{-12} Ci/l.

The data obtained on the ratio between ^{89}Sr and ^{90}Sr indicated that the ratio was steady and during the period from 14 to 20 May it was within the range 7–14.

The survey of the radioactive contamination of the Dnieper water reservoirs from 14 to 20 May showed that water radioactivity was principally caused by ^{131}I dissolved in water. A 'wave' of radioactive iodine was observed with its southern boundary crossing this time the Kremenchug water reservoir. Below the Kremenchug water reservoir ^{131}I concentration stabilized at $(2-4) \times 10^{-10}$ ci/l.

Further studies of the contaminated waters in rivers and water reservoirs indicated that contamination did not exceed permissible limits (even during the spring flooding in 1987).

7.4. Systems of Environmental Monitoring in the Soviet Union and Elsewhere

The general monitoring (observation, assessment, prediction, elements of regulation) of the state of the environment in the USSR is carried out mainly by various geophysical services, most of which are under the authority of the USSR State Committee on Hydrometeorology and Environmental Monitoring (Goskomgidromet).

Such services collect and process information on the natural state of the environment, and on changes in this state due to natural and anthropogenic factors. In addition, they study the regularities of these changes, forecast future changes in the state, and provide all interested organizations with information to be used in the interests of nature conservation and economic development.

Such services include: meteorological and hydrological services, a service for observing the state of the seas and oceans (the above services also provide essential information for the data service connected with climate), an agrometeorological service, a tsunami warning service, ionospheric and magnetometric services, a service for observing and monitoring the radiation situation in outer space, a service for observing and monitoring the level of environmental pollution, as well as a service for satellite observations of renewable natural resources. Goskomgidromet is responsible for the operation of all these services. The solution of hydrometeorological problems and the procedure for monitoring the envirionment within the framework of the above services are described in [10, 11].

The seismic service, which is directed by the USSR Academy of Sciences, is a somewhat special case.

With regard to anthropogenic changes in the state of the environment, both man and the environment can be affected severely by the anthropogenic pollutants now being produced in substantial amounts.

The All-Union Service for Observing and Monitoring the Pollution of the Environment is operating effectively in the USSR [14]. As already mentioned, Goskomgidromet is responsible for the operation of this service (which operates in conjunction with other interested ministries and departments).

The All-Union Service, which is principally a national system for monitoring environmental pollution, was based on the experience gained earlier by the USSR Hydrometeorological Service, as well as on the results of an analysis of available information on environmental pollution; the scientific establishment of the above system is based on the results of a comprehensive analysis of the state of the environment as a whole, which makes it possible to determine the optimum nature of the All-Union Service being organized.

Thus, during the organization of the above service, the following were used as much as possible:

(1) experience gained by the hydrometeorological service when organizing the representative network of hydrometeorological observations, data collection, and data processing;
(2) experience from field studies performed by the hydrometeorological service (regular observations of the background chemical composition of the surface waters, dating back to the 1920s);
(3) the existing operational network of hydrometeorological stations located throughout the country.

For a rational distribution of the stations of the All-Union Service, the determination of priority pollutants took into consideration the following:

(1) general information about existing and potential pollution sources (large cities, industrial regions, river mouths with industrial plants, large livestock farms, individual enterprises in a zone of unique natural projects, like the Baikal Pulp and Paper Works, etc.);
(2) results of past observations of pollution levels (mostly under field conditions), which are tentative in nature;
(3) data on the levels of environmental pollution in other countries and large cities (this made it possible to attack the most serious pollution problems).

The use of the network of hydrometeorological observations when organizing the All-Union Service made it possible to ensure comprehensiveness of the observations, whereby hydrometeorological and biological observations are made simultaneously with observations of the level of environmental pollution. This provides an integrated approach to monitoring environmental pollution (this being the approach which is the most broad, efficient, and at the same time economical). Such an approach forms a basis for transforming the All-Union Service for Observing and Monitoring Environmental Pollution into an ecological service designed to collect information and to assess and forecast the state of the biosphere, in the widest sense of the term. The system should be scientifically established on the basis of a comprehensive analysis of the environment. This will make it possible to determine the most vulnerable elements of the biosphere, which should be of particular concern.

The establishment of an ecological service is simplified somewhat by the fact that the observation and assessment of the state of the environment according to biological indicators (for water objects) have been already carried out within the framework of the All-Union Service, while a service for carrying out observations and for the collection of satellite data on renewable natural resources is being organized by Goskomgidromet.

The All-Union Service provides information on the natural (background) state of the environment and on anthropogenic changes at present and in the future at various impact levels (high, intermediate, and low (background) impacts).

The organization of investigations and observations according to a complex program on the basis of the biosphere reserves is an example of the complex, integrated approach to monitoring the natural environment at the background level [13]. These elements are already the building blocks of the future ecological service for monitoring the state of the environment.

The transformation of the pollution monitoring service into an ecological service should also provide for detection and monitoring of sources of various impacts, and especially for the monitoring of pollution sources. Ecological and public-health standards, established allowing for the technological and economical capabilities of the economy, should be used as a basis for limiting such pollution sources and regulating the environmental quality.

The All-Union Service for Observing and Monitoring Environmental Pollution (as well as the future ecological service) is built on a hierarchical principle (the primary point for monitoring pollution or any other state of the environment is a territorial center, that is, a regional or main center for gathering information). The service already uses existing Goskomgidromet automated systems (and will incorporate systems being developed) for data collection, processing, and transmission, using the latest techniques (nuclear-physical and biological methods; remote sensing techniques, including the use of satellites).

The described service provides, first of all, the data on the environment which are needed to ensure human health. In this respect, the All-Union Service should be closely allied with the corresponding services in the USSR Ministry of Health.

The structure and elements of the All-Union Service for Observing and Monitoring Environmental Pollution are described in detail [14, 23]. Below we will give only that information about the functioning of the service which reflects and promotes the development of the concept of environmental monitoring considered in this book, and which illustrates the scientific approaches used when organizing and developing a pollution-monitoring system on the scale of the entire Soviet Union.

The above service can be divided into three large subsystems:

(1) *A subsystem for monitoring pollution in zones of possible significant anthropogenic impact* (zones of strong impact). In the case of atmospheric pollution, these will be urban and industrial regions. In the case of surface-water pollution, these will be river mouths and estuaries, reaches of large rivers and small rivers, as well as lakes and reservoirs, in industrial regions, particularly places where wastewaters are discharged, and the vicinities of livestock farms and large cities (taking into account municipal sewage). In the case of seawater pollution, these will be coastal areas in industrial regions, near large cities, particularly ports, and areas of off-shore oil drilling. In the case of soil pollution, these will be areas near large cities, freeways and highways, and agricultural areas where toxic chemicals have been applied intensively.

(2) *A subsystem for monitoring pollution at the regional level* (a zone of general monitoring of intermediate pollution levels): in the atmosphere of small towns, and in regions adjoining highly industrialized zones; pollution measurements in a large number of places on rivers, lakes, and seas, and in cultivated soil. These measurements should be performed everywhere but they need not be very frequent (in space and time). In the case of

substances with a relatively regular distribution, a field survey can be made (for instance, a survey of the background global radioactive fallout). The monitoring of long-range pollutant transport (for instance, transboundary transport) can be referred to this as well.

(3) *A subsystem for monitoring pollution at the background level* (observations are carried out in zones far from any local sources): an example is a complex of measurements and investigations carried out in biosphere reserves. This subsystem includes monitoring the state of the environment and the pollution levels in endangered natural preserves.

Let us consider in somewhat more detail how the state and the pollution of the environment are monitored in the Soviet Union with the aid of these three main subsystems.

Pollution monitoring in zones of possible strong impact includes monitoring of air pollution, the pollution of water on land and in the sea, and soil pollution [23].

In 1977 the air pollution in cities and industrial regions was monitored by a stationary observation network in 210 Soviet cities, or, taking into account en-route surveys, in 350 cities [14]. By 1983 the number of cities monitored had increased to 450 (1100 stationary observation points and 500 en-route points) [19].

The monitoring system receives and transmits on an operational basis data on sudden changes in pollution levels due to accidents or unfavorable weather conditions.

The content of dust, sulfur gas, soot, carbon monoxide, and nitrogen dioxide in the air is measured in all cities.

The contents of ammonia, sulfuric acid, hydrogen sulfide, carbon disulfide, chlorine, methylmercaptan, phenol, heavy metals, and hydrogen fluoride are measured in cities where the relevant industries are operating. In some cities hydrocarbons and ozone are measured (which is important for detecting smog situations).

Observations and monitoring of the surface-water pollution level are performed according to physical, chemical, and hydrobiological indicators. The observation points are combined with hydrological stations providing hydrological computations. In a region of wastewater discharge, two (or several) observation points are established: one above the pollution source and the other below it.

The stationary observation network in the USSR consists at present of 3300 observation stations located on 1900 water bodies [19].

The following ingredients and indicators are measured: mineralization, suspended solids, pH, dissolved oxygen, BOD_5, COD, major ions, biogenic substances, petroleum products, detergents, phenols, pesticides, heavy metals, and some specific ingredients. The temperature and color of the water are determined as well.

All the sites of observation of the pollution level of land water are divided into four categories, according to the frequency and amount of detail of the program of observations. First-category sites are located at places where wastewaters are discharged and at regions of the spawning and wintering of valuable fish. Second-category sites are located in the vicinity of industrial cities, and at river mouths and other zones subjected to a significant impact.

These two categories of sites are referred to the monitoring group for zones of severe anthropogenic impact.

Observations at first-category sites are performed every few days according to an extended program. At second-category sites observations are performed once a month according to the same program.

The introduction of hydrobiological methods into the system for monitoring the water pollution level is of importance: the results of the analyses are a direct indication of the state of the ecosystem of the water bodies. These observations thus already form a part of the ecological service.

Observations of such elements of aquatic ecosystems as zoobenthos, zooplankton, phytoplankton, and macrophytes form the basis of hydrobiological monitoring. For instance, the abundance of benthic organisms, the total number of their species, and the number of species in the major groups are determined. When zooplankton and phytoplankton are monitored, their total biomass, and the number of phytoplankton and zooplankton have to be determined.

Hydrobiological monitoring of the state of macrophytes (a higher aquatic plant) includes observations of the species composition, the phytomass, and anomalies in development. Even a visual assessment of the state of the air-water macrophytes yields a conclusion about the extent of pollution of the water body. A detailed assessment of the water quality, in the case of hydrobiological observations, is made on the basis of a set of hydrobiological indicators (indices of saprobity, diversity, toxicity, etc.).

When monitoring sea pollution, particular attention is given to organizing measurements in the coastal zone, at health resorts and fishery zones, and in areas of strong impact.

During marine observations, measurements are made simultaneously at various depths, including the near-bottom layer and the layer of discontinuity.

Stations for marine pollution measurements are divided into three categories (60–70 stations of category I, 570–600 stations of category II, and 1000–1100 stations of category III). Stations of categories I and II pertain to zones of increased pollution, and stations of category III pertain to background regions. Measurements at stations of category I are made once every 10–20 days, and at category II they are made once a month. The reduced program (once every ten days) includes measurements of dissolved oxygen, petroleum products, and 1 or 2 specific ingredients. The complete program (once every 30 days) includes measurements of petroleum products, organochlorine pesticides, heavy metals, phenols, detergents, and specific ingredients, as well as the main indicators of the environment and the elements of the hydrometeorological regime. Monitoring of soil pollution in agricultural regions includes measurements of the residual amount of DDT and its metabolites, hexachlorocyclohexane (HCCH), granozol, metaphos, and other pesticides, heavy metals, and polycyclic hydrocarbons, as well as observations of the soil-quality indicators. In soils close to factories, power plants, and roads various metals, benzo(a)pyrene, and PCBs are determined. The above soil pollutants are measured in all zones, and the influxes of these pollutants from the atmosphere are measured as well (for this purpose monthly precipitation samples are collected and snow samples are taken at the end of winter.)

The subsystem of pollution monitoring at the regional level differs from the one described (for zones of severe impact) by its wider observation network (for surface water and seawater); however, the measurements in this network are made less frequently and according to a less broad program. For instance, in order to monitor the land water pollution at this level, stations of category III are used (monthly observations according to the reduced program and measurements according to the complete program, combined with the principal hydrological phases).

Monitoring the pollution of the atmosphere, seawater, and soil at this level of impact complements monitoring in zones of severe impact, and in some cases monitoring at the background level.

Universal pollution monitoring at the regional level is particularly important for widespread pollutants: for water these are petroleum products; for the air and other parts of the environment they are sulfur dioxide, its transformation products, and products of global radioactive fallout distribution over vast territories.

Determination of the 'reserve' of long-lived radionuclides, such as ^{137}Cs and ^{90}Sr, over Soviet territory, as well as of the dose rates (and dose loads) near the Earth's surface due to natural radioactivity and the global fallout of artificial radioactive isotopes (produced by nuclear weapons tests) is an example of the regional approach to pollution monitoring. This type of monitoring is closely associated with background monitoring.

The implementation of the above activities is based on regular (but not frequent) single measurements, which can be provided by aircraft gamma surveys (measurements from the air of the gamma radiation of the indicated radionuclides). The survey frequency will depend on the possible rate of changes in the situation (radionuclide concentrations and gamma dose rates). Taking into account that the reserves of such isotopes vary rather slowly (due to decay, low-intensity stratospheric fallout, and migration), at present no more than a few percent per year, it is sufficient to make such surveys once every few years. One survey, made over Soviet territory for several months, practically constitutes a single survey, and the data can be extrapolated without substantial error to the median time for the survey time interval.

The results of the last such gamma survey are given in [4]. Figure 7.5 shows the distribution of the global reserve of ^{137}Cs, i.e., the amount of radioactive isotope ^{137}Cs per unit area (mCi/km^2). The mean ^{137}Cs concentration over Soviet territory is 92 mCi/km^2. The cesium-isotope distribution over this territory is principally zonal (latitudinal), with maximum contamination levels of 100–175 mCi/km^2 in the latitude belt 50°–60°N. In mountain regions the contamination levels increase, due to the higher fallout rates in these regions. The contamination over the territory as a whole is rather patchy: the ^{137}Cs fallout from the atmosphere correlates with the precipitation amount in the region.

A laboratory analysis revealed that there is everywhere a correlation between the concentrations of long-lived ^{137}Cs and ^{90}Sr isotopes (^{90}Sr is not detected from aircraft). The ^{90}Sr reserve can be determined using the relation

$$qs_r = 0.54qc_s, \tag{7.1}$$

with an error no higher than 10% (the coefficient in this relation varies slightly for different soils). The calculated mean reserve of ^{90}Sr for Soviet territory amounts to 50 mCi/km^2.

The dose rate of ^{137}Cs at the Earth's surface is on the average 0.6 μR/hr. It can be estimated for any region using the relation

$$p = 6.5 \times 10^{-3} q \tag{7.2}$$

(where p is in μR/hr and q is in mCi/km^2).

The change in the coefficient is connected with the penetration of ^{137}Cs into the soil, which is most pronounced in arable land (in ploughed land the dose rate can be less by a factor of 2.5).

Maps of the dose-rate distribution over Soviet territory, produced by natural radioactive elements, indicate that this distribution depends on the types of soils and rocks present, which contain different amounts of radioactive elements (uranium, thorium, and potassium).

The contribution of the ^{137}Cs dose rate to the total dose rate is insignificant at present: it amounts to approximately 10% of the dose rate produced by natural radioactive elements in rocks and soils.

The external annual exposure over Soviet territory (at the Earth's surface), produced by natural and artificial radionuclides and cosmic radiation, varies from 50 (tundra) to 90–95 mrad/year (steppes and deserts). In some

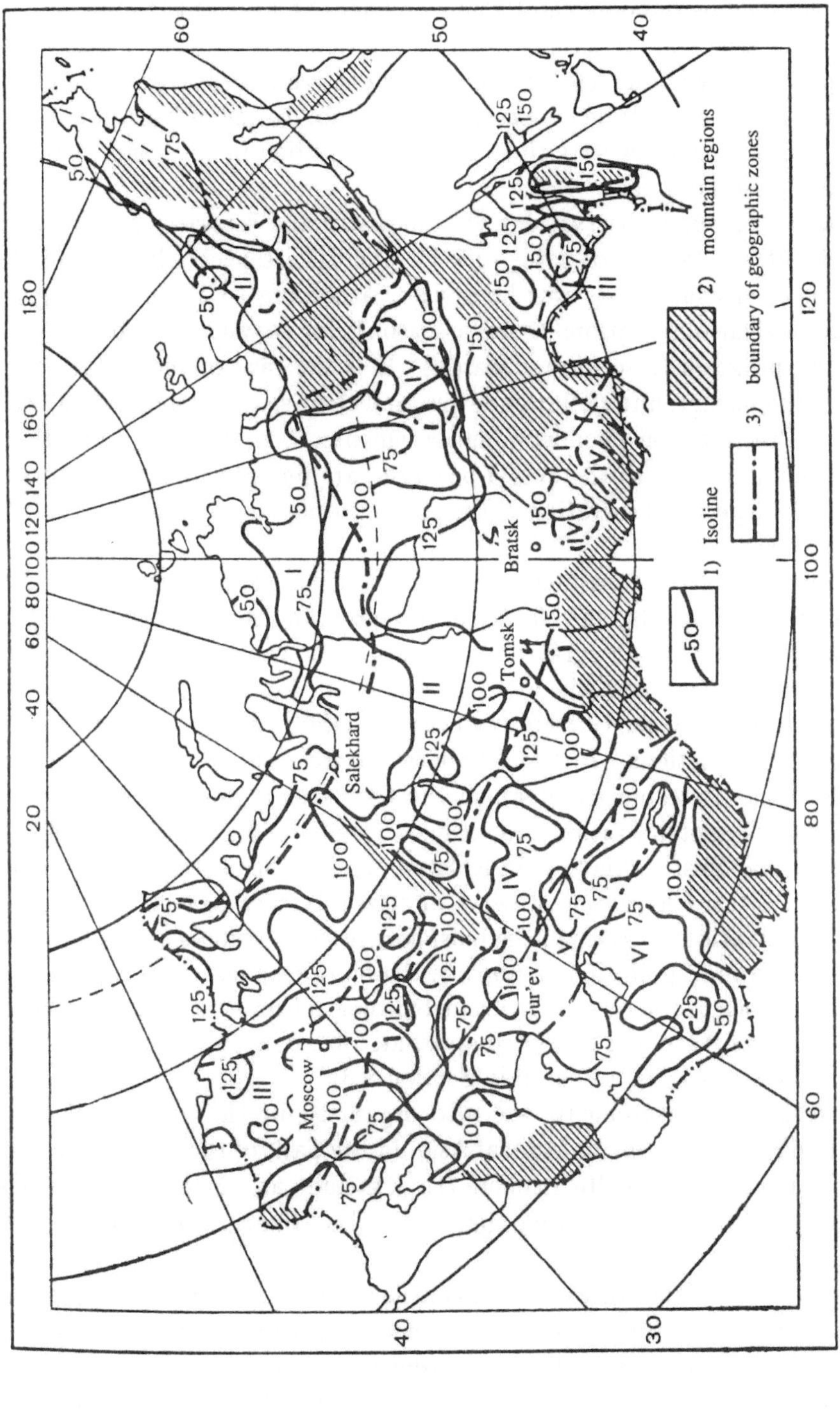

Figure 7.5. Distribution of reserves of ^{137}Cs (mCi/km^2) over Soviet territory: (I) tundra and forest-tundra; (II) coniferous forests; (III) mixed forest; (IV) forest-steppe and steppe; (V) semidesert; (VI) desert.

mountain regions the dose can be much higher (because of the radioactivity of the rocks and the increase of cosmic radiation with altitude). The above example of determining the amount of radioactive isotopes from the global fallout is at the same time an example of the monitoring of background pollution, since it provides us with data on the pollution in any zone, including the least polluted areas.

Mass background observations of land water and seawater pollution are carried out at stations of categories IV and III, respectively. These stations make measurements: for land water according to a general program of hydrological observations (once a season); for seawaters according to a complete program (also once a season). It is the task of these stations to observe the background pollution levels, and their seasonal and annual variability. Precipitation chemistry measurements at more than 70 stations, performed according to the WMO program, are referred to the background level (this type of observation corresponds to measurements made at the regional level).

For an integrated assessment of the state of an environment exposed to anthropogenic impacts, and, first and foremost, to the pollution impact on the background level, integrated investigations in areas specially chosen for these purposes were needed.

Here biosphere reserves were organized on the basis of existing reserves or permanent stations. In the USSR biosphere reserves were established in Belorussia (Berezina Reserve), in the Caucasus (Caucasus Reserve), in Turkmenia (Repetek Station), in Kirgizia (region of Lake Sary Chelek), in the Soviet Far East (Sikhote-Alin Reserve), and in the central part of Soviet Europe (Central Chernozem Reserve and Oka-Terrace Reserve). The integrated observation program in the biosphere reserves includes measurements of the background pollution, a study of the biota response to this pollution, and also associated necessary hydrometeorological observations. The latter thus also constitute one of the components of ecological monitoring. The program was described in detail in Section 5.3.

Background measurements are also organized in the vicinity of high-mountain glaciers in Soviet Central Asia, in the path of the steering current encircling the Northern Hemisphere (they include an analysis of snow and ice samples) [22], and they are organized in the Atlantic, on Soviet weather ships.

Observations of the state of the ozonosphere (measurements of the integral ozone content) are carried out at ozonometric stations; these observations are also background observations. In order to extend these observations, we have to organize measurements of ozone profiles and of the content in the upper atmosphere of minor constituents which affect the ozone layer.

All the information (emergency information as well as information summarized for certain time intervals) from areas exposed to various impact levels is sent from the pollution-monitoring systems to interested institutions. Priority is given, of course, to information from areas of strong impact.

As already mentioned, the pollution monitoring system is based on the hierarchy principle. The primary step (level) is the point (or points) of observation, from which the information is transmitted to a center for the collection and processing of information. In the case of the automated system now under development in the USSR, the primary step is a local system serving a separate region or city. This step consists of two parts: monitor-measurement stations and a computing center.

In the information-collection centre (computing center) data are processed, sorted, and transmitted to the next level, the regional (territorial) computing center (RCC, TCC). From these centers the information on environmental pollution is transmitted to local interested organizations.

The third level of the system is the main data center, at which information about environmental pollution on a national scale is collected and generalized.

An intermediate level is also possible providing for the collection and generalization of data on the pollution of various parts of the environment at computing centers of research institutes responsible for related activities (such data, which are collected simultaneously, must be transmitted to the main data center without delay).

At the above centers the information obtained is analyzed thoroughly, and any observed peculiarities of the environmental pollution are studied. The measurements are automated, in order to provide monitoring of the pollution of the air, the land water, and the seawater. The first local automated systems for monitoring air pollution are being developed in Moscow, Leningrad, and Kiev. The first pilot system for automated monitoring of water quality has been established on the Moskva River.

The automated stations for monitoring water pollution measure temperature, pH, eH, electrical conductivity, water turbidity, and amount of dissolved oxygen.

The detectors employ ionselective electrodes and other methods, providing a continuous determination of ions of chlorine, fluorine, copper, nitrates, and cyanides, as well as ammonia, hydrogen sulfide and primary biological production.

In the future automated continuous observations (in situ) are to be combined with laboratory analyses of water samples. The automated systems are to provide complete automation of the data processing and transmission.

The USSR State Committee for Hydrometeorology and Environmental Monitoring, together with the USSR Ministry of Health and the USSR Ministry of Water Resource Management and Reclamation, are responsible for organizing the system for monitoring environmental pollution control. All the information on environmental pollution is to be transmitted to the main data center (where it is stored).

Along with monitoring environmental pollution, Goskomgidromet also keeps track of the water and its use, and it is responsible for the State Water Inventory. These activities ensure that water pollution is monitored. The work is carried out jointly with the USSR Ministry of Water Resource Management and Reclamation (monitoring water use) and the USSR Ministry of Geology (monitoring groundwater).

All the water and all the water bodies which form the total state water stock of the USSR, as well as the water consumption by all users and enterprises, organizations, and institutions, are subject to the State Inventory. The above system includes a network of observations, which receives, processes, transmits, and stores information about the state of all water objects, and about water use and water quality. The information is sent to any interested organizations.

The Goskomgidromet bodies provide the State Inventory of Water Resources (except for groundwater), i.e., an inventory of the natural component of the surface-water inflow. In addition, they organize observations and monitoring of the water-pollution level according to physical, chemical, and hydrobiological indicators.

Goskomgidromet, together with the USSR Minsitry of Water Resource Management and Reclamation, inventories the water use by enterprises and institutions, and it determines the water users (those who use water and discharge wastewater); the Ministry determines the quality of the wastewater, and it notes the state of the equipment which records water intake and discharge.

The USSR Ministry of Geology is responsible for inventorying the groundwater resources and observing their state and quality.

The State Water Inventory involved with checking 'water quality' is carried out within the framework of the All-Union Service for Observing and Monitoring Environmental Pollution.

The State Water Inventory and the direction of the State Water Survey of the USSR lie outside the framework of the pollution monitoring system, but they are closely connected with the monitoring system as well as with the wider system for monitoring anthropogenic changes in the state of the environment.

A timely warning about high pollution levels, and also about unfavorable hydrological and meteorological conditions promoting the appearance of such levels, can lead to measures designed to eliminate the pollution or at least to draw attention to it.

For instance, a warning about unfavorable meteorological conditions enables the authorities to take efficient organizational and technological steps (change to low sulfur fuel, reduction of power-plant output, limiting the number of cars in urban areas) to halt the occurrence of dangerous pollution levels.

Such a warning service (along with an in-place system of preventive measures) is already operating in 122 cities of the Soviet Union [19]. For example, in Dzerzhinsk it enabled the authorities under unfavorable meteorological conditions to reduce several times the levels of air pollution by chlorine and nitrogen oxides. The validity of such warnings is quite high (86%) [7].

The most important step in regulating the state of the environment is to detect pollution sources and control them by limiting pollutant emissions and introducing rigid standards for such emissions. The large-scale development and introduction of standards for the maximum permissible discharge into the environment began in 1980 after adoption of the Soviet Law of Protection of the Atmosphere, followed by the Decree of the USSR Supreme Soviet and the Decree of the USSR Council of Ministers on state control of the state of the atmosphere.

Goskomgidromet is responsible for monitoring sources of atmospheric pollutants, the USSR Ministry of Water Resource Management and Reclamation is responsible for monitoring water-pollution sources. By now, all the pollution sources have been identified: an all-union inventory was carried out of sources polluting the atmosphere and other parts of the environment. This inventory yielded the emission characteristics of the main enterprises in various branches of the national economy for each Soviet Republic, for the major economic regions, and for almost 500 towns and cities [19].

Inspections by Goskomgidromet and the USSR Ministry of Health ensure that ministries, departments, enterprises, and organizations are obeying the Law of Protection of the Atmosphere. If the law is violated, then sanctions are to be applied.

The Law of Protection of the Atmosphere provides for the introduction of standards limiting the emission of any pollution source and any harmful impact, including impacts affecting adversely the atmosphere itself.

When developing such standards, we should begin by determining the ecologically permissible impact and the permissible impact on man; we should set up standards limiting such an impact, for instance, MPC standards for man in the environment, fishery MPC and MPEL standards for ecosystems. Then criteria are worked out, so as to reduce pollutant emissions from sources: MPE norms for pollutants in the environment, which are to be established while meeting the MPC and MPEL standards.

Many standards for monitoring the air-pollution level have been developed and introduced in the USSR. MPC standards (public-health standards, designed to protect human health) have been set up for 263 chemical substances and for 47 combinations of these; maximum permissible loads for ecosystems are being established. When developing the MPE, we have to take many factors into account: the hydrometeorological characteristics; the relief; the distribution (in space and time) of pollution sources and contingencies of populations and ecosystems susceptible to the impact; and the sensitivity of various organisms to impacts.

If a valid standard (based on health or ecological criteria) cannot be set up at a given moment, because of economic or scientific-technological factors, then temporary agreed-upon emissions (TAE) are introduced, followed by a step-by-step transition some time later to an established standard. However, this is possible only on condition that the impact levels for man are not dangerous and the critical levels for the ecosystems are not exceeded. Otherwise, 'prohibiting' decisions are needed, with respect to certain pollution sources.

Standards for MPEs are introduced simultaneously for all enterprises which pollute urban or rural air. Therefore, the permissible emissions from the aggregate of sources are optimized. For the solution of this problem, several versions are considered, and the version whose implementation seems the simplest, costs the least, and provides the best results is selected.

The state standard determining the maximum permissible emissions (MPE) has been in effect since January 1, 1980. As a first step, it is planned in the near future to set up standards for the maximum permissible emissions from major facilities, located in 250 cities and towns and contributing most (more than 80%) of the environmental pollution. Much has already been done, and by January 1984 standards were set up for the maximum permissible emissions of several thousand of the largest facilities [19].

Goskomgidromet is also responsible for relocating and reconstructing large facilities of the national economy, with a view toward preventing atmospheric pollution. The organizations under the USSR Ministry of Health, the USSR Academy of Sciences, the USSR Ministry of Fisheries, and Goskomgidromet measure and study the impact $\mathcal{E}_{\text{im}}$ of pollutants on man and different organisms and on the biota as a whole.

Prediction of the amounts of pollution $\sigma_i(\mathbf{R}, t)$ in different parts of the environment and of the effects of man's activities on the biosphere is the job of the USSR State Committee on Science and Technology, the USSR Academy of Sciences, Goskomgidromet, and other organizations.

The institutes of the respective ministries and departments and the USSR Academy of Sciences are developing methods of decontamination or of preventing pollutants from entering the environment with discharges (emissions), and they are also selecting the appropriate optimal technological processes.

The institutes of the USSR Academy of Sciences assess the ecological and economic effects of environmental pollution, and a comprehensive analysis of changes in the environment is performed by the Laboratory for Monitoring the Environment and Climate under both Goskomgidromet and the USSR Academy of Sciences and by the Fedorov Institute of Applied Geophysics under Goskomgidromet in cooperation with other institutes.

The described integrated organization of activities, with allowance for a possible broadening of the functions of a number of ministries and departments (and, in particular, setting up and monitoring standards limiting the emissions from different pollution sources to permissible values) will ensure the effective protection of the environment in the Soviet Union against anthropogenic impacts, and especially protection against pollutants so as to maintain a high quality of the environment here.

Many countries consider it necessary first of all to organize at a national level (and to participate in at the global level) a system monitoring the anthropogenic impacts, especially pollution. In the author's opinion, the principal characteristics of the monitoring systems in different countries should be the same. There should be a single concept of monitoring, which includes observations, assessments, and forecasts of the environmental state, unification of measurements, and a unified determination of priority when observing, working out criteria estimates, assessing the environmental state, etc. However, it is evident that in each country such a system is based on the existing observation systems, which often reflect to some extent the organizational structure of the management of the particular national economy. Thus, it is of interest to consider the scheme of organization of the monitoring system in some other country, in addition to describing the way the system is realized on a broad scale in the Soviet Union.

The approach used to organize the environmental monitoring system in Great Britain will be discussed below [20, 24, 26, 29, 32, 66].

The organization of environmental protection in Great Britain does not provide for a unified system planned in advance (with regard to the duration of its formation) [32].

The problems of environmental protection which are of paramount importance are those connected with human health and survival, with the protection of animals and plants used by man, with recreation, and with the reduction of any unfavorable effects on the environment.

The battle against pollution is essential in order to prevent damage to human health and also to protect the natural environment. Special attention is given to investigating those aftereffects, which show up only after a long time.

In Great Britain there is a system of nature reserves, national parks, and green belts for protecting and conserving biological systems. In addition to existing measures, a system for monitoring pollution has been set up.

The British monitoring system is organized in accordance with existing needs, and the results obtained are used for decision making. This program is designed to provide tracking of both the cause and the effect of the impact; the purpose of the monitoring is to determine the relation between the change in pollution level and the consequences of such pollution for biological systems. The authors of [32] note that the Soviet approaches to creating a scientific basis for the monitoring program are shared in Great Britain.

The economic aspect is recognized as being of great importance for organizing a monitoring system. Thus, in order to strike a balance between the costs of pollution control and monitoring and other programs for nature protection, we have to have precise and comprehensive information.

Monitoring in Great Britain is implemented in two directions:

(1) quality control; an operational assessment of 'what's going on', whereby known pollutants are measured, and standards are set for such pollutants in order to protect the environment and human health;
(2) 'alertness and foresight'; the danger of new pollutants, new poisons, new responses of biota to different substances, and new diseases. The assessment of new effects is of great significance for this part of monitoring.

It is characteristic of the development of various systems in Great Britain that local authorities are responsible for the systems and that special attention is paid to close cooperation between the population and private companies.

Therefore, though there are some national programs in Great Britain, the main activities in the field of monitoring are realized at a local level. The two main supervision programs are: The National Service for Monitoring Atmospheric Pollution and The Service for Monitoring River Pollution [26]. Both programs represent their results in the form of maps of the pollution levels.

The measurements showed a good correlation between the observed concentrations of sulfur dioxide and smoke and the calculated (according to the amount and types of fuel burned) emissions (and sources) of pollutants, as well as the morbidity rate of the population. This regularity makes it possible to simplify the monitoring somewhat.

It is stressed in [26] that the monitoring of pollution effects is the most significant type of monitoring, since it provides a check of pollution sources, pollutant concentrations in 'hot spots', and the state of health of 'critical' groups of people, who are the most vulnerable to harmful effects of their surroundings. For instance, special observations were carried out of the citizens of Birmingham, who were exposed to increased lead concentrations [26].

Biological monitoring is also being developed, with a view toward monitoring the state of the atmosphere (by observing lichens and moss) and the surface water (by observing algae and certain fish species).

However, as noted by British scientists, efficient operation requires the establishment of a centralized information service; attempts are being made to establish a network to supply 'early warning' data on the most dangerous chemical compounds polluting the environment. Toxicological investigations, epidemiological surveys, and clinical tests are carried out on a regular basis.

The idea of developing a comprehensive, unified monitoring system found reflection in the report of the Royal Commission on Environmental Pollution (1974).

In order to harmonize (coordinate) monitoring programs at various levels, several groups (committees) on monitoring management have been established in Great Britain; they deal with the physical aspects of various parts of the environment (air, fresh water, sea, and land), human health, and the state of biological systems.

British specialists believe that, in order to evaluate the pollution damage to human health and to natural ecosystems, it is necessary to identify pollution sources, to determine the pollutant concentration and characteristics, their migration paths, and the nature of their interaction with other substances. In this connection, there is a tendency in Great Britain to set up coordinated indicators of the environment quality, which take into account the ability of individual parts of the environment to bear the loads put on them (this ability is considered to be a natural resource, while an indicator of environmental quality is considered to be a control lever) [24].

It is asserted that measures to regulate the environmental quality (by changing the intensity and location of sources) need not be the same everywhere. The specific capacity of the environment to dilute and remove pollutants should be taken into account. This requires a comprehensive approach to the analysis of the environment and the use of various models, i.e., it leads directly to the method of comprehensive analysis of the environment which has been in use in the Soviet Union since 1974 (see Chapter 1).

Thus, in Great Britain the approach based on environmental quality indicators is preferable to an approach based on monitoring the pollution sources. Such an approach places rather strict demands on the monitoring system, and it requires the inclusion in the observation program of a wide range of substances. However, the observation of substances on some priority basis is not considered in Great Britain to be the sole basis of action. The integrated analysis of impacts, taking into account the actual situation, and the identification of the possible effects of various substances even before their production (particularly medical preparations and pesticides) are dominant with regard to the present scientific basis of monitoring.

This is why a centralized network of data on chemical compounds is being developed in Great Britain. It will collect basic data on pollutants, the results of testing these, and estimates of the state of the environment. An integrated pollution data bank is being organized.

The state of the environment in Great Britain is not homogeneous. In some places the air and surface waters are very polluted. This calls for the organization of a detailed monitoring system. It is difficult to establish unique criteria for environmental quality, for river-water quality in all places. Thus the quality of each stream or river is considered individually. According to [8], half of all the data on river-water quality (more than 30 rivers are mentioned), judged by the BOD (biological oxygen demand), indicate questionable or poor water quality (5–7 mg/l or higher). In order to use such water as drinking water, it should be chlorinated up to the clarification point, coagulated, deposited, filtrated, cleaned with activated charcoal, and treated with ozone. Some portion of this water should be biologically treated or desalinated. In accordance with [26], the upper reaches of a river are usually not polluted; the tendency toward pollution increases with the approach to the estuaries.

Since British rivers have a low water content, the varying hydrological conditions should be taken into account when making rules for the discharge of waste.

The river water in Great Britain is thoroughly analyzed with regard to a number of variables, first of all pH, temperature, electrical conductivity or total amount of dissolved solids, suspended solids, ash, chlorides, ammonia, COD (chemical oxygen demand), BOD and dissolved oxygen.

The measurement accuracy is, as a rule, ± 20% of the measured value; when sampling, the homogeneity criterion (at the river cross-section) of the given parameter is observed.

In Great Britain much attention is given to monitoring with the aid of hydrological indicators [21].

It is very important to refer to air pollution monitoring data (background data), as well as modeling of the possible pollution of the neighboring zone, before new facilities (power plants, etc.) are designed and constructed [20, 26].

Recently in Great Britain an approach has begun to be widely used whereby environmental monitoring is based on ecological analysis (see, for instance, [25]).

PART III

Global Ecological Problems. Critical Anthropogenic Effects

Anthropogenic effects on the atmosphere and the climate

The transport of pollutants over long distances. The ecotoxicology of acid rain

Stressing the atmosphere and the climate

The geophysical and biological consequences of nuclear war

Anthropogenic effects on the world ocean

CHAPTER 8

Anthropogenic Effects on the Atmosphere and Climate

8.1. The Assessment of Anthropogenic Climatic Changes

Of the global problems of the biosphere arising due to the anthropogenic impact, the most significant one, which demands particular attention, is the possible man-made change in the climate of our planet.

The urgency of this problem is due to at least two factors. First, even a slight climatic change can influence substantially man's economic activity, especially food production. Second, slight anthropogenic climatic changes, which develop against the background of natural climatic variability, can disturb the stability of the climate and can develop into catastrophic changes. Paleoclimatic research indicates a succession of glacial periods and other climatic cataclysms in the past, which took place without any anthropogenic impact whatsoever.

In order to comprehend the possible climatic changes, it is necessary to study the state of the climate system (atmosphere – ocean – land surface (including rivers and lakes) – cryosphere – biota), the interactions between the elements of this system, and the effects of external and internal factors (particularly anthropogenic factors) on this system. These are the goals of climate monitoring (see Chapter 6).

The major anthropogenic factors affecting the climate system are:

A. The direct impact on the state of the atmosphere (heating, change in humidity, etc.).

B. I. Impacts which alter the physical and chemical properties of the atmosphere, and in particular its radiation and electrical characteristics. These factors include the change in the troposphere composition, the increase in CO_2 concentration (due to the burning of fossil-fuel and forest fires), the increase in the concentration (or generation) of minor gaseous pollutants such as nitrous oxide (N_2O), chlorofluorocarbons (freons), methane, and ozone, produced by industry and agriculture, the increased amounts of pollutants, affecting the electrical state of the atmosphere and its ionization (krypton-85), and the production of a great amount of tropospheric aerosols of anthropogenic origin.

B. II. The impact on the upper atmosphere, which alters its characteristics and state. This category includes factors affecting the ozone layer (freons, nitrous oxide) and factors promoting the production of stratospheric aerosol (volcanic eruptions, etc.).

C. Factors affecting the characteristics of the underlying surface and changing its reflectivity (albedo change), and also affecting the interaction between the elements of the climatic system (for example, gas exchange between ocean and atmosphere, a change in atmospheric humidity due to a change in evaporation, etc).

Some effects can pertain simultaneously to several categories of impact, and they should be analyzed allowing for this. For example, forest fires lead simultaneously to a direct heating of the atmosphere (A), an increase in the content of CO_2 and other gases in the atmosphere (B.I), an increase in the content of stratospheric aerosols (B.II), and a change in the albedo of the underlying surface (C). The first two factors promote an increase in the tropospheric temperature, while the last two factors lead, as a rule, to a decrease in this temperature.

The above factors affect not only the state of the climate system but also all the elements of the biosphere, leading both to climatic changes and to the most diverse ecological (biological) consequences, altering the nature of the biogeocenoses, affecting the state of living organisms, and, what is very important, affecting directly or indirectly human health and welfare.

The above factors influencing the climate system will be considered briefly below, and their possible effect on the present and future climate will be analyzed. Of course, the possible climatic changes produced by anthropogenic impacts can be assessed properly only after the natural climatic fluctuations have been studied thoroughly.

THE CHANGE IN THE ATMOSPHERIC CONCENTRATIONS OF CO_2, AEROSOL PARTICLES, AND VARIOUS MINOR POLLUTANTS

CO_2 is generally recognized to be the most significant (albeit not the only) component of the atmosphere, an increase in the concentration of which leads to the absorption of long-wave radiation, the appearance of the 'greenhouse' effect, and a significant increase in the temperature of the troposphere, and consequently also to a possible climatic change.

Based on an analysis of the change in the CO_2 content of the atmosphere, a 'preindustrial' (end of 19th century) concentration equal to 290 ppm [5] was determined (with a scatter of values in different sources ranging from 265 to 298 ppm) [40].

From 1958 to 1980 the average CO_2 concentration in the air increased

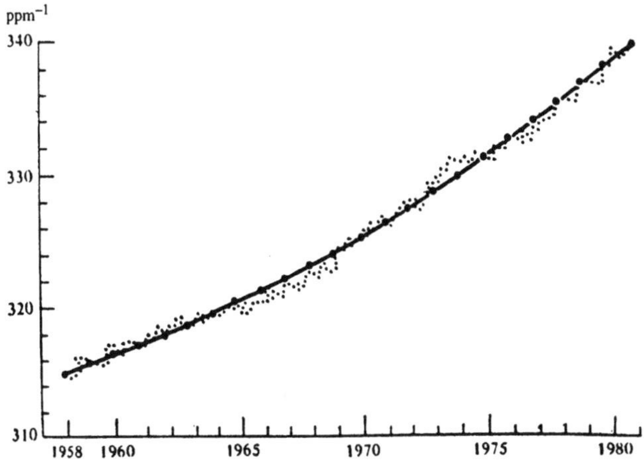

Figure 8.1. Change in CO_2 concentration in time, according to measurements at Mauna-Loa station.

by 22 ppm (according to data of the Mauna-Loa Observatory, in Hawaii). The increase in concentration estimated from industrial carbon emissions (about 82 Gtons) amounts to 39 ppm. The difference between these figures is explained by CO_2 absorption by the ocean (around 40%) owing to an increase in the partial pressure of CO_2 [5].

At present the average CO_2 concentration in the atmospheric air is estimated to be 338 ppm, and its change in time (according to measurements at the Mauna-Loa station) is shown in Figure 8.1. A very similar curve was obtained from measurements at the South Pole (the difference between the average concentrations in different zones of the world does not exceed 3 ppm).

Measurements of the CO_2 concentrations at different latitudes revealed two maxima: one at the equator, apparently caused by the release of CO_2 by tropical waters in upwelling regions, and the other (which increased by approximately 2 ppm over the last 20 years) in the latitude zone between 30° and 50°N. The latter maximum is apparently due to the ever-increasing burning of fossil fuels [5, 32]. This is confirmed by an increase in the amount of atmospheric CO_2 which does not contain the long-lived radioactive isotope ^{14}C (fossil fuel contains practically no ^{14}C, whose halflife $T_{1/2} = 5000$ years).

For instance, it was shown in [42] (on the basis of tree rings) that the amount of ^{14}C in the atmosphere has decreased by 20 ± 1.2% since the middle of the 19th century.

Modeling of the carbon cycle in accordance with existing scenarios of power-production development (and, consequently, various scenarios for the entry of 'anthropogenic' CO_2 into the atmosphere) makes it possible to forecast the CO_2 concentration in the atmosphere. For the year 1990, it has

Table 8.1. Change of CO_2 concentration with time.

Forecast period	CO_2 concentration, ppm
preindustrial period, 1880	275–297
1980	338
1990*	360 ± 6
2000*	394 ± 9
2030–2060	double
2050 (according to different scenarios of development)	700 ± 100

* Supposing that the present rate of power-production development will remain constant during the next 20 years.

been set at 360 ± 6 ppm, and for the year 2000 at 394 ± 9 ppm (see Table 8.1) [5] (assuming that the part of the CO_2 which remains in the atmosphere after the interaction with the sea and the biota remains unchanged during this time interval).

Of the minor constituents (gaseous pollutants) of the atmosphere whose concentrations are sufficient (or may become sufficient in the near future) to impact significantly the tropospheric temperature regime, we should mention: nitrous oxide (N_2O), halocarbons (freons such as $CFCl_3$, CF_2Cl, $CHCl_3$, CH_3Cl, and $CHClF_2$, and methylchloroform CH_3CCl_3), methane (CH_4), and tropospheric ozone (O_3). According to different measurement data, the concentrations of ozone, methane, and several other pollutants, as well as NO_2, which is mostly produced by the breakdown of mineral nitrogen fertilizer, continue to increase, while the rate of increase of freons has decreased somewhat.

At present the concentrations of minor gases in the troposphere (in background regions) are, on the average, as follows: tropospheric ozone 30×10^{-9} to 68×10^{-9} [10], nitrous oxide 290×10^{-9} to 350×10^{-9}, methane 1.4×10^{-6} to 1.6×10^{-6}, carbon tetrachloride 100×10^{-12} to 180×10^{-12}, freons: CH_3Cl 0.5×10^{-9} to 2.0×10^{-9}, CH_2Cl_2 30×10^{-12}, $CHCl_3$ 20×10^{-12}, CF_2Cl_2 200×10^{-12} to 250×10^{-12}, $CFCl_2$ 160×10^{-12} [24, 25], and methylchloroform CH_3CCl_3 120×10^{-12} [24]. Many of the above pollutants, by participating in photochemical reactions with gases, can influence greatly the radiation fluxes, and in turn the tropospheric and stratospheric temperature distributions. Ozone and other minor pollutants tend to the 'greenhouse' effect [12].

An analysis of the aerosol component of the atmosphere, particularly the stratospheric aerosol, is of great importance when assessing the state of the atmosphere (from the viewpoint of possible climatic changes). A great amount of aerosol particles, produced both by natural processes (weathering, volcanic activity, forest fires) and by human activity, enter the atmosphere. The total annual emission into the atmosphere of man-made particles amounts to several hundred million tons; many anthropogenic gaseous emissions (e.g. sulphur dioxide emissions) promote the formation of aerosol particles in the atmosphere.

There exist some quantitative estimates of the stratospheric aerosol, including aerosols produced by volcanic eruptions. More complicated are estimates of the possible effect of the tropospheric aerosol on the climate. An enormous amount of man-made aerosol is ejected annually into the troposhere (industrial SO_2 emissions alone produce around 150 million tons of aerosol); anthropogenic forest fires and wood burning contribute significantly as well. Although the region of spread of this aerosol is limited to the land zone, particularly in industrialized countries, while its residence time in the atmosphere amounts to only a few days, still the effect of the tropospheric aerosol on the state of the atmosphere can be quite substantial. It is noteworthy that existing models of possible anthropogenic climatic changes do not pay sufficient attention to this factor.

Let us assess roughly the relative role of the tropospheric aerosol in scattering the incident solar radiation. The assessment will be performed in comparison with the stratospheric aerosol, the effect of which is approximately known. For instance, it is estimated that the stratospheric aerosol layer (produced mostly by volcanic activity) reduces the mean temperature at the earth's surface by 0.1 to 0.3°C (sometimes to 0.58°C) [49].

The stratospheric aerosol layer has increased severalfold during recent years, due to volcanic activity and other sources [49].

It is known that this layer, because of the effect of local natural and man-made sources, has the form of a latitude belt (maximum concentrations) between 20° and 40°. Therefore, in the Northern Hemisphere the effective area of the Earth's surface 'covered' by this stratospheric aerosol layer is, at the latitude of the industrialized countries, about 20% of the entire surface of the world.

As already noted, the annual amount of tropospheric aerosol formed additionally from SO_2 emissions is around 150 million tons; if it is assumed that a like amount of particles is produced from other anthropogenic sources (man-caused forest fires, various industrial emissions) which contain a substantial amount of minute fractions, fly ash, and solid products of incomplete combustion (ash, hydrocarbons, etc.) [24], then the total amount will be 300 million tons. If the lifetime of these particles in the troposphere is taken to be 4 days, and if the amount of high-disperse aerosol (whose reflecting characteristics are similar to the stratospheric aerosol) is equal to half the amount of aerosol produced in the troposphere (here it must be kept in mind that mainly high-disperse aerosol particles are formed from gaseous products), we then find that about 1.5 million tons of high-disperse anthropogenic aerosol are permanently present in the troposphere. It is noteworthy that, according to the data of [30], the introduction of 2 million tons of dust into the Earth's atmosphere will result in a temperature drop of 0.4°C. The area of the Earth's surface covered by this aerosol layer is apparently somewhat less than that covered by the tropospheric aerosol layer, but for the corresponding estimates the total aerosol mass was taken. It should be kept in mind that the tropospheric aerosol also includes aerosols absorbing solar radiation (these aerosols, which according to estimates are fewer, will attenuate the above effect).

On the basis of recent measurements of the concentration of the stratospheric aerosol, it is estimated that the relatively constant amount (mass) of particles in the stratosphere (their lifetime is rather long, being measured in years) is 0.25 million tons, that is, less than the permanent amount of tropospheric aerosol. Only after major volcanic eruptions can the amount of stratospheric aerosol particles exceed several million tons [16].

Even if we introduce a correction to the ratio of the areas covered by these layers, their sizes will be comparable.

The area under the tropospheric aerosol layer (projected onto the earth surface), if we base ourselves on the areas of the industrialized countries in North America, Europe, and Asia and the area of spread of the aerosol particles during their lifetime, amounts to approximately 2×10^7 to 3×10^7 km^2, i.e., it is only 1/3 to 1/5 of the area under the stratospheric aerosol layer.

A very important conclusion can now be drawn with regard to the possible effect of the tropospheric aerosol on the Earth's climate, namely that it can be substantial and that this factor has to be taken into account in the corresponding models.

PERSPECTIVES FOR POWER-PRODUCTION DEVELOPMENT AND FOR EMISSIONS OF CO_2 AND OTHER GASEOUS COMPONENTS AND AEROSOL PARTICLES INTO THE ATMOSPHERE

In order to forecast possible changes in the state of the atmosphere and subsequent climatic changes, we have to know the dynamics of future emissions into the atmosphere of various pollutants, especially CO_2. Thus it is important to know the prospects for the development of energy. With regard to this, in [5] scenarios of energy development are considered in accordance with [18, 36, 41].

Legasov and Kuz'min [18] single out three phases of energy development. In the first (near future) phase, up to the year 2000, power production is based on fossil fuel, hydroenergy, and nuclear power produced in thermal reactors. The second phase (up to the year 2100) is the phase of energy optimization, based on a broad application of nuclear energy and renewable power resources. In the third phase (after the year 2100) the energy con-

Table 8.2. Natural and anthropogenic energy fluxes.

Source	Energy flux, W
Solar radiation	$\alpha = 1.8 \times 10^{17}$ (solar constant is equal to 1367 W/m^2)
Anthropogenic (annual) energy rate	
1975	7×10^{12}
2000	2×10^{13}
2100	$2 \times 10^{14} (0.1\beta)$
Energy release causing a 1 – 1.5°C temperature increase (under equilibrium conditions)	$\beta = 1.8 \times 10^{15} (\beta = 0.01\alpha)$

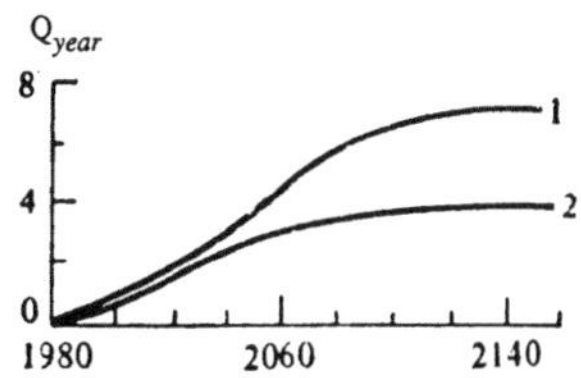

Figure 8.2. Growth of worldwide energy consumption (for the stabilization level of per capita energy consumption)
1) 20 (kW year)/year, 2) 10 (kW year)/year; $Q = 33.5$ (TW year)/year

sumption per capita (up to 20 kW year/year) will be stabilized, and the total energy consumption will range from 1550 to 2450 TW year/year. In the distant future nuclear energy will play the dominant role.

Proceeding from these premises, the following conclusions are drawn in [18]: between 1975 and 2000 the total power generation will triple, and the consumption of conventional fuel will increase correspondingly by a factor 3; by the year 2025 the total energy production will still increase by a factor two and will amount to 50 or 60 (TW year)/year mainly due to nuclear power; it is assumed that during this period there will be no significant increment in the rate of fossil-fuel consumption. The increase in energy consumption with time according to the scenario in [18] is shown in Figure 8.2.

Table 8.2 can be used to compare the existing and predicted fluxes of produced energy with the natural energy flux and with the energy release which would produce an appreciable warming effect in the atmosphere due to direct heating.

According to Styrikovich and Vnukov [24], fossil-fuel consumption will increase until the middle of the next century, when the annual consumption may reach 20 billion tons of comparison fuel (in 1975 it was around 8 billion tons). Then fossil-fuel consumption will start to drop, giving way to physical types of energy resources which do not pollute the environment.

Table 8.3. **Increase of industrial CO_2 emission into atmosphere Gt/year (according to carbon) [5].**

Source	1975	2000	2025	2050
Legasov and Kuz'min [18]				
maximum variant	5	175	18	23
minimum variant	5	14.5	15	16
Rotty [41]	5	12	26	–
Hafele [36]	5	9	14	21
Styrikovich [24]	–	–	–	20

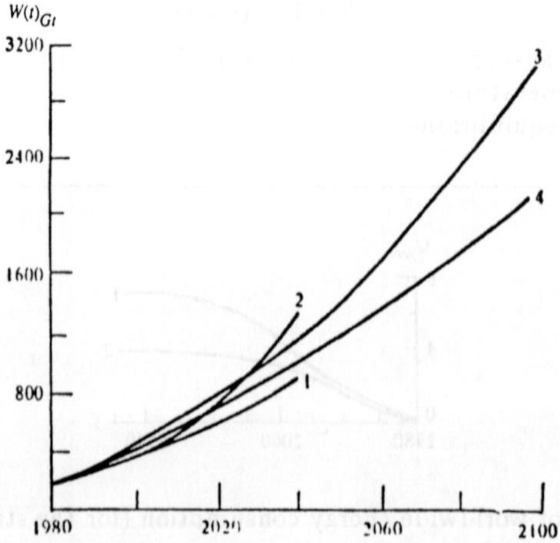

Figure 8.3. Total industrial CO_2 emission, accumulated by given year (Gt). (1) after Hafele [36]; (2) after Rotty [41]; (3, 4) after Legasov and Kuz'min [18]; maximum variant (3) and minimum variant (4).

In [36, 41] the predicted energy development is somewhat more gradual than in [17], although the CO_2 emissions into the atmosphere are close (Table 8.3 and Figure 8.3).

The predicted change (increase) in the CO_2 concentration of the atmosphere until the middle of the next century is given in Table 8.1.

It should be noted that, according to all scenarios using different models of the carbon cycle, the mean CO_2 concentration should reach about 700 ppm by the year 2050. However, all the models did not take into account the interaction between the atmosphere and the deep layers of the ocean, and the processes of CO_2 transport to these depths. Moreover, the fraction of the CO_2 remaining in the atmosphere during the whole period under consideration was assumed to be constant (about 60%).

FLUCTUATIONS AND CHANGES IN PRESENT-DAY CLIMATE. POSSIBILITIES OF PREDICTION

A detailed analysis of regularly measured (over a long period) meteorological parameters makes it possible to evaluate fluctuations and changes in the

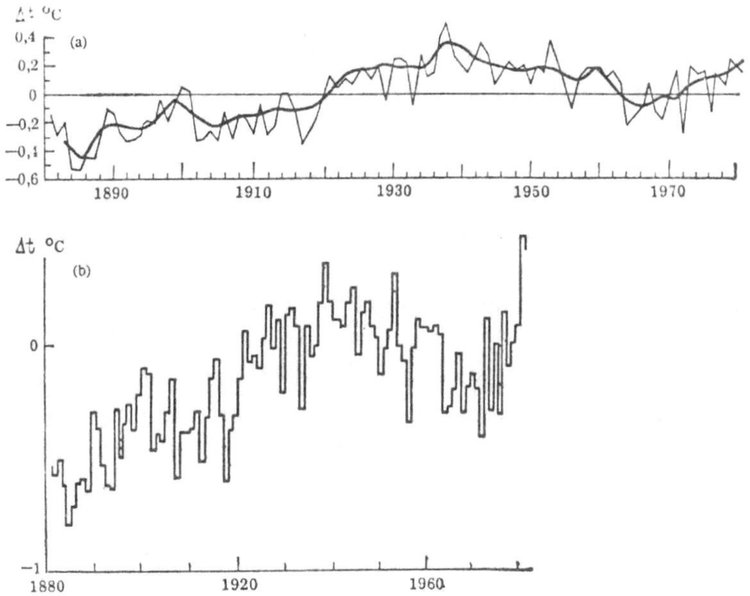

Figure 8.4. Variation of average surface temperature in Northern Hemisphere since 1880. (a) data of [4], (b) data of [35] (agrees with [40]).

present-day climate.

Long series of observations of the air temperature and precipitation are at our disposal. By averaging these data over a large area (e.g., over the whole Northern Hemisphere or a large part of it), and by considering their changes over a long time interval (e.g., over a hundred years), we can draw some conclusions about climatic changes (during this period).

Figure 8.4 shows the change in the mean surface temperature of the air in the Northern Hemisphere in the belt from 17.5° to 87.5°N [4] and over the whole Northern Hemisphere [35]. It is evident from the figure that, beginning at the end of the last century and up until the 1940s, there was a rather sharp cooling, while at the end of the 1960s a certain stabilization occurred. In the middle of the 1970s a persistent discussion took place on the tendency of the subsequent temperature change. However, accumulated new data made it clear that the cooling is being replaced by warming. This was the conclusion drawn by scientists in 1980–1981 [5, 21, 47], on the basis of empirical data (see Figure 8.4) and data in [4, 35, 40], etc. Of course, this process may slow down (even the sign of the derivative can change), for various reasons. For instance, a major volcanic eruption can change the trend of the temperature curve for several years. Such a change was observed after the El-Chichón eruption (Mexico) in 1982 [16].

Naturally, however, just changes in the global temperature characteristics of the surface air layer do not yet constitute climatic changes.

At present aerological measurement data are being collected and generalized for the entire thickness of the troposphere (see, e.g. [5]). Data are also needed on the dynamic characteristics of the atmosphere and their changes (in addition to mean values). The frequencies of various anomalous phenomena can indicate that a climatic change has occurred.

Recently many scientists have tried to forecast possible climatic fluctuations and variations in the future on the basis of existing empirical data and models of varying complexity. It is obvious, of course, that estimates of the obtained trends in the climatic changes cannot be used to forecast by simple extrapolation.

As already noted, scientists are inclined to agree on the possibility of climatic changes in the near future and a general warming caused by the anthropogenic impact. An interesting summary of the data on this is contained in [5, 40].

Most scientists consider that the ongoing accumulation of anthropogenic CO_2 in the atmosphere is the reason for the global warming [3, 15, 38, 40], although other reasons are cited as well.

The analysis performed gives reason to assume that the warming in the Nothern Hemisphere between 1880 and 1970 occurred due to an increase in CO_2 concentration by 0.4 to 0.6°C [5]; according to [40], the increase was 0.2 to 0.62 K. Some scientists have questioned whether the effect of the changed CO_2 concentration on the tropospheric air temperature is really noticeable. Even in [5], which definitely supports the idea that the increased CO_2 concentration in the atmosphere affects the climate, it is nevertheless stated that: "it is impossible to be sure at present that warming due to the CO_2 effect is being observed, although certain existing data seem to indicate this".

In [38] it is shown that a doubling of the CO_2 concentration in the atmosphere will result in a global temperature increase of, on the average, 3°C. In [5] it is stated that the most probable change will be 2–3°C, and also that, in conformity with most scenarios, the CO_2 concentration will most probably double between the years 2030 and 2060.

It is obvious that the increase of the temperature of the Earth's surface will occur irregularly; for instance, the greatest warming will take place at high latitudes [5]. In [39] data are given on changes in the midlatitude surface air temperature with an increase of 3°C in the mean global temperature. In areas north of 70°N the changes will reach 6.8 to 7.6°C. The size of the variation in total precipitation will vary as well (in some areas it will reach +10 to +13 cm/year). If we compare the various scenarios of possible changes in the mean global temperature at the Earth's surface, then we can expect an increase in temperature (compared with climate data of the last century) of 1 to 2°C by the year 2000, 2 to 3°C by the year 2025, and 3 to

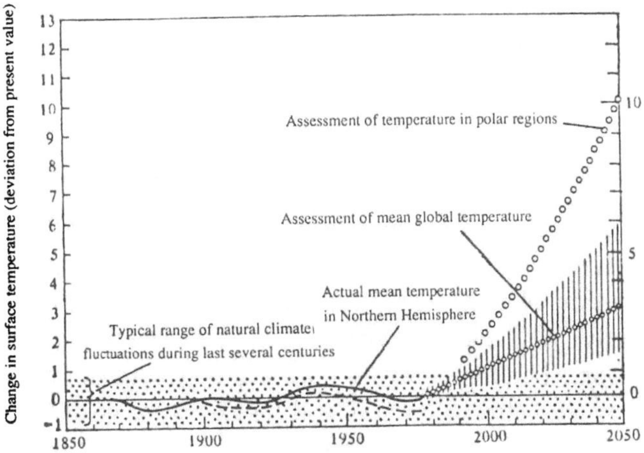

Figure 8.5. Actual changes of mean surface temperature in Northern Hemisphere since 1860 (solid line) and values of these which would exist in the absence of carbon dioxide (dashed line). Dotted region covers almost entire range of temperature fluctuations for last thousand years or more. Line of circles indicates future changes in mean global temperature. Vertical shading indicates region of uncertainty for model assessments, delineated by multipliers 2 and 0.5. Temperature changes in polar region are expected to be 3 to 5 times greater than global mean.

5°C by the year 2050 [5].

The possible temperature change up to the year 2050, according to the data in [15], is shown in Figure 8.5.

The possible temperature increase because of the minor gaseous components has been estimated in a number of works. For instance, in [13] it was shown that doubling the amount of N_2O in the atmosphere, compared with the present amount, can lead to an increase in the mean surface air temperature of 0.72°C, doubling the CH_4 content can lead to an increase of 0.37°C, and doubling the tropospheric ozone content can lead to an increase of 0.67°.

The participants of a Soviet–American meeting of experts, held in Leningrad in July 1983, pointed out the particular role of the minor gaseous components in the appearance of the greenhouse effect. The author of [14] used data of model assessments and data from measurements of the increase in the concentrations of CCl_3F, CCl_2F_2, CH_4, and N_2O in the 1970s to establish that the greenhouse effect produced by these gases is comparable with the contribution of CO_2 during this same period. If we take into account the increased amount of tropospheric ozone, then the role played by the minor gaseous pollutants in climatic change becomes comparable with the effect of CO_2 (or even exceeds it [14]).

As already mentioned, most modelers of the future climate take the effect of the stratospheric aerosol into account (although the role of the tropospheric aerosol is not given sufficient attention), and in a number of works the possible change in the albedo of the Earth's surface is taken into account (in particular, the change due to forest cutting).

THE EFFECT OF POSSIBLE CLIMATIC FLUCTUATIONS AND VARIATIONS ON THE STATE OF THE BIOSPHERE AND ON HUMAN ACTIVITY

Any climatic fluctuations and changes will affect the state of the biosphere and human activity.

The distribution of water resources, the development of the biota, and the bioproductivity all change with a change in the fields of temperature and precipitation. The temperature rise, together with other anthropogenic factors, leads to desertification (the soil loss because of desertification and other factors is now estimated to be 60,000 km^2/year [46]). It is believed that an increase of 3°C in the mean air temperature in the Northern Hemisphere would cause the perennial ice of the Arctic Ocean to disappear.

Climatic fluctuations and changes have the greatest effect on agriculture [21a, 28]. The anticipated warming of the climate should lengthen the growing season (by approximately 10 days per degree of temperature increase [15]), which can in general have a positive effect on agricultural development. A direct increase of the CO_2 concentration will also promote an increased agricultural productivity. On the other hand, in areas where there is a precipitation deficit, the frequency and intensity of droughts can increase and serious damage can take place. As mentioned above, the worldwide pattern of precipitation variation with an increase in mean temperature will be quite diverse, and it is hardly possible to predict it with sufficient accuracy, although serious preventive measures should be taken in arid zones, so as to reduce the possible damage.

Agricultural zones in the Soviet Union and other countries are subject to droughts of different intensity. In the USSR, according to data for the last hundred years, droughts occurred in the Lower and Middle Volga Region, in the Northern Caucasus, in the Southern Urals, in Western Siberia, and in Kazakhstan once in 3 or 4 years, and in the Ukraine in the Central Chernozem Region, in the Central Urals, south of Volgo-Vyjatsk and the Central Regions, once in 5 or 6 years.

The greatest damage to the Soviet grain economy is caused by spring-summer droughts, which are typical of a number of major grain-growing regions of the European USSR, Siberia, and Kazakhstan. The most severe droughts occurred in the years: 1891, 1901, 1920, 1921, 1924, 1936, 1946, 1954, 1955, 1963, 1965, 1972, 1975, 1979, 1981. Particular attention should be given to the droughts of 1975 and 1981, which spread over the main grain-growing regions of the European USSR, Siberia, and Kazakhstan and which did not have analogs since the year 1891 [21a, 28, 29].

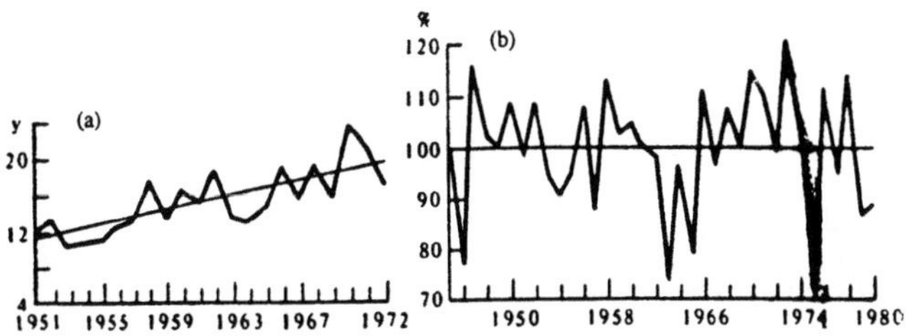

Figure 8.6. Change in yield of winter wheat for USSR [28, 29].

An analysis of the recurrence of droughts by years shows that droughts over large territories can sometimes recur two years in succession. Examples are: the droughts of 1891 and 1892; 1920 and 1921; 1938 and 1939; 1948 and 1949; 1954 and 1955. Local droughts can even recur 3 or more years in a row (see [21a]).

An analysis of droughts by decades shows that from 1891 to 1980 three massive droughts in a row, covering five to ten grain-growing regions, were characteristic of most decades. However, between 1931 and 1940 there were five such droughts, while between 1901 and 1910, as well as between 1921 and 1930, there were only two droughts in each decade.

Figure 8.6 shows the change in the average grain yield in the USSR between 1945 and 1980. The smooth curve indicates the trend characterizing the change in average yield due to improved agrotechnology. Inspection of the figure shows a substantial deviation of the actual yield in certain years, sometimes upward and sometimes downward. The droughts of 1946, 1954, 1963, 1965, and 1975 show up distinctly in Figure 8.6b.

We see that, against the general background of a significant increase in the productivity and gross yield of grain, their variation from year to year (interannual variability of yield) is not only not weakening but rather is becoming more pronounced.

The given curve indicates convincingly the effect of climatic (weather) fluctuations and variations on one of the most significant fields of human activity. It should be emphasized that the influence of climatic (weather) fluctuations on the grain yield is evident, although it is very difficult to single out the impact of any directional climate change from the data given.

The following example is connected with the distribution (and redistribution) of water resources in Eastern Siberia (in the upper reaches of the Yenisei and Angara Rivers). Figure 8.7 shows the change in the water content of these rivers in the vicinity of the Krasnoyarsk and Bratsk reservoirs (according to the data of Shiklomanov). The figure shows that a substantial change in the water content of these rivers is connected with climatic fluctuations during various time periods and also that the low-water periods lasted for 5 to 10 years, although it does not appear to be possible to discern

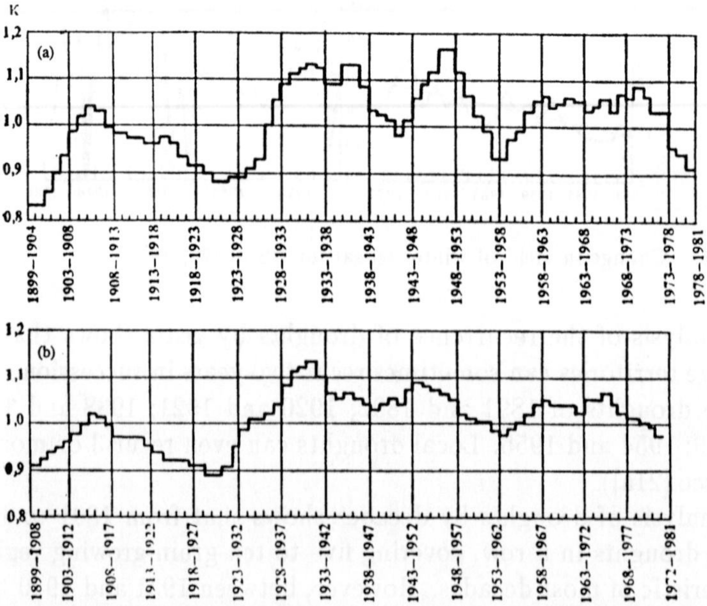

Figure 8.7. Chronological variation of inflow (in relative units) to Bratsk water reservoir (or in its channel) according to six-year (a) and ten-year (b) periods.

any constant climatic changes in this zone on the basis of these data.

In order to optimize the national economy, we must know about the climatic fluctuations in various zones. It was the regularities in the water content of the above rivers which led to the construction in this zone of reservoirs with long-term regulation and to the formulation of the problem of optimization of the regulation regime for these reservoirs.

This is a good example of the need to use information about the state of the environment (climatic fluctuations and the associated changes in the water content of rivers) to organize the management of human activity on a regional scale.

8.2. The Anthropogenic Disturbance of the Earth's Ozone Layer

Although the amount of ozone in the atmosphere is not great, it is the most significant component of the atmosphere. Ozone actively absorbs ultraviolet solar radiation, and thus it determines the stratospheric temperature distribution and also creates a protective shield against the hard UV radiation, which is dangerous for all life on Earth. The ozone in the atmosphere forms the ozonosphere. The atmospheric ozone content amounts to $6 \times 10^{-5}\%$ (by mass); its total amount reaches 3.3×10^9 tons (1.16×10^8 tons of which is in the troposphere). The mean lifetime of ozone in the atmosphere is about

50 days [1].

The beginning of ozone formation in the upper stratosphere is connected with the photodissociation of molecular oxygen (by UV solar radiation with a wavelength of 242 nm):

$$O_2 + h\nu \rightarrow O + O.$$

The interaction between oxygen atoms and oxygen molecules (in the presence of a third body) then leads to ozone formation:

$$O + O_2 + M \rightarrow O_3 + M.$$

The first reaction proceeds with a higher probability than the second, at high altitudes. Most of the ozone formed with the help of this mechanism appears at altitudes of 35 to 45 km. In the lower stratosphere and troposphere the beginning of the reactions which lead to ozone formation produces nitrogen oxides, which dissociate under the influence of UV and visible radiation.

The main amount of ozone is distributed at altitudes between 15 and 45 km, with a maximum somewhere between 15 and 25 km (at lower altitudes in the polar latitudes and at higher altitudes in the tropics). The ozone content reaches a maximum in spring in the Northern Hemisphere, being on the average 446 DU, or Dobson Units (1 DU = 0.001 cm of ozone at normal pressure and temperature); in autumn it is only half as great.

The mean annual ozone amount in the ozonosphere is 300 DU (i.e., only about 3 mm under normal conditions), while at the equator it is still less (up to 200 DU). The ozone concentration in the atmosphere is subject to a sizable natural variability, due to ozone transport by both orderly and turbulent flows (primarily at altitudes below 25 km) and also due to the fact that different photochemical reactions (above 25 km) proceed at different rates. Just the daytime variations of the ozone concentration can reach $\pm 25\%$.

The mean rate of ozone destruction (due to photodissociation and other natural processes) is 6.6×10^{10} molecules/(cm^2 s). Although the anthropogenic impact on the ozone layer is real and dangerous, it is not so easy to distinguish the anthropogenic effects against the background of the high natural variability of the ozone concentrations.

Let us consider various ways in which the ozone is destroyed (by both natural processes and anthropogenic impacts).

The atmospheric ozone can be destroyed as a result of the reaction (oxygen cycle, or Chapman cycle):

$$O_3 + O \rightarrow O_2 + O_2.$$

This reaction proceeds slowly. However, it can be accelerated catalytically in the presence of compounds of nitrogen, hydrogen, and chlorine (the nitrogen, hydrogen, and chlorine cycles, respectively).

Examples of the reactions of the corresponding cycles are given below [23, 33]:

the nitrogen cycle

$$O + NO_2 \rightarrow O_2 + NO$$
$$NO + O_3 \rightarrow NO_2 + O_2$$

$$O + O_3 \rightarrow O_2 + O_2$$

the hydrogen cycle

$$O + HO_2 \rightarrow OH + O_2$$
$$OH + O_3 \rightarrow HO_2 + O_2$$

$$O + O_3 \rightarrow O_2 + O_2$$

the chlorine cycle

$$Cl + O_3 \rightarrow ClO + O_2$$
$$O + ClO \rightarrow Cl + O_2$$

$$O + O_3 \rightarrow O_2 + O_2$$

The above reactions are most probable at altitudes of 20 to 40 km.

The catalyst concentrations are thousands of times lower than the ozone concentrations, and the number of destruction cycles per catalyst molecule ranges from 10^2 to 10^7 [25]. The contributions of the various catalyst cycles depend on the altitude; the chlorine cycle predominates above 35 km, and the nitrogen cycle predominates below 35 km.

It is interesting to note the possible existence of an ion cycle of ozone destruction, either due to ion-molecular reactions leading directly to ozone loss (ion cycle) [23]:

$$O_2^- + O_3 \rightarrow O_2 + O_3^-$$
$$O_3^- + O_3 \rightarrow O_2^- + 2O_2,$$

or due to ion-molecular reactions leading to the formation of catalyst particles which destroy ozone.

For instance, the authors of [26] consider the formation of hydroxyl radicals OH in the lower stratosphere (due to ion-molecular reactions of charged particles, produced by galactic cosmic rays) and the subsequent ozone destruction by this radical in the familiar hydrogen cyle. This is apparently a unique process, which leads to ozone loss under nocturnal conditions. It is of great importance during the intensive ionization of the atmosphere at high latitudes which accompanies strong solar flares.

The arrival of various natural and anthropogenic substances into the troposphere has a substantial effect on the ozone layer. For instance, volcanic eruptions alone annually inject into the stratosphere from 10,000 to 100,000 tons of chlorine.

The main anthropogenic processes producing substances that can destroy the operation of the ozone layer are: refrigerators and aerosol units (and blocks), which release halocarbons (chlorofluoromethanes or freons); the decomposition of mineral fertilizers (releasing nitrous oxide); flights of high-altitude aircraft (emitting nitrogen oxides and water vapor); nuclear explosions (producing significant amounts of nitrogen oxides). Other chlorine compounds (of anthropogenic origin) are also observed to reach the stratosphere: methylchloroform (CH_3CCl_3), carbontetrachloride (CCl_4), and methylchloride (CH_3Cl).

Only (chemically) inert substances, such as freons and nitrous oxide, can pass from the troposphere to the stratosphere. In the stratosphere, when acted upon by UV radiation, freons release chlorine, which leads to destruction of the stratospheric ozone.

In 1974 the production of chlorofluorocarbons (mainly F-11 (CCl_3F) and F-12 (CCl_2F_2)) amounted to 850,000 tons. The total production of freons before 1976 amounted to 8.1 million tons. The freon production by the beginning of the next century is expected to reach 1.7 to 3.7 million tons per year. The lifetime of freons in the atmosphere is estimated to range from 29 to 205 years [33].

Calculations performed using models of different complexities [1, 20, 33] indicate a decrease in the ozone content due to anthropogenic emissions of halocarbons and nitrous oxide.

According to estimates of the UNEP Coordinating Committee [20] in 1981, the continuing emission of freons at the 1977 level will finally lead to a decrease in the ozone content by 5 to 10% (and an equilibrium steady state will be achieved).

According to the estimates, the ozone content has so far decreased by less than 1% (due to freons), which is below the present limit of detectability.

It is of interest to compare these values with other estimates. An earlier study [33] (1976) showed that, if the freon production (F-11 and F-12) is kept at the 1973 level, then in approximately 100 years the ozone content can decrease by 8% (and, by the year 2000, by 3%); estimates of the decrease taken from other sources are also given in [33], and they vary from 6.5 – 7% to 10 – 18%. Similar results are given in [37]: an 8 to 10% reduction by the year 2040.

It is also noted in [33] that the ozone decrease due to released freons and other chlorocarbons will be 1.8% (0,6% due to freons, 0.8% due to methyl

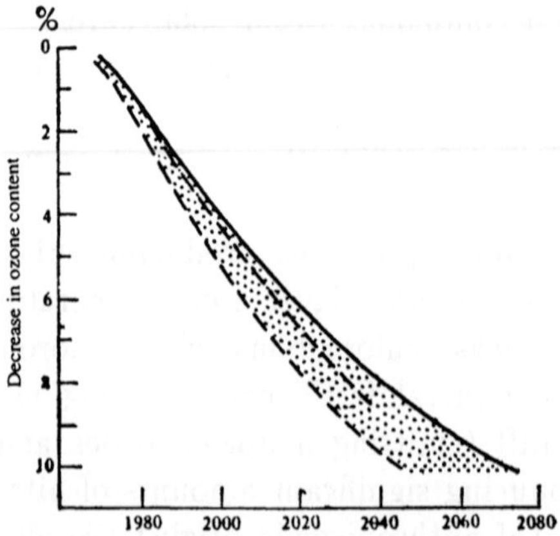

Figure 8.8. **Ozone decrease (%) for continuing use of freons at level of 1973–1978 (about 8×10^5 tons/year), according to data of various authors [37].**

chloride, and 0.4% due to carbon tetrachloride).

The model computation given in a later work [34] (1979) indicated that the decrease in ozone content due to freon emissions by the time of the computation had reached 0.7 to 1.6%, the average being around 1%.

It was shown that, if freon production had ceased completely by 1979, then the decrease would have reached 2% in 5 to 15 years, and if it were stopped by 1983 then the maximum decrease would be 2.5%.

If freons continue to be emitted at the 1975 level, then the decrease will be 11 to 16% (reaching half this value in 50 years) (Figure 8.8). The decrease in ozone content will be especially great at high altitudes. Note that Figure 8.8 shows only the effect of freons. If freon emissions had stopped in 1979, then the decrease would have been around 2% (see Figure 8.9).

It is very important to determine the combined effects produced by the increase in the concentrations and the emissions into the atmosphere of other halocarbons, CO, NO_x, CO_2, and bromides. Models [34] demonstrate that, when the annual production of nitrogen fertilizers (releasing N_2O) amounts to 200 million tons, the ozone destruction will increase (in combination with freons) by a factor of 1.1–1.2, compared with a constant level of N_2O.

It was also calculated that, with an increase of the CO_2 content in the atmosphere, the stratospheric temperature decreases. On the whole, a drop in the stratospheric temperature is associated with an *increase* in the amount of ozone. It is calculated that due to the CO_2 increase the ozone content should have already increased by 1–2% compared with the preindustrial period, and by the year 2030 it will increase by 6–7%.

The stratospheric temperature decrease and the change in ozone content (in percent) at different altitudes (and in different seasons) with an increase

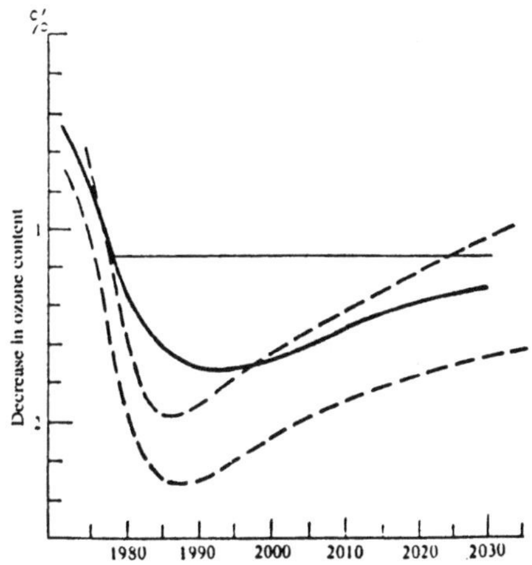

Figure 8.9. Variation of ozone content (%) if freon emissions had stopped in 1979, for intensive use of freons in 1973–1978 of about 10^6 tons/year, according to data of various authors [37].

in CO_2 concentration from 275 to 600 ppm are considered in [34].

Calculations of the combined effect of freons and CO_2 (for steady-state freon levels and a CO_2 concentration of 600 ppm) demonstrate that the ozone concentration in the upper atmosphere above 40 km decreases by 40% in winter, while the temperature drops by 18 K at an altitude of 48 km in summer.

The increase in CO_2 concentration to 600 ppm smooths over the general effect of a decrease in ozone content due to freon emissions from 10% to 7%. Therefore, the effect of an increased CO_2 concentration in the atmosphere on the ozone content in the upper stratosphere is opposite to the effect of freons [34]. It is also noted in [34] that, when 200 million tons of nitrogen fertilizers are produced per year, the ozone destruction will amount to 1.4% (if the NO_2 lifetime is assumed to be 12 years) and 4.1% (if the NO_2 lifetime is 40 years).

It is shown that the effect of all the other anthropogenic halocarbons (except freons) will reinforce the effect of the freon emissions by approximately one third.

The impact on the ozone layer of anthropogenic factors in their various combinations (using in the calculations improved values of the rates of the photochemical reactions) was considered in [20].

The freon emissions were assumed to remain at the 1980 level; the annual NO_2 increase is 0.2%; the number of flights of subsonic aircraft will increase tenfold (from 1975 to 1990, with a subsequent constant value). In this case the calculations indicate that the total ozone content in the Northern Hemisphere will change by less than 0.5% during the following few decades.

However, the stabilized ozone concentration at an altitude of 40 km will decrease by 40%, and it will increase by 25% at an altitude of 10 km. Such a change in the vertical ozone distribution can, of course, affect the climate significantly.

Ground-based and satellite measurements of the ozone content in the atmosphere enable us to analyze the existing trends. At present the accuracy of measurements of the mean annual ozone content using Dobson instruments is estimated to be 1.5%, although in practice the accuracy of the measuring system is worse than the value given (apparently it is around 4% [20]).

The total ozone content and its vertical distribution are measured with satellites (by measuring the backscattered UV solar radiation or IR radiation in the 9.6 μm band).

Because of the great natural variability, it is very difficult to analyze and identify any trends in the variation of the ozone content.

Data of the ground-based measurement system show that between 1965 and 1980 there were several episodes of increase and decrease in the total ozone content (within limits from one to several %) [34]. There are some indications (though not proven) of a possible connection between the changes in the ozone content and the solar cycle.

A significant (4%) ozone decrease in the layer from 32 to 48 km was observed just after the Fuego volcanic eruption in 1974 (the original values were reached again only in 1980).

In the tropospheric layer from 2 to 8 km ozonosondes have revealed an increase in ozone concentration by approximately 20% between 1967 and 1980 [20].

On the whole, observations and calculations show that since 1970 the total change in ozone content which can be attributed to anthropogenic impacts amounts to no more than 2%. This value is below the threshold of confident detectability.

Thus, although in the years to come anthropogenic effects will cause the total ozone content to change slightly, there may occur a significant ozone redistribution with height, which may result in significant climatic changes and other negative consequences. This indicates the sensitivity of the ozonosphere to the already existing anthropogenic impact. It is noteworthy that the maximum decrease in the ozone content should be expected at the end of winter and at high latitudes (this effect is 3 to 4 times stronger than the effect at low latitudes in summer) [34].

It should be emphasized that in general the upper atmosphere is very vulnerable. In addition to the already considered effects connected with the ozonosphere, we should note the occurrence in the upper atmosphere, in the ionosphere region, of zones with a lower electron concentration (so-called 'ionospheric holes'). There is typically the appearance or intensification of emissions which are not characteristic of the natural luminescence of the upper atmosphere. This effect is due to:

(1) the accumulation of various substances in the upper atmosphere because of diffusion (this was discussed above);
(2) the accumulation in the upper atmosphere of various substances from the launching of powerful rockets;
(3) the effect of electromagnetic radiation from transmitters.

Particular attention should be given to the emission (ejection) of water and hydrogen-bearing compounds (and other substances) during the launching of rockets. This can alter substantially the state of the ionosphere and its ability to transmit radiosignals over long distances, and, consequently, it can cause significant economical damage.

The extent of the anthropogenic changes in the hydrogen content at high altitudes is so great that it constitutes a disturbance of the global hydrogen balance in the upper atmosphere.

All this goes to show that the anthropogenic impacts are not only 'disturbing', but are also close to the 'critical' limit.

As already mentioned, the disturbance of the ozone layer leads to an increased influx of UV solar radiation. This increase is nonlinear and depends on the radiation wavelength.

Ultraviolet radiation covers a wide spectrum of wavelengths; the following three ranges of wavelengths are of particular importance for biological systems:

UV-A: 400 – 315 nm
UV-B: 315 – 280 nm
UV-C: $\leq$ 280 nm

UV-B radiation has a significant (and dangerous) biological effect, and at the same time it is very sensitive to changes in the ozone content of the atmosphere. The most important components of all living organisms, nucleid acids, absorb radiation in precisely this range.

It was found that a 1% change in the total ozone content results in a 1.6 to 3% change of the diurnal dose in the spectral range of active biological impact for most latitudes and seasons.

Figure 8.10 shows the dose increase (in %) of ultraviolet (UV-B) radiation with a decrease in the total ozone content at various latitudes. We note that the total increase in the dose relative to the equatorial dose is 15% in the middle latitudes.

Experiments [20] have demonstrated the harmful effect of UV-B radiation on the productivity of such agricultural crops as wheat, rice, soy beans, barley, and potatoes. Plants in their early growth stages are particularly vulnerable.

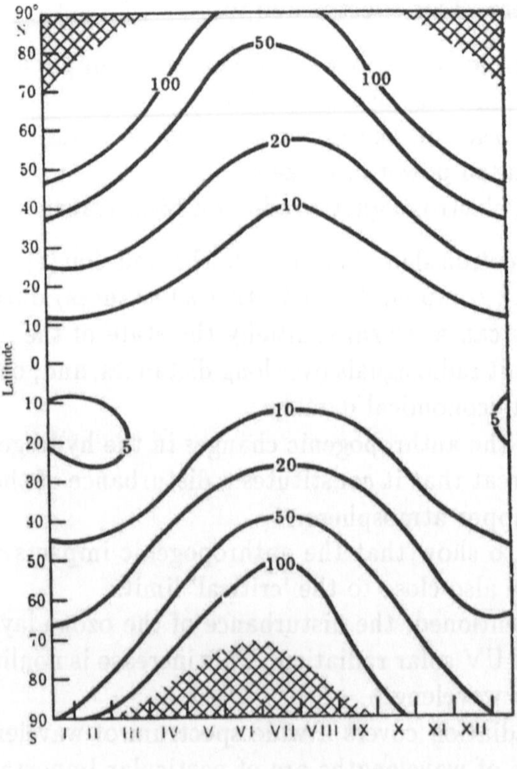

Figure 8.10. Increase in % dose of UV-B radiation at various latitudes during different seasons with decrease in total ozone content due to freon use at 1973–1976 level.

UV-B radiation penetrates into deep water (down to 20 m in clean water). A large number of marine organisms which are very sensitive to this radiation (eggs, larvae, and young fish, shrimps, crabs, many plants) inhabit the upper 20-meter layer of the ocean. Even the present levels of UV-B radiation are sufficient to suppress to phytoplankton productivity. An increase in UV-B radiation alters the composition of the ecosystems.

UV radiation affects human health significantly, promoting the formation of vitamin D_3 and producing sunburn, eye diseases, allergic reactions, and skin diseases, including cancer. Except for the formation of vitamin D, all these things are harmful to humans. A 1% increase in UV-B radiation increases the rate of nonmelanomic skin cancer by more than 2% (this has been statistically verified). There is evidence that an increase in hard UV radiation leads to the appearance of malignant melanomas. Particulary vulnerable to such diseases are light-skinned people living at low latitudes. UV radiation changes the response of the immunological system in both humans and animals.

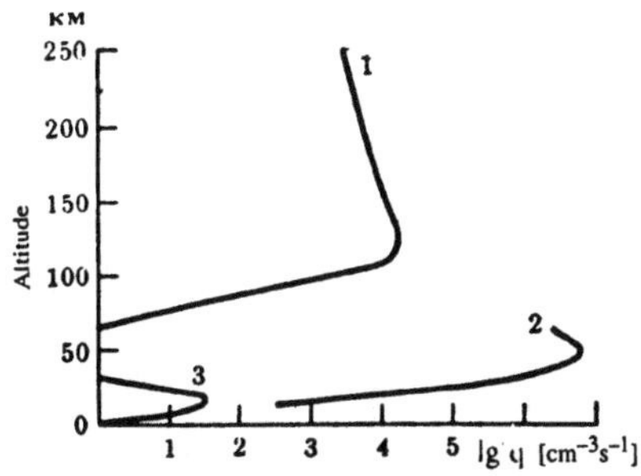

Figure 8.11. Rate of ion formation at various altitudes: (1) rate of ionization by hard ultraviolet solar radiation; (2) rate of oxygen photodissociation; (3) rate of ionization by galactic cosmic rays.

8.3. Ionization in the Atmosphere (Troposphere) and the Variation of Its Electrical Properties

The ionization of the air, observed against the background of active natural processes, is the most significant aspect of the anthropogenic impact on the atmosphere.

The ionization determines the electrical properties of the atmosphere, particularly the troposphere, and to a great extent the composition of the upper atmosphere as well, especially the ozonosphere.

The ionization of the air is produced by the ultraviolet and hard electromagnetic radiation of the Sun and by galactic cosmic rays. It develops as a result of oxygen photodissociation (the main process of ozone production in the stratosphere) and the emission of radio-isotopes (both natural and anthropogenic).

The importance of these radiation sources varies for different atmospheric layers. Figure 8.11 [26] indicates the rate of ion formation at various altitudes, the ions being produced by the ionization associated with the processes mentioned above.

The ionization produced by radioisotope radiation takes place primarily in the surface air layer, although individual long-lived isotopes can spread throughout the whole troposphere and can even penetrate into the stratosphere.

It is seen from the figure that galactic cosmic rays contribute substantially to the ionization of the air in the ozonosphere; this source also contributes significantly to the ionization of the underlying layers of the atmosphere.

Let us consider the effects of the ionization (including the anthropogenic impact) in the troposphere; the problem of anthropogenic changes in the ozonosphere was dealt with in Section 8.2.

It stands to reason that the ionization of the air in the troposphere determines its properties and affects its electrical conductivity. So far, however, the effects produced by a variation in the atmospheric electrical conductivity have not been sufficiently studied. This property of the atmosphere determines to a great extent thunderstorm activity, and it influences the formation of condensation nuclei and the growth of cloud droplets, and consequently also the development of cloud systems, the formation of aerosol particles (in particular sulfate particles), and their growth rate. In any case we can agree with the opinion of the author of [31] that a change in the electrical properties of the atmosphere by more than 10% leads to noticeable and even undesirable effects. This level can be considered to be the level of 'disturbing' impact.

The effect of natural and anthropogenic radiation on the extent of ionization and the electrical properties of the atmosphere has been discussed in detail in a number of recent papers [9, 26, 31].

The ionization under natural conditions is attributable primarily to cosmic radiation and depends considerably on the height: it decreases with the approach to the Earth's surface. The mean rate of ion production by cosmic radiation at sea level in the equatorial zone is 1.5×10^6 ion pairs/(m^3s). At high altitudes (approximately 50 km or higher) the ionization is so great that the electrical conductivity of the air is sufficient to equalize the potential of the entire Earth. This equalizing layer of the atmosphere can be considered to be one plate of a giant spherical condenser, with the Earth's surface as the other plate. The condenser, which becomes charged during electrical storms, constantly (and gradually) discharges due to the electrical conductivity of the air.

Besides cosmic radiation, ionization is produced by radioisotopes contained in the atmosphere itself. Natural radon-222 and its short-lived decay products (including lead-210) have the greatest effect on the ionization. Radon-222 enters the atmosphere from uranium-bearing rocks. Since the halflife of radon is 3.8 days, its maximum concentrations are observed within a limited distance from the source (no more than several hundred to a thousand kilometers). The air over the open ocean is practically free of short-lived natural radioisotopes; the ionization in these regions is mainly produced by cosmic radiation. The radon-222 concentration over ocean areas remote from the continents is from 0.1 to 0.5 pCi/m^3, whereas over the land it ranges from 50 to 200 pCi/m^3. The energy of all types of radiation from the decay of atoms in the chain from radon-222 to lead-210 is 2.1×10^7 eV per decay of radon-222. We find (assuming the energy of production of one ion pair to be 32.5 eV) that, due to radon-222 and its decay products, 2.4×10^3 to 12×10^3 ion pairs/(m^3s) are formed in the atmosphere over the ocean, while 1.2×10^6 to 4.8×10^6 ion pairs/(m^3s) are formed over the continent. Consequently, the ionization rate over the oceans is more than 99% determined by cosmic radiation, while over the land at sea level its contribution is only 20 to 50%.

Radioactive products from nuclear explosions and from nuclear power plants and enterprises are examples of man-made radioisotopes.

Most of these products settle on aerosol particles and are rapidly washed out of the troposphere by rain (the lifetime (residence time) of aerosol particles in the troposphere can be several days, or a maximum of 1 or 2 weeks).

Particularly important are long-lived isotopes of inert gases, which are difficult to remove from the troposhere; the major one of these is the radioisotope of krypton ^{85}Kr, with a halflife of 10.75 years. Krypton-85 is a fission product of fissile nuclei (uranium, plutonium, etc.), and it can be produced from nuclear power plants and nuclear weapons tests.

99.6% of the nuclear decays of Kr-85 are accompanied by β-emission, with a mean energy of 251 keV, 0.4% with an energy of 47.5 keV, and 0.4% of the decays are followed by γ-radiation with an energy of 514 keV.

Krypton-85 is one of the radioisotopes of the inert gas krypton, and its chemical properties do not differ from those of stable krypton isotopes. The concentration of stable krypton in the atmosphere is 3.29×10^{-4}% (by mass). Although recently the ability of inert gases to react chemically with highly active substances has been proven, under real atmospheric conditions such reactions practically do not occur. Krypton is slightly water-soluble (1.85×10^{-10} at equilibrium); it can dissolve in some organic substances, forming chèmical compounds known as clathrates.

There are various possible mechanisms of removal of Kr-85 from the atmosphere (excluding decay). The authors of [17] considered that krypton dissolves in seawater, is sorbed on the land surface, and absorbed by aerosol particles. The data obtained indicate that around 3% of the atmospheric krypton becomes dissolved in the World Ocean. Taking into account the gas-exchange rates at the sea-air interface and between the surface water and the deep water, it is estimated that no more than 0.1% of the annually produced krypton-85 is dissolved in the ocean.

Since the rates of the other mechanisms of krypton removal from the atmosphere are even lower, radioactive decay should be considered the single practical possibility for reducing the Kr-85 content in the atmosphere.

The boiling point of krypton is very low (-156.6°C), so it is present in the atmosphere only in the form of a gas.

Obviously, under conditions of a stationary equilibrium in the atmosphere, the accumulated amount of krypton-85 will be

$$Q = \tau q,$$

where q is the rate of entry of krypton-85 into the atmosphere; τ is the mean lifetime of krypton-85 in the atmosphere; $\tau = T_{1/2}/\ln 2$, and $T_{1/2}$ is the halflife of krypton-85, i.e., $\tau = 15.4$ years.

There are two mechanisms of krypton-85 formation: fission of heavy nuclei and neutron capture by a stable krypton-85 nucleus. The most probable

Table 8.4. Krypton-85 yields for various fissile materials and neutron energies.

Fissionable nucleus	Neutron characteristics	Yield per fission
Uranium-235	Thermal neutrons	0.0029
	Fission neutrons	0.0032
Uranium-238	Fission neutrons	0.0017
Plutonium-239	Thermal neutrons	0.0014
	Fission neutrons	0.00145

natural means of krypton-85 formation is spontaneous fission of natural uranium and induced fission via the interaction between natural uranium and the neutron component of cosmic radiation. Natural sources ensure the constant presence in the atmosphere of a very slight amount of krypton-85 with a total activity of only 10 to 15 Ci.

Beginning in 1945, krypton-85 has been entering the atmosphere as a result of nuclear tests. Estimates indicate that in all 3 to 5 MCi of krypton-85 from nuclear tests have entered the atmosphere. Of this amount, 1.5 to 2 MCi still remain in the atmosphere [11,17].

A significant quantity of krypton-85 has accumulated in the atmosphere from the operation of nuclear reactors. In order to estimate the krypton-85 content in the atmosphere, we have to know its yield from nuclear fission in various sources.

Table 8.4 gives the yields of krypton-85 for various fissile materials and neutron energies.

During the operation of nuclear reactors practically all the krypton-85 remains in the fuel elements, and only less than 1% of the krypton-85 produced diffuses through the element walls and enters the atmosphere [11, 17]. As a rule, the depleted fuel elements are kept for a long time before being reprocessed (half a year or longer), so as to reduce the total activity of the fission products. At the plants where the fuel elements are reprocessed the depleted fuel is dissolved and krypton-85 enters the atmosphere. Rough estimates have shown that, just due to the plutonium production in the USA before 1964, 15 MCi of krypton-85 were released into the atmosphere [17]. Allowing for plutonium production in other countries, we can conditionally estimate the worldwide krypton-85 emission (due to this source) to be twice the above amount, i.e., 30 MCi. Of this, about 15 MCi is still present in the atmosphere.

Since the early 1970s, and the beginning of the intensive development of nuclear power engineering and, accordingly, the reprocessing of nuclear power-plant fuel, the worldwide output of nuclear power plants (NPPs) has increased from 24 GW(e) in 1970 [17] to 80 GW(e) in 1976 [11].

The release of krypton-85 into the atmosphere from NPPs depends on their power output and mean loading factor, on the neutron spectrum of the

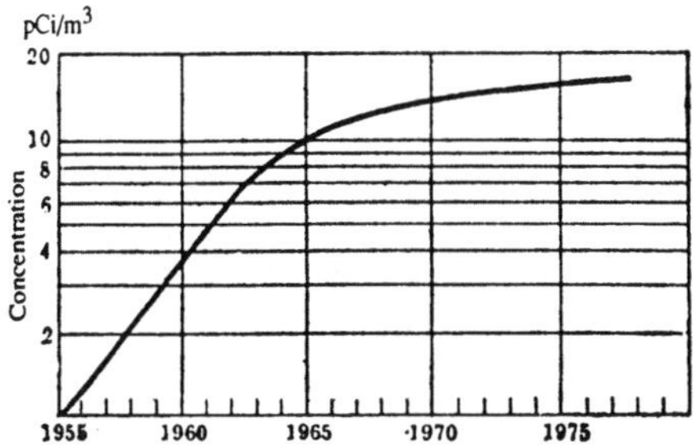

Figure 8.12. Variation of krypton-85 concentration in atmosphere.

reactors (i.e., the reactor type), on the type of fuel used, on the regeneration cycle of the depleted fuel, and on the efficiency of the traps. It is noteworthy that in practice neither at the NPPs nor at the plants reprocessing the fuel elements was the krypton-85 trapped [11], so all the krypton-85 left in the fuel elements after the holding time entered the atmosphere.

At present practically all the NPPs in the world are based on water-cooled water-moderated reactors (water-water reactors) using a mixture of uranium-235, uranium-238, and plutonium-239. According to the data in [11], the plutonium contribution is around 40%.

If the mean rate of krypton-85 formation during the operation of a water-water reactor is assumed to be 300 Ci/MW(e) per year, for a mean loading factor of 63%, then the krypton production in 1976 can be estimated as 15 MCi (19). Such an approach makes it possible to estimate the krypton-85 concentrations and to forecast future concentrations.

Since the late 1950s, the krypton-85 concentrations have been measured on a regular basis. These data, obtained by different investigators between 1954 and 1978 [6, 17, 27, 42] (as an averaged curve), are shown in Figure 8.12. Inspection of the figure shows that during the period before 1965 a drastic increase in activity was observed, mainly due to plants producing plutonium and to a lesser extent due to nuclear weapons tests. After 1965 the growth rate dropped significantly, which can be ascribed to the reduced production of plutonium. For 1965 the concentration can be estimated as 10 pCi/m^3, and for 1976 as 16 pCi/m^3. According to the data in [50], the concentration in 1977 was 25.1 pCi/m^3. Thus, over a period of 11 years, the atmospheric krypton content changed on the whole by 24 MCi (assuming the volume of the whole atmosphere under normal conditions to be 3.96×10^{18}m^3).

Practically all of the krypton-85 produced is released into the atmosphere of the Northern Hemisphere. This results in some irregularity of its distribution over the global atmosphere. According to data in [6], the krypton-85 concentration in the Southern Hemisphere is lower by a factor of 1.3 to 1.4 than in the Northern Hemisphere. The height distribution of krypton-85 is

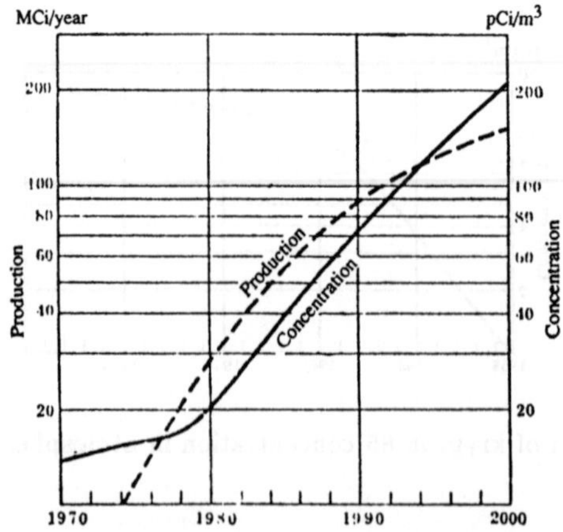

Figure 8.13. Predicted krypton-85 production (emission into atmosphere) and increase in its concentration.

practically uniform up to altitudes of 20 to 25 km [42].

The release of krypton-85 from NPPs since the 1960s will presumably determine its content in the atmosphere in the future. Obviously, the concentration increase will be closely related to the worldwide development of nuclear power engineering. In order to provide sufficiently reliable forecasts of the krypton-85 concentrations, we must have information about the predicted growth rates of NPP outputs and about the types of reactors which will be used in the near future, as well as the tempos and the technology of fuel-element reprocessing.

The perspectives for the development of nuclear power engineering have been discussed in a number of works [7, 18, 19, 22]. However, the actual development of this field differs somewhat from some of the predictions.

According to the data of [2], water-cooled water-moderated reactors using uranium fuel will predominate up to the year 2000 (80% of the power by 2000). In the 1990s high-temperature gas-cooled reactors and fast breeder reactors with liquid-metal coolants will begin to be widely used. Their contribution to the total power output in the year 2000 will be about 10%. After 2000, the role of breeders will become more significant.

The output of the nuclear power plants in 1975 was 120 GW(e). By the year 2000 the total output should reach 1270 to 1700 GW(e) [48]. Here the lower estimate seems to be more realistic.

Figure 8.13 shows the assumed relationship between the rate of krypton-85 production and its concentration in the atmosphere for the period up to the year 2000. To plot this curve, it was assumed that the specific rate of krypton-85 production was 300 Ci/MW(e) per year, and that the reactor loading factor will gradually decrease from 0.63 in 1975 to 0.55 in the year 2000. This computation was based on the development of nuclear power engineering predicted in [19, 22]. Note that, according to [17, 45], the krypton-85 concentration in the atmosphere will reach 3 nCI/m^3 by the beginning of the next century.

The relationship in Figure 8.13 is based on the assumption that the atmospheric krypton-85 is completely mixed and that the portion of the reprocessed fuel elements will increase from 50% in 1980 to 60% in 1990, and then to 80% by the year 2000. It also follows from Figure 8.13 that, compared with 1975, the krypton-85 concentration will increase by a factor of 13 by the year 2000 and will then be approximately 200 pCi/m^3. Obviously, however, measures may be taken in the future to limit, or to stop completely, the release of krypton-85 into the atmosphere.

Let us estimate the probable effect of the increasing krypton-85 concentration in the atmosphere on the ionization of the air and on the electrical properties of the atmosphere.

A peculiarity of the ionization produced by the krypton-85 radiation is its relative uniformity per unit air mass, in accordance with the concentration distribution in the atmosphere, whereas the ionization produced by cosmic rays decreases with decreasing height and the ionization produced by natural radioactivity decreases with an increase in height. Krypton-85 already begins to contribute significantly to the ionization of the air (compared with the natural ion production) when its concentration at sea level is around 1 nCi/m^3 (1000 pCi/m^3). As shown in [31], when the krypton-85 concentration is 3 nCi/m^3, a total of 8.55×10^5 ion pairs/(m^3s) are formed. This increases the electrical conductivity of the air over remote ocean regions by 43%. Integration of the possible effects over the land and ocean indicates that the total resistance of the Earth's entire air column is reduced by almost 10%. As already noted, precisely such a change can be considered the threshold of the 'disturbing' impact.

It should be kept in mind here that the effect of krypton-85 will not be limited to the given spatial region. Its effect will be felt everywhere in the world. The produced effects will persist and will develop for decades to come, because of the long halflife of Kr-85.

If the distribution is uniform, then the total content of Kr-85 in the atmosphere (in this case) will be 1.2×10^4 MCi. To keep the concentration constant at a level of 3 nCi/m^3, 780 MCi have to enter the atmosphere annually, i.e., 6 times as much as we estimated will enter in the year 2000.

The results of the possible effect of Kr-85 on the atmosphere should be compared with the other negative impacts of Kr-85, with its radiation effect on man.

When krypton-85 radiation affects the human organism, the critical organ is the skin. This is because of the slight penetrating ability of this radiation: the mean depth of penetration of beta-radiation into the tissues is around 0.2 mm. The present norm for the maximum permissible concentration of krypton-85 is 2600 nCi/m^3 (radiation safety standard). This value was obtained from calculations for a limited part of the population living or working in the vicinity of a nuclear power plant. For the global population as a whole, one tenth of the above value, i.e., 260 nCi/m^3, can be taken as the maximum permissible level. This value is almost two orders of magnitude higher than the level of the 'disturbing' effect associated with ionization and practically three orders of magnitude higher than the concentration predicted for the year 2000.

It follows from the foregoing that the 'traditional' way of estimating the danger of the radioactive isotope Kr-85 by its radiation effect is by no means the most 'critical' link subjected to this effect ([8] and others). Only the ecological approach, involving a comprehensive analysis of the environment, makes it possible to estimate the real damage correctly.

CHAPTER 9

The Transport of Pollutants over Long Distances. The Ecotoxicology of Acid Rain

The long-range (transboundary) transport of significant quantities of various manmade pollutants has made it necessary to solve a number of complicated ecological and technological problems [34].

These problems first arose in connection with the long-range transport of radioactive substances in the atmosphere and the appearance of remote and global radioactive fallout [6, 10]. Such transport was distinctly latitudinal in nature, although meridional transport was observed as well. The exchange of radioactive substances between the Northern and Southern Hemisphere was even observed.

The experience gained in various countries in organizing a network for the observation of global radioactive fallout, together with research on the long-range transport of radioactive substances in the atmosphere and the migration of radionuclides in different parts of the environment [3, 10, 28], were useful for organizing a global system for monitoring any anthropogenic pollutants.

At present the atmospheric spreading of many pollutants over long distances is observed. The possibility of this is determined by the amount of pollutants emitted into the atmosphere and by the lifetimes of these substances and their transformation products. Of prime importance are substances possessing a high toxicity.

In view of these factors, priority should be given to the long-range transport of sulphur dioxide and its transformation products, nitrogen oxides and their transformation products, heavy metals (particularly mercury), pesticides, and radioactive substances.

The first of these substances, when they interact with cloud water droplets and falling rain drops, produce acids and then acid salts, which often strengthen their toxic properties. The washing of pollutants out of the atmosphere is accompanied by significant negative ecological effects [7]. For instance, the washing out of sulfur dioxide and nitrogen oxides leads to the production of sulfuric and nitric acids (and their salts), so that acid rain falls. At present this phenomenon is widespread, and it has led to a considerable acidification of the environment and to significant ecological changes over enormous territories. Thus, in recent years the problem of acid rain and acidification of the environment has arisen [11, 21, 34]. Acid rains can fall at points very far (many hundreds or thousands of kilometers) away from the source of the primary pollutant emission.

However, it should be stressed that this problem is part of the problem of long-range transport, of the spreading of anthropogenic pollutants throughout the world.

The largest-scale anthropogenic pollution sources producing acid rain and the long-range transport of pollutants are discharges into the atmosphere of sulfur dioxide (sulfurous anhydride) and nitrogen oxides. The worldwide anthropogenic emissions of sulfur dioxide now reach 140 to 200 million tons per year [33, 34, 53]. The wide (and ever-increasing) use of coal for energy production, taking into account the high cost of intercepting the sulfur dioxide in the outgoing flue gases or desulfurizing the fuel, presupposes a further increase in the worldwide emission of sulfur. According to some forecasts, anthropogenic SO_2 emissions into the atmosphere should reach 250 to 300 million tons by the year 2000.

The present global sulfur cycle is, according to [33], produced in the following way.

Around 120 million tons of sulfur are extracted annually by man from the lithosphere, together with fossil fuels and other sulfur-bearing materials. Seventy million tons of this are then emitted into the atmosphere during the burning of fuel and processing of ore. The total amount of sulfur washed away from the continents into the ocean via rivers is 224 million tons, of which 109 million tons are of anthropogenic origin. The total transfer of sulfur from the ocean water to the bottom sediments is about 130 million tons per annum. Seventy nine million tons of sulfur enter the atmosphere from the ocean (the atmospheric transport from the continents to the ocean is 27 million tons), while 90 million tons pass from the atmosphere to the ocean (about 10 million tons of which are anthropogenic sulfur). Furthermore, the annual transfer of ocean biomass, via the mineralized remains of organisms (most of which is assimilated again), amounts to 324 million tons.

Thus, the total amount of anthropogenic sulfur in the annual flux from the lithosphere and ocean into the atmosphere amounts to about 30% (i.e., 70 million tons out of 203 million tons). The total sulfur flux from the continents and from the atmosphere into the ocean amounts to 334 million tons (119 million tons of which are anthropogenic sulfur.)

Figure 9.1 indicates schematically the major sulfur fluxes in nature [33].

Since there are many common aspects of the long-range transport of different substances (both gaseous substances and substances transported by highly dispersed aerosols) and their deposition onto the Earth's surface, let us consider in more detail the theory of transport of such substances, taking

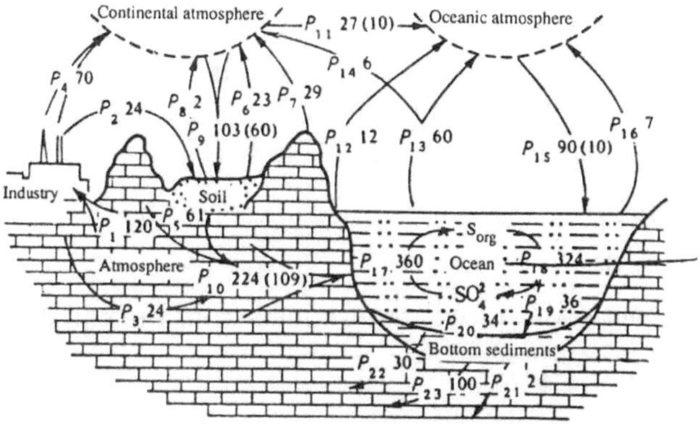

Figure 9.1. Major sulfur fluxes in nature (in millions of tons per year, anthropogenic fluxes are given in brackets).
P_1 flux from lithosphere for all means of extraction; P_2 flux into soil with fertilizers; P_3 flux with wastewater; P_4 anthropogenic flux into atmosphere; P_5 flux produced by water erosion; P_6 flux of biogenic sulfur; P_7 volcanic sulfur; P_8 flux of sulfur with dust (weathering); P_9 flux onto Earth's surface produced by settling from atmosphere; P_{10} flux with river water; P_{11} atmospheric flux from continents into ocean; P_{12} flux of biogenic H_2S from shelf shores; P_{13} flux into atmosphere with seaspray; P_{14} flux of marine sulfur to continents; P_{15} flux from oceanic atmosphere into ocean; P_{16} sulfur emission from ocean surface; P_{17} sulfur in biomass of marine organisms; P_{18} mineralized sulfur in dying organisms; P_{19} flux of organic sulfur to ocean bottom; P_{20} sulfur oxidation to sulfates when reentering seawater; P_{21} organic substances buried in marine bottom sediments; $P_{22,\,23}$ sulfates of sulfur buried in marine bottom sediments.

into account their peculiarities (first, radioactive substances, and then sulfur and nitrogen oxides).

9.1. Transport Theory

Let us consider the regularities in the long-range atmospheric transport of radioactive products from a single source, for instance, from a nuclear explosion, as well as the regularities in their depositions.

The pollutant concentration in the surface air layer at a large distance from the source can be calculated using the semiempirical equation of eddy diffusion [5, 6]

$$\frac{\partial q}{\partial t} - v_g\frac{\partial q}{\partial z} + u\frac{\partial q}{\partial x} = w_x\frac{\partial^2 q}{\partial x^2} + w_y\frac{\partial^2 q}{\partial y^2} + w_z\frac{\partial^2 q}{\partial z^2} \tag{9.1}$$

for the boundary condition

$$w_z\frac{\partial q}{\partial z} + v_g q = \beta q. \tag{9.2}$$

Here q is the pollutant concentration in the air; w_x, w_y, w_z are the respective eddy-diffusion coefficients; β is a coefficient characterizing the interaction between the pollutant and the underlying surface, v_g is the particle deposition rate (at a great distance $v_g \to 0$; in the calculations it is assumed to be 1 cm/s, the so-called rate of 'dry' deposition), and u is the wind velocity. The order of magnitude of w_i is 10^4 to 10^5 cm^2/s.

The eddy-diffusion coefficients are defined as

$$w_{x,y} = \frac{1}{2}\frac{\mathrm{d}\sigma^2_{x,y}(t)}{\mathrm{d}t}, \tag{9.3}$$

where the horizontal variance is determined from the equation

$$\sigma^2_{x,y}(t) = \sigma_0^2 + \alpha^2 u^2 t^2 \tag{9.4}$$

(σ_0^2 is the initial dispersion of products in the cloud, u is the wind velocity, and α is a coefficient depending on the cloud height).

The vertical flux for a weightless pollutant is

$$w_z = \frac{\partial q}{\partial z} = \beta q. \tag{9.5}$$

The density of radioactive fallout for the ith substance (dry, i.e., without precipitation) is determined using the expression

$$P_d(x,y) = Q_{0_i} e^{-\lambda_i x/u} \beta \int_0^\infty q(x,y,0,t;h)\,\mathrm{d}t, \tag{9.6}$$

where Q_{0_i} is the amount of the substance deposited from the cloud at a distance.

The 'wet' (with precipitation) fallout is calculated allowing for the washout duration, the precipitation intensity, and the coefficient of capture of various particles by precipitation. Simplified (somewhat overestimated for $P_w(x)$) calculations yield

$$\frac{P_w(x)}{P_d(x)} = \frac{h_0 u}{\sqrt{2\pi}\beta\sigma_x} e^{h^2 u/4w_z x}, \tag{9.7}$$

where h_0 is the height of the washout layer.

Figure 9.2 shows the densities of the 'dry' and 'wet' radioactive fallout, calculated using the above formulas for a single source, corresponding to 1 kton for the debris radioactive products [6]. During the calculation the values $w_x = w_y = x$ were used, where x is the distance from the explosion site, $\gamma = 0.06$ cm/s; the mean wind velocity $u = 40$ km/hr, and the height of ascent of a cloud $h = 1$ km. During the calculation of the wet fallout, the coefficient of capture of 1 μm particles by rain droplets was taken to be 0.01.

Figure 9.3 gives some more detailed calculation data for a source equal to q arbitrary units, for various cloud heights h and washout coefficients λ [6].

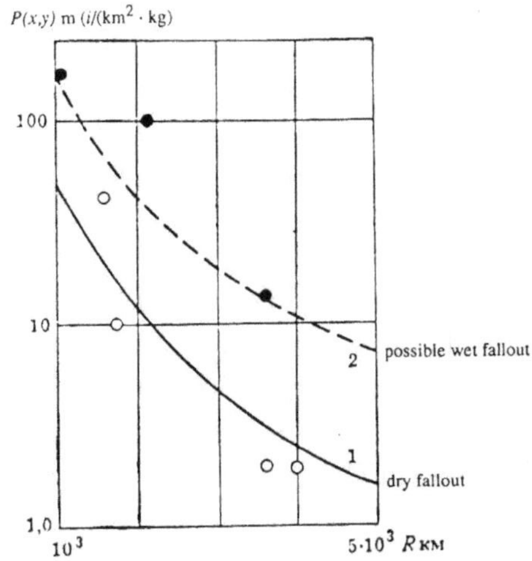

Figure 9.2. Density variation $P_{(x,y)}$ of radioactive fallout with distance ($h = 1$ km, $w_z = 10^5$ cm^2/s).

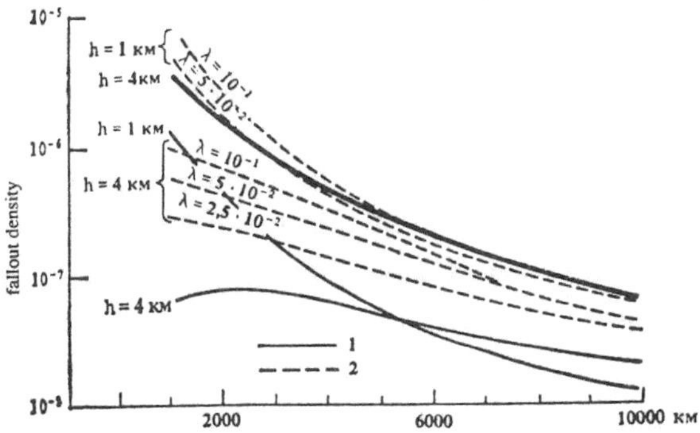

Figure 9.3. Variation of density (q km^{-2}) of dry (1) and wet (2) fallout with distance for $h_0 = 2$ km, $u = 40$ km/hr, $w_z = 10^5$ cm^2/s, and various values of λ.

These figures show that, at a considerable distance from the explosion site, the wet fallout (with precipitation) can exceed the dry fallout considerably.

The calculated data in Figure 9.2 agree with the experimental data [1].

In order to understand the peculiarities of the long-range atmospheric transport of sulfur (or nitrogen) compounds and the formation of acid rain, it is necessary to keep track of their 'fate'. A number of studies have been made of this problem, and in particular in [4, 35].

The horizontal transport of sulfur (or nitrogen) compounds is determined by wind currents, and it can be calculated from the available information on the structure of the baric fields at various levels. The existing European network of meteorological stations and aerological sounding stations can, in principle, provide such information. In the USSR the transport of polluted air masses is observed at three baric levels: 1000, 925, and 850 hpa (mb). We should emphasize the importance of using the 925 mb level (which is practically not used outside the USSR), since the main mass of pollutants is transported at precisely that level [34].

Let us consider the theory of the atmospheric transport of sulfur dioxide and its transformation products, and their washing down to the Earth's surface. We will use a model constructed by us for the operational assessment of transboundary flows of anthropogenic pollutants [4].

The concentrations of SO_2 and its transformation products are calculated from the following formula (in the coordinates of the spreading 'jet'):

$$c(r,l,z,t) = \frac{Q}{\sqrt{2\pi}\sigma_l u}\exp(-l^2/2\sigma_l^2)\psi(z,t)A(t)B(t), \tag{9.8}$$

where $c(r,l,z,t)$ is the concentration; Q is the source strength; r is the distance along the jet axis; σ_l^2 is the pollutant dispersion in a direction z perpendicular to r; u is the rate of the jet 'elongation'; $t = r/u$ is the 'age' of the fraction of pollutant in the jet; $\psi(z,t)$ is the distribution of the pollutant concentration in the vertical direction; $A(t)$ is the chemical transformation coefficient; $B(t)$ is the coefficient of washout by precipitation and dry deposition of pollutant.

In the model used by the Eastern European Synthesizing Meteorological Center (in the Soviet Union) to calculate the transboundary transport in Europe, the pollutant concentration is integrated over the height:

$$p(r,l,t) = \int_0^\infty c(r,l,z,t)\,\mathrm{d}z. \tag{9.9}$$

For the calculations we will use about 600 sources, located at points on a grid 150 × 150 km in size. The distance covered by the jet in four days is assumed to be its 'length'. The flows are calculated through an ideal boundary, consisting of rectilinear segments 150 km long.

The $\psi(z,t)$ distribution in formula (9.8) is obtained for certain boundary conditions, the first of which determines the nature of the interaction between the pollutant and the underlying surface ($z = 0$)

$$\left[w_z\frac{\partial\psi}{\partial z} + v\psi\right]_{z=0} = v_g[\psi]_{z=0}, \tag{9.10}$$

where w_z is the coefficient of vertical eddy diffusion, v is the velocity of fall of pollutant particles in the atmosphere ($v \to 0$); v_g is the velocity of dry deposition (accomodation) of a weightless pollutant onto the underlying surface.

The second boundary condition is $\psi \to 0$ at $z \to \infty$.

If we integrate function $\psi(z,t)$ over the entire height of the layer being considered, then the function obtained

$$\zeta(t) = \int_0^\infty \psi(z,t)\,dz \tag{9.11}$$

will represent the fraction of the substance of age t which remained in the atmosphere, while the fraction of the pollutant absorbed on the underlying surface will be determined by the difference $1 - \zeta(t)$.

Dry depositions are determined by the turbulence and by the nature of the underlying surface.

Of course, individual molecules and highly dispersed particles (fractions of a μm or several μm in size) are practically not deposited (because of sedimentation); however, when they collide with the Earth's surface, they are absorbed by it (adsorption, chemosorption). Therefore, the ratio between the pollutant flow deposited in this manner on the Earth's surface and its volume concentration in the surface air layer (which has dimensions of velocity) can be considered to be the effective rate of dry deposition of a weightless pollutant. This quantity was widely used to calculate the radioactive fallout far away from a source [1, 6, 10].

In [1, 6] the value of v_g was assumed to be 7 to 10 mm/s; in [10] the range for v_g was different: from fractions of a mm/s to 14 mm/s, while the effective rates of removal of atmospheric pollutants via dry and wet deposition ranged from several mm/s to 30–40 mm/s.

Obviously, the linear velocity of absorption (deposition) v_g thus introduced to describe the process of dry deposition (uptake) at the underlying surface will depend on the nature and the physicochemical characteristics of the underlying surface, as well as on the degree of turbulence of the lower layer of the atmosphere and the properties of the absorbed gases and particles.

The value of v_g is different for different surfaces: for soils it is 5 to 10 mm/s, for fresh water 5 mm/s, for snow 1 mm/s, for dry vegetation 2 to 5 mm/s, and for wet vegetation (grasses and bushes) it can be up to 30 mm/s [6, 10, 26].

The mean height of the distribution layer of SO_2 and sulfates for European conditions is 1200–1500 m [20, 34] (and no more than 2000 m [31]).

Figure 9.4 shows the functions of dry removal of atmospheric sulfur, calculated for the following parameter values: $w_z = 0.05$ km^2/h; $h = 0.1$ km; $v_g = 0.03$ km/h (0.8 cm/s) for SO_2; $v_g = 0.0036$ km/h (0.1 cm/s) for sulfates and sulfuric acid. Here, by 'sulfates' we mean sulfate ions coupled with hydrogen or any other cation; the removal of sulfuric acid is not specially singled out, since it is represented in the atmosphere by aerosol particles of the same size as neutralized sulfates.

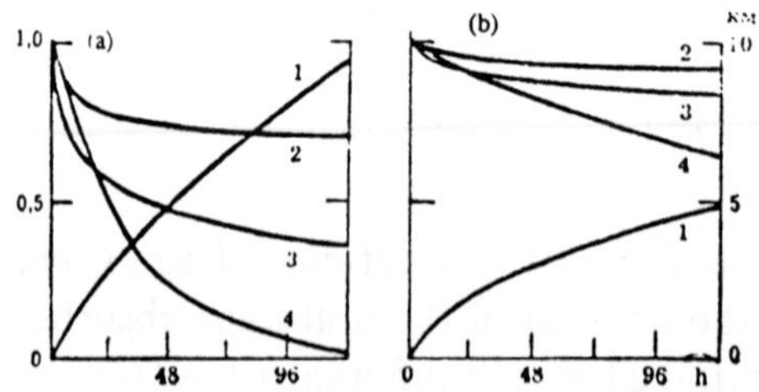

Figure 9.4. Functions of dry removal of sulfur compounds from atmosphere. (a) SO_2, $v_g = 0.8$ cm/s; (b) SO_4^{2-}, $v = 0.1$ cm/s; (1) $H_l(t) = \xi(t)/\psi(0,t)$; (2) $\exp(-v_g t H_l)$; (3) $\zeta(t)$; (4) $\exp(-v_g t/lk_m)$.

Figure 9.4 also gives values of the effective height of the layer of travel of the pollutants $H_l(t) = \xi(t)\psi(0,t)$, calculated using Equation (9.2) and the plots of two other functions of dry removal $\exp(-v_g t/l)$ and $\exp(-v_g t/H_l)$.

9.2. Chemical Transformation in the Atmosphere and the Formation of Acid Rain

The acidic and alkaline characteristics of rainwater and the aerosol are determined by the presence in a solution of free ions of hydrogen H^+ or OH^-, produced by dissociations of acids and bases. Under natural conditions rain precipitation is neutral or weakly acidic in nature, depending on the amount of dissolved salts of natural origin and on the amount of dissolved carbon dioxide.

Very slightly mineralized rain precipitation, in a state of equilibrium with CO_2 at its mean atmospheric concentrations, has a pH of about 5.6, and since $\text{pH} = -\log[H^+]$ the above pH value corresponds to a concentration of free hydrogen ions 2.5×10^6 g-ion/l.

In addition to CO_2, an unpolluted atmosphere can contain a number of natural acid-producing compounds: nitrogen oxides (and nitric acid), hydrogen sulfide, sulfur dioxide (and sulfuric acid), and various organic acids.

The presence of these compounds reduces the pH of the precipitation. At the same time, the presence of atmospheric dust, particularly in areas with alkaline rocks (soils), increases the pH. The sea spray over the ocean also increases the pH of the atmospheric precipitation.

The acidic characteristics of precipitation (rain) are due both to natural reactions inside the atmosphere and to the anthropogenic emission of acids (sulfur and nitrogen oxides, producing acids and their salts, and sometimes hydrogen chloride and hydrogen fluoride, etc.).

Let us cite some examples of (natural) acid reactions in nature [34]:

oxidation

$$NH_4^+ + 2O_2 = NO_3^- + H_2O + 2H^+ \tag{9.12}$$

hydrolysis

$$Al_2(SO_4)_3 + 6H_2O = 2Al(OH)_3 + 6H^+ + 3SO_4^{2-} \tag{9.13}$$

ion-exchange reactions

$$Ca^{2+} + x\begin{cases} H^+ \\ H^+ = 2H^+ + xCa^{2+}. \end{cases} \tag{9.14}$$

An example of an alkaline reaction observed in nature is

$$CaCO_3 + H_2O + CO_2 = Ca^{2+} + 2HCO_3^-. \tag{9.15}$$

Emissions of SO_2 into the atmosphere are produced by the burning of sulfur-bearing fossil fuel (coal, oil), as well as by the industrial processing of sulfur-bearing raw materials, etc.

The emission of SO_2 into the atmosphere can give rise to the following reactions:

(1) gas phase molecular oxidation;
(2) gas phase oxidation by radicals;
(3) oxidation on the surface of solid particles;
(4) oxidation of liquid-droplets particles.

Reactions of the first type progress slowly (the reaction rate $K_1 = 10^{-3}$ h^{-1}, which corresponds to a lifetime $\tau = 1/K = 1000$ h).

Reactions of the second type depend on the time of day, the season, and the latitude (radicals are formed as a result of photochemical reactions) $K_2 = 0.015\ h^{-1}$, $\tau \approx 70$ h during the day and in summer at a latitude of 45°).

The gas-phase oxidation of sulfur compounds at night can apparently be supported only by molecular reactions. Such gas-phase oxidation proceeds much more slowly than oxidation by radicals.

Reactions of the third type (heterogeneous reactions) depend on the number of particles, on their sizes, and on the physicochemical characteristics of the surface. For an industrial region $K_3 \approx 0.003$ to $0.01\ h^{-1}$ and $\tau \approx 100$ to 300 h. This reaction is sharply accelerated under conditions of increased humidity.

Reactions of the fourth type depend on the probability that a molecule will enter the liquid phase (droplets in clouds and fog); this reaction can be catalytic; taking into account the alternation of dry and rainy (or foggy) time periods, $K_{4,\text{eff}} = 0.007\ h^{-1}$ and $\tau \approx 140$ h.

The following reactions are examples of gas-phase molecular oxidation reactions (first type):

$$SO_2 + O_2 \rightarrow SO_3 + O^{\cdot}, \qquad (9.16)$$

$$SO_2 + O_3 \rightarrow SO_3 + O_2. \qquad (9.17)$$

Reactions of gas-phase oxidation by radicals (second type):

$$SO_2 + OH^{\cdot} \rightarrow HSO_3^-, \qquad (9.18)$$

$$SO_2 + O^{\cdot} + M \rightarrow SO_3 + M. \qquad (9.19)$$

Reactions of oxidation at a solid surface (third type):

$$(SO_2)_{ads} + 2(OH^{\cdot})_{ads} \rightarrow [(OH)_2SO_2]_{ads} \rightarrow (SO_4^{2-})_{ads}. \qquad (9.20)$$

Reaction of oxidation in the liquid phase (fourth type):

$$SO_2 + H_2O + O_3 \rightarrow H^+ + HSO_4^- + O_2. \qquad (9.21)$$

All these reactions lead to the formation of sulfuric acid (SO_3 is almost instantly transformed into sulfuric acid).

It is noteworthy that the above K values are not constant; they depend on the distance from the emission source. The rate of oxidation of sulfur decreases with an increase in the distance from the source.

Generalizing the above reaction coefficients (and oxidation mechanisms), we see that the total effective rate of chemical transformations of SO_2 in the atmosphere is characterized by a mean value $K_\Sigma = 0.027\ h^{-1}$ or $\tau_\Sigma = 37$ h [11].

The value is typical of summer and daytime conditions (at night the oxidation rate drops sharply). Sulfuric acid is still neutralized significantly in the atmosphere (in the liquid-droplet phase or on the surface of particles) mainly by ammonia. For the European region the corresponding values are $K = 0.003\ h^{-1}$ and $\tau \approx 33$ h (for neutralization).

For sulfate removal the determining factors are the concentration and surface area of the water droplets. For SO_2, in addition to the above, the pH of the rainwater and the presence of ammonia and other pollutants in the air are of great importance.

The process of removal can take place either inside the cloud (intracloud scavenging) or under the cloud (subcloud scavenging). In the former case a particle either serves as a condensation nucleus for a developing droplet or else is captured by droplets in the cloud. In the latter case a particle (substance) is captured by a falling drop.

The portion of the pollutant $\alpha(t)$ scavenged by precipitation can be estimated using the relation

$$\alpha(t) = \alpha_0 e^{-\Lambda\sqrt{I} f_t}, \qquad (9.22)$$

Table 9.1. Rates of transformation and scavenging of SO_2, H_2SO_4, and sulfates from atmosphere.

Substance	Process	Factor K_i, h^{-1}	Lifetime, h
SO_2	Oxidation	0.027	37
	Removal by precipitation	0.01	100
	Absorption by underlying surface	0.015	70
	Scavenging from atmosphere as a whole	0.052	19
H_2SO_4	Neutralization	0.03	33
	Removal by precipitation	0.0125	80
	Absorption by underlying surface	0.005	200
	Scavenging from atmosphere as a whole	0.0475*	21
Sulfuric-	Removal by precipitation	0.0125	80
acid salts	Absorption by underlying surface	0.005	200
	Scavenging from atmosphere as a whole	0.0175	60

* The value given in [11] is 0.037.

where: $\Lambda(SO_2)$ is a proportionality factor (Langmuir coefficient) equal to 10^{-5} s^{-1} (for SO_2), $\Lambda(SO_4^{2-}) = 10^{-4}$ s^{-1}; I is the precipitation intensity (mm/h), f_t is the fraction of the time required for a pollutant of age t to be scavenged by precipitation, and α_0 is the initial quantity of pollutant.

The quantities f_t and I are climatological characteristics of the region.

The rate of scavenging depends substantially on the dispersion composition of the particles.

The sulfate-particle sizes vary from 0.1 to 0.5 μm; the rates of scavenging (removal) from the atmosphere are nonlinear in this interval.

For the European region the scavenging factor K_{scav} for SO_2 is assumed to be 0.01 h^{-1}, and for sulfates $K_{scav} = 0.0125$ to 0.015 h^{-1}.

As already mentioned, the intensity of dry scavenging of SO_2 and sulfates from the atmosphere depends mainly on the turbulence in the lower layer of the atmosphere and on the roughness and the physico-chemical characteristics of the underlying surface.

The rates of transformation and scavenging of SO_2 and sulfates are generalized in Table 9.1.

It is more complicated to estimate for nitrogen oxides and nitric acid the rate of removal by precipitation and the rate of uptake by the underlying surface.

Taking into account that, in the atmosphere, nitric acid molecules are easily captured by liquid or solid particles, we assume the scavenging rate for nitric acid to be the same as for sulfuric acid, while the rate of uptake by the underlying surface is the same as for SO_2.

The rates of the major processes of transformation and scavenging of NO_x

Table 9.2. Rates of transformation and scavenging of NO_x and HNO_3 from atmosphere.

Substance	Process	Factor K_i, h^{-1}	Lifetime, h
NO_x	Oxidation to HNO_3	0.12	8.3
	Interaction with organic compounds	0.025	40
	Removal by precipitation	0.01	100
	Uptake by underlying surface	0.01	100
	Scavenging as a whole	0.165	6
HNO_3	Neutralization	0.035	30
	Removal by precipitation	0.0125	80
	Uptake by underlying surface	0.015	70
	Scavenging as a whole	0.0625*	16

* The value given in [11] is 0.042.

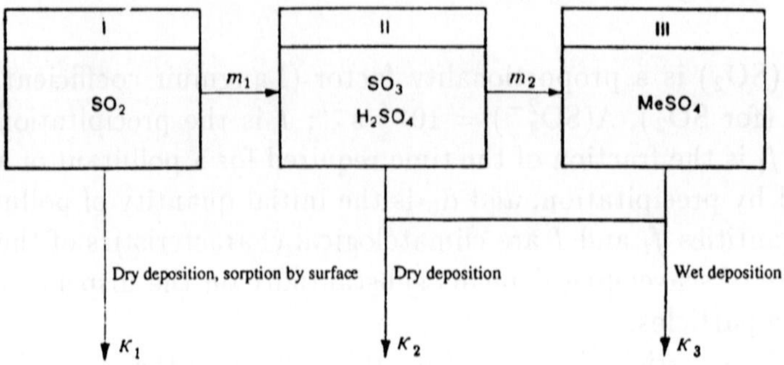

Figure 9.5. Block diagram of transformations and depositions of sulfur compounds from atmosphere.

and HNO_3 from the atmosphere are given in Table 9.2 [11,34].

The paths of transformation of sulfur dioxide and of the deposition of its transformation products onto the Earth's surface are shown schematically in Figure 9.5.

For an equilibrium system (the emission of sulfur compounds is compensated by sink mechanisms, edge effects being neglected) we have

$$N_{\text{I}} m_1 = N_{\text{II}}(k_2 + k_3 + m_2)\frac{M_{\text{I}}}{M_{\text{II}}} \tag{9.23}$$

$$N_{\text{I}} m_1 = N_{\text{III}}(k_2 + k_3 + m_2)(k_2 + k_3)\frac{M_{\text{II}}}{m_2 M_{\text{III}}}, \tag{9.24}$$

where $N_{\text{I-III}}$ is the content of sulfur compounds in the atmosphere (indices I, II, III refer to sulfur dioxide, sulfuric acid, and sulfates, respectively), and $M_{\text{I-III}}$ is the molecular mass of the corresponding compound (k_i, m_i are

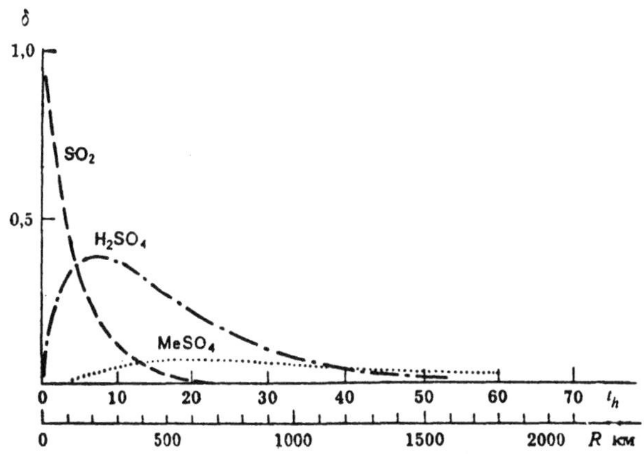

Figure 9.6. Relative variation in content of sulfur compounds with distance and with time required for pollutant to spread from emission site.

shown in Figure 9.5; their values are given in Table 9.1).

The above formulas made it possible to determine the ratios N_{II}/N_I, N_{III}/N_I. It was found that for the permissible SO_2 concentrations (in the air) the H_2SO_4 concentration will be close to the permissible value, whereas the sulfate concentration will be 5 to 10 times higher than the threshold value of the impact on man [28].

If the SO_2 concentrations are assumed to be constant and equal to the MPCs, then the H_2SO_4 flux will reach 50–80 g/(m^2 year), while the pH will drop to 2.7, which is absolutely unacceptable from the viewpoint of ecology [28].

If SO_2 spreads from a single source or from several sources (absence of equilibrium), then the changes in its concentration (and transformation products) are described by the following system of equations:

$$N_I^{\cdot} = -(K_1 + K_3 + m_1)N_I, \tag{9.25}$$

$$N_{II}^{\cdot} = m_1 N_I - (K_2 + K_3 + m_2)N_{II}, \tag{9.26}$$

$$N_{III}^{\cdot} = m_2 N_{II} - (K_2 + K_3)N_{III}. \tag{9.27}$$

The relative change in concentrations with distance is shown for this case in Figure 9.6 (for a transport rate of 30 km/h and the maximum values of factors K_i and m_i [35]).

It is seen from the figure that the maximum H_2SO_4 concentration in the atmosphere is reached at a distance of 200–250 km from the source, while the sulfate maximum is reached at about 600 km.

Therefore, the ratio $H_2SO_4/MeSO_4$ on various-sized particles demonstrates the analog of fractionation of the SO_2 transformation products in the atmosphere, so characteristic of radioactive products of nuclear explosions

[3]. Ultimately this also determines the H_2SO_4 and $MeSO_4$ distribution over the terrain (in Figure 9.6 it is seen that the maxima for these compounds do not coincide). In the case of distant fallout, this effect is smoothed out.

When the SO_2 concentration near the source is equal to the MPC for humans, the H_2SO_4 concentration in the zone of maximum is only 0.3 MPC, but at the same time the pH of the precipitation will be very low, close to 4.0, which can have negative ecological consequences.

A very similar picture is observed when NO_x spreads from individual sources: the maximum content of nitric acid is reached at a distance of about 200 km from the source.

Anthropogenic acids (either emitted or formed in the atmosphere) and their transformation products, falling in the form of acid rain, affect the surface water, the soil, terrestrial and aquatic ecosystems, and different constructions and buildings. These substances also have a negative effect on human health. Numerous studies have been made of the injurious effects of the above chemicals (see, for instance, [20, 21, 26, 28, 35, 37, 38, 44]).

Sulfates affect humans through inhalation already at concentrations of 6–10 $\mu g/m^3$, while the maximum permissible concentration is 50 $\mu g/m^3$ for SO_2 and 100 $\mu g/m^3$ for H_2SO_4.

Increased concentrations of the above substances lead to an increased incidence of respiratory diseases, chronic bronchitis, emphysema, asthma, and allergic diseases.

It is mainly vegetation which is exposed to acidic sulfur-bearing substances. In many cases the impact is indirect: the fertility of the soil and its other characteristics deteriorate. The washing out of metals from the soil, especially aluminum and magnesium, is dangerous for the vegetation.

The effects of SO_2 transformation products can turn out to be dangerous for the atmosphere as a whole, and for the climate in particular. The accumulation of sulfate aerosol can affect the Earth's radiation balance, leading to an accelerated cloud formation (due to the increased number of condensation nuclei); the appearance of a stratospheric sulfate layer can result in a rise in the stratospheric temperature and a drop in the tropospheric temperature.

Therefore, it is obvious that the anthropogenic emission of sulfur dioxide and the acid rain produced by it will be accompanied by substantial ecological, economic, and esthetic damage.

The total emission of SO_2 in Europe is 65 to 70 million tons per annum, whereas in the member countries of the UN Economic Commission for Europe (which includes the USA and Canada) it is about 100 million tons a year [20, 27]. The improvement of systems for trapping particles during emission (usually alkaline in nature) with the help of filters promotes the transport of acid-producing substances over long distances.

The acidity of precipitation has increased rapidly during the last 20 to 30

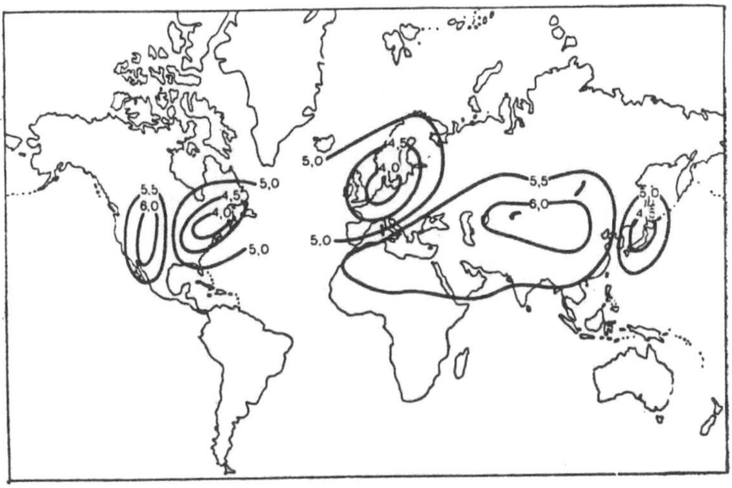

Figure 9.7. pH values in precipitation for Northern Hemisphere at the end of the 1970s.

years, particularly in industrial regions. In the north-western USA [40], for instance, even the average annual pH of the precipitation has been very low (4.1 to 4.3) during recent years. The content of free hydrogen ions increased threefold to fivefold over that period, and by a factor of 10 to 30 since the preindustrial period.

In individual rains or fogs the pH can be significantly lower; for instance, in the state of New York the pH values of rains vary from 2.66 to 4.66 [30].

In Europe a quite severe situation is observed in Scandinavian countries, where the pH of the precipitation dropped from 5.5–6.0 in 1955 to 4.0–4.2 (in recent years) [27, 28]. The acid precipitation in this case is mainly due to SO_2 emissions in other countries (Central Europe and Great Britain). The sharpest increase in the acidity in Scandinavian countries was from 1960 to 1965, during the period when high smoke stacks were put into operation in Central Europe [11].

The vast area of precipitation acidification in Europe includes Great Britain, West Germany, Holland, Denmark, part of East Germany, and southern Scandinavia, and it stretches toward the north-western USSR. In these regions and countries the pH is, on the average, 4.1 to 4.3.

The mean value of the pH in the Oka-Terrace biosphere reserve in the Moscow region decreased from 6 or 7 in 1975 to 4 or 5.1 in 1978 [16]. Figure 9.7 gives the average values of the pH in the precipitation in the Northern Hemisphere at the end of the 1970s [56].

Figure 9.8 shows a map of the sulfate (sulfuric acid neutralization products) depositions in Europe in 1980 (in terms of sulfur). Figures 9.9–9.11 indicate the densities of the depositions of free hydrogen ion, sulfates, and

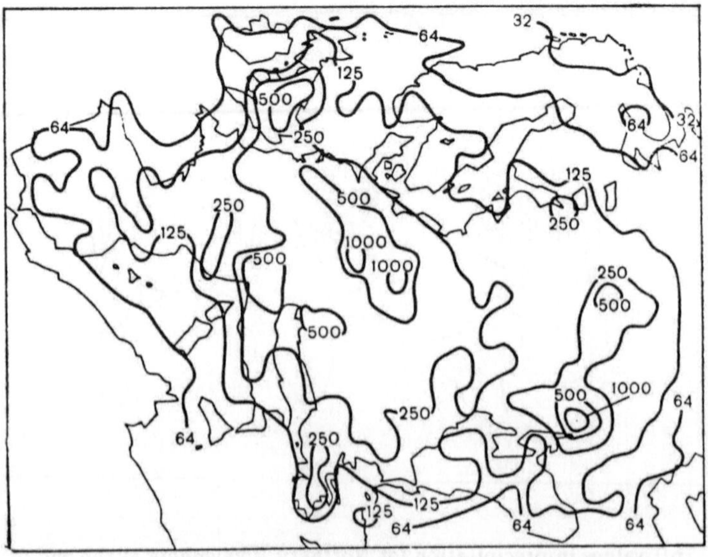

Figure 9.8. **Density of sulfur depositions (mg/m²) in Europe in 1980.**

nitrates in the USA and Canada in 1980 [42].

Average data are given here. On certain days the SO_4^{2-} depositions at a number of stations exceeded 100 mg/(m² day) [45]. A high correlation between the precipitation acidity and the sulfate deposition is everywhere observed.

This is because the sulfuric acid entering a particle leaches out a portion of the inorganic compounds, which are usually insoluble, and then is transformed into sulfates. A similar pattern of precipitation acidification is observed in Japan, where the free-hydrogen-ion content in rain precipitation has increased by more than 100 times during the last 20 or 30 years [46].

Recently in a number of industrial countries the concentration of hydrogen ions in the precipitation has increased greatly, due to the presence of nitric acid (the relative contribution of HNO_3 to the precipitation acidification has increased) (see Figure 9.11).

The median values of the mean annual ion concentration in the precipitation for 15 stations in southern Scandinavia [21] (in μg eq/liter) are:

Ion	1955–1959	1960–1964	1965–1969	1970–1974	1975–1979
SO_4^{2-}	55	90	90	90	85
NO_3^-	15	20	25	35	40

In southern Sweden the sulfur deposition in recent years has been 23.6 to 23.8 kg/(ha year) and the nitrogen deposition has been 13.1 to 13.9 kg/(ha year) [31].

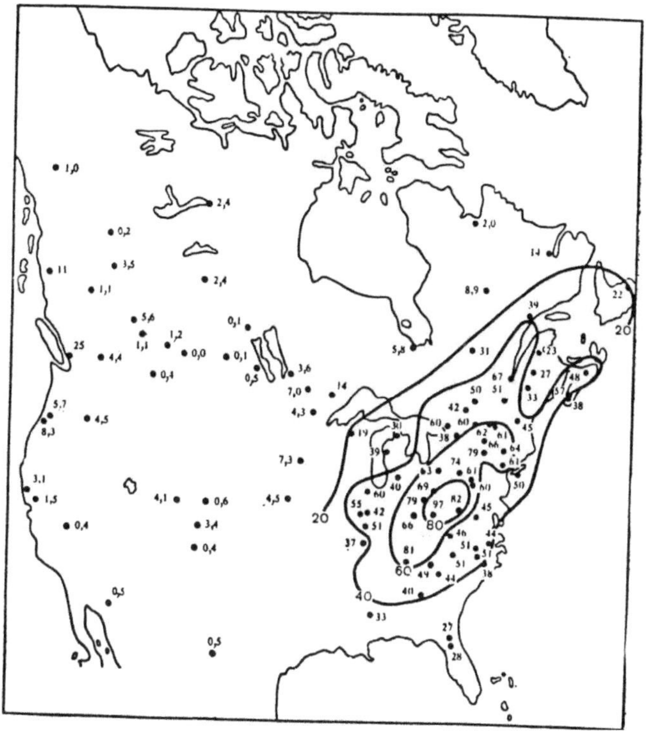

Figure 9.9. Annual deposition of free hydrogen ion in precipitation (mmole/m^2) in United States and Canada in 1980 (1 mmole/m^2 = 0.01 kg/hectare).

In a substantial number of cases the pH can be a good indicator of the anthropogenic acidification of the precipitation (particularly if the pH value is considered together with other indicators).

Therefore, an analysis of the pH values measured in the North Atlantic from Soviet research vessels, together with the air-mass trajectories, showed that for pH < 5 in practically all cases the trajectories passed over industrial zones of the USA (in other cases the pH values were much higher).

Therefore, in the case described the pH was a specific indicator of the anthropogenic effect on the precipitation.

It is noteworthy, however, that a decrease in the pH of the precipitation to very low values, for example, to 4.0, is not necessarily connected with anthropogenic acidification [51].

Such a decrease may well be related to a decreased content of alkalis in the precipitation (for example, of the Ca^{2+} ion), compared with the acid content.

At the same time, in regions with a high content of alkaline dust (particles) in the atmosphere, the pH value in the precipitation can be much higher

Figure 9.10. Annual deposition of sulfates (in terms of SO_4^{2-} ion, mmole/m^2) in United States and Canada in 1980 (1 mmole/m^2 = 0.961 kg/ha).

than 5.6 (even 7.0; for example, in internal regions of India the pH of the precipitation varies from 5.8 to 8.9 for a median value of 7.5 [51]).

This, in particular, means that an appreciable amount of alkaline dust in the air can substantially mitigate the problem of precipitation acidification in the internal regions of India.

Substances of natural origin transported through the atmosphere, which were produced by bioprocesses, forest fires, desertification, volcanic activity, and movement of sea salt, can be acidic, alkaline, or neutral. The anthropogenic impact considerably changes not only the ratio between acids and alkalis in the precipitation but also the role played by various compounds in the process of precipitation acidification.

Therefore, whereas at the background level the main contribution to the precipitation acidification is that of carbon dioxide (80%), the main contribution (more than 90%) to the atmosphere of industrial cities is made by sulfuric and nitric acids, only about 2% being due to dissolved CO_2 [11].

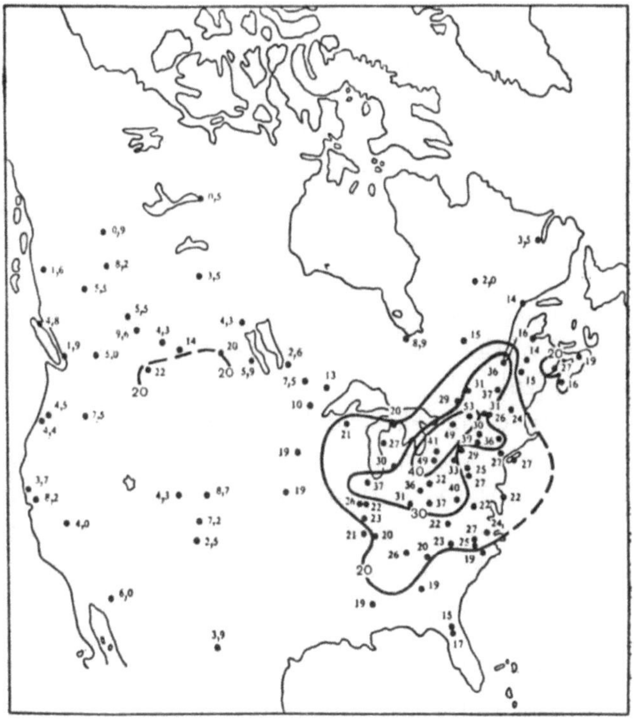

Figure 9.11. Annual deposition of nitrates (in terms of NO_3^- ion, mmole/m^2) in United States and Canada in 1980 (1 mmole/m^2 = 0.62 kg/ha).

9.3. The Acidification of Lakes, Streams, and Soils. The Ecotoxicology of Acid Rain

The acidification of precipitation led to a substantial acidification of lake water and had many other ecological consequences (in Scandinavia, Canada, and the USA). In the lakes of southern Norway, for instance, the pH value dropped, on the average, by 1.25 over 20 years. From 1965 to 1975 the pH value of a number of Norwegian lakes dropped from 5.9 to 5.4, and in some lakes it fell below 5.0 [25]. In lakes with an area of 13.000 km^2 there are already no fish, and in lakes of 20.000 km^2 in area the fish population has been substantially reduced.

In nine rivers in Nova Scotia the pH value is 4.7 (salmon and trout do not reproduce there), in eleven rivers pH = 4.7 to 5.0, and in seven rivers pH = 5.1 to 5.3.

Lakes in many regions of the USA and Canada are exposed to acidification: New York, Minnesota, Wisconsin, Michigan, Idaho; the Canadian provinces of Ontario, Quebec, etc. [54] (see also [21, 22, 39]).

The pH in more than 50% of the mountain lakes in the USA (at altitudes higher than 600 m above sea level) does not exceed 5.0, whereas back in 1929–1937 only 4% of the lakes had pH values as low as this [30, 40].

In western Sweden the pH in most lakes has dropped from 6.5–7.5 (1930–1940) to 4.5–5.5 (1971) [24], and in lake Gardzön the pH value dropped from 6.0 to 4.5 between 1960 and 1979 (historically the pH in this lake was 6.5–7.0 for many thousands of years) [47].

A direct relation has been established between the acidity of the precipitation, the sulfate content in the precipitation, and the sulfate content in the lake water.

For instance, in Lapland, as a result of the melting of polluted snow with pH = 4.4, the pH of the water in small rivers dropped from 7.0 to 4.9 (in about two weeks) [26].

Of prime importance for lake acidification is the composition of the rocks in the catchment area and in the vicinity of the lake. Maximum acidification is observed in the vicinity of igneous rocks and quartz sand (since the chemical composition is not greatly affected by runoff water). On the other hand, sedimentary rocks with their high leaching ability can greatly influence the chemical composition of lake water. For instance, in regions with a prevalence of limestone or dolomite no lake acidification is observed (because of the high buffering capacity of mountain rocks).

Conditions promoting the acidification of lake water are observed in northern Europe (particularly in Scandinavia, in the Murmansk region, and in the Karelian Autonomous Republic), in the northeastern USA and southeastern Canada, and in Siberia [11].

The leaching ouf of heavy metals from rocks and the entry of these into lake and river water present a serious problem [21]. Figure 9.12 indicates the extent of leaching out of various metals from the soil, as a function of the pH and Figure 9.13 shows the accumulation of heavy metals in the bottom sediment of a Swedish lake: it is especially pronounced in the top layers of sediment, where the pH is lower.

It should be noted that in water of increased acidity toxic monomethyl mercury is produced more intensively (when inorganic mercury compounds enter the water bodies) [34].

Acidification affects the physicochemical properties of the soil and its structure (the granulation is reduced, and the air permeability of the soil drops sharply). When the pH < 5.0 the soil fertility is reduced noticeably. The optimum soil acidity appears to be from 5.0 to 7.0.

For instance, the pH of forest soil (under natural conditions) is above 5.0. A six-year experiment was conducted, in which 100 kg of H_2SO_4 were added every year to each ha of forest soil with pH = 5.3 (this corresponds to an average acid concentration in the air which is 5 to 10 times less than the MPC). By the end of the experiment the soil pH was 4.75 [10].

If during the acidification the pH in the surface layer of soil decrases to

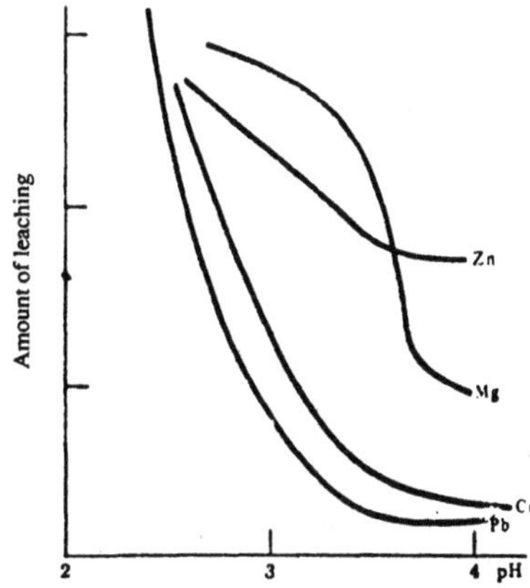

Figure 9.12. Leaching out of heavy metals from podzol forest soil at various pH values.

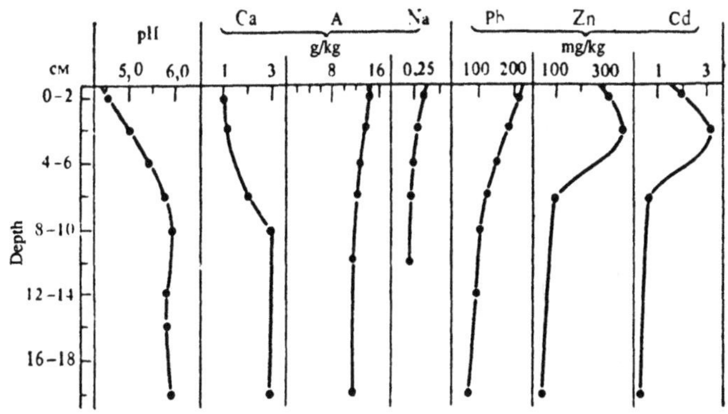

Figure 9.13. Content of heavy metals and aluminum in bottom sediment of Swedish lake Horsikan (western shore) in 1977–1979.

3.5–4.0, then at a depth of 50 cm it will be 0.7 to 2.0 higher than this. Under different trees the acidity in the humus layer (other conditions being the same) was consistently different. For example, the pH under pine trees was 0.5 lower than the pH under birch trees [21].

Various fertilizers added to the soil can affect its acidity, too: the greatest decrease in pH (by almost 1.5), from 6.4 to 5.1, was observed when ammonium sulfate was added to the soil for almost 5 years [21].

Let us note that the buffer action of some soils is quite great (for example, Central European and Eastern European limy soils), so that acid rain is not dangerous for such soils for a certain time (some soils even need some additional sulfur). It should be emphasized, however, that this buffer action is not unlimited and can be used up rather quickly. Podzol soils, which have a high natural acidity (the pH is 4.0–5.0) are very sensitive, and acid rains can have serious ecological consequences.

In a number of regions a great deal of money has been spent on introducing lime, in an attempt to reduce the acidification of the soil and lake water caused by acid rain.

Just due to the acid rain moving in across the western boundary of the USSR, the damage done to the agriculture in the northwestern part of the country is approximately 100 million roubles (annually). To reduce the soil acidification there, up to 3.5 million tons of lime should be introduced annually.

Lime is introduced into Swedish lakes on a large scale, in an attempt to deacidify the lake water. Carrying out this operation at 1500 lakes cost a total of 100 million Swedish crowns [21].

In different soils, depending on their buffer action, the free hydrogen ion is absorbed (neutralized), either due to the dissolution of calcium carbonate or due to the leaching out of AlOOH from the silicates and the formation of an Al^{3+} ion or during the formation of iron from its oxides [30].

It should be noted that the mobile aluminum ions produced by leaching out AlOOH lead to a reduced soil productivity. The alkali reserves of the soil can obviously become exhausted by the falling of acid precipitation.

The soil acidity determines the accessibility of nutrients in the soil and the soil fertility: an increase in acidity reduces the amounts of calcium, manganese, and magnesium in the soil.

Acidification reduces the rate of decomposition of organic compounds, and this retards their cycle and reduces the productivity of nitrogen-fixing bacteria. Soil acidification leads to disturbances of the functional and structural characteristics of the soil biota.

The hydrogen ion retards nitrogen uptake by plants, suppresses the synthesis processes, and disturbs the carbohydrate-protein metabolism.

The nature of the plant nutrition changes: the concentrations of various elements in the soil solution increase (when passing from the exchange phase to the water-soluble phase): Mn, Co, Zn, Fe.

The phosphorus is bound into an almost insoluble salt, and thus it becomes practically inaccessible for plants.

An excess of the soluble forms of Al and Mn has a very negative effect.

A calcium deficit appears, since Ca is bound into the almost insoluble form $CaSO_4$, so that the root system suffers damage.

For acidified podsolic soil there can occur a molybdenum deficit, because the Mb is bound by Al, Fe, and other cations.

SO_2 affects the vegetation quite significantly. The impact on transpiration, respiration, and photosynthesis alters the plant metabolism.

The leaves of higher plants suffer acute necrosis: a reduction in the yield of wood increment. Morphological changes of plants associated with the SO_2 impact are described in detail in [2].

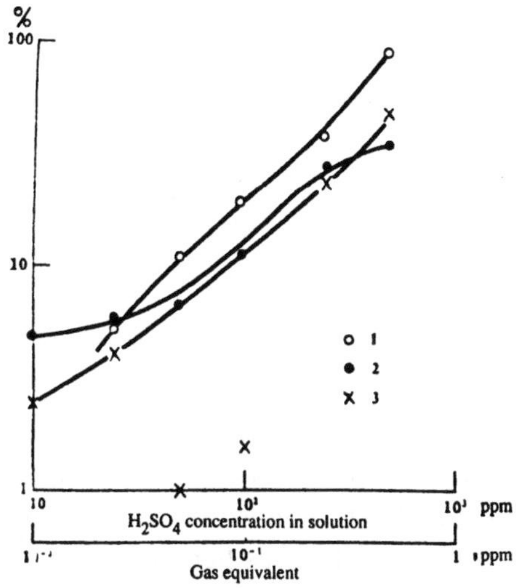

Figure 9.14. Rate of reduction of CO_2 assimilation by plant (1), rate of loss of Mg^{2+} ions (2), and rate of reduction of oxygen release (3), as functions of SO_4^{2-} concentration (in solution).

The SO_2 effect reduces the plant resistance to external factors (frost, drought) and diseases, and it disturbs the physiological processes: the cell permeability and ion balance change, and unbound water is lost.

The negative impact of SO_2 on the vegetation increases with an increase in its concentration (to a considerably lesser extent it depends on the total dose), since the increased rate of uptake makes detoxification difficult (in connection with the transformation of sulfur dioxide and sulfate to a less toxic sulfate) [32].

Among higher plants conifers are very sensitive to the impact of SO_2 and acids.

However, lower plants (lichen, moss, blue-green algae, some fungi) are most susceptible to the effect of SO_2 and H_2SO_4. The photosynthetic processes of lichens are already retarded sharply when the SO_2 concentration in the air is 0.01 ppm [43]. Lichens die at acid concentrations of 0.005 to 0.03 ppm, while conifers are damaged at 0.07 to 0.08 ppm [7, 9]. In many areas adjacent to industrial cities lichens have already completely disappeared. The chloroplasts are the main target when SO_2 affects a cell, since that is where sulfur-bearing ions accumulate. SO_2 affects the enzymes which catalyze the initial stages in the synthesis of fatty acids [8, 19].

Data in [12] indicate that an increased SO_4^{2-}, SO_3^{2-} (HSO_3^-) concentration in solution (in gaseous equivalent SO_2 from 0.01 to 0.5 ppm) results in a decreased rate of CO_2 assimilation by plants and production of total chlorophyll, in a loss of Mg^{2+} ions, and in decreased chlorophyll activity and decreased oxygen release (Figure 9.14). Experiments with conifers and

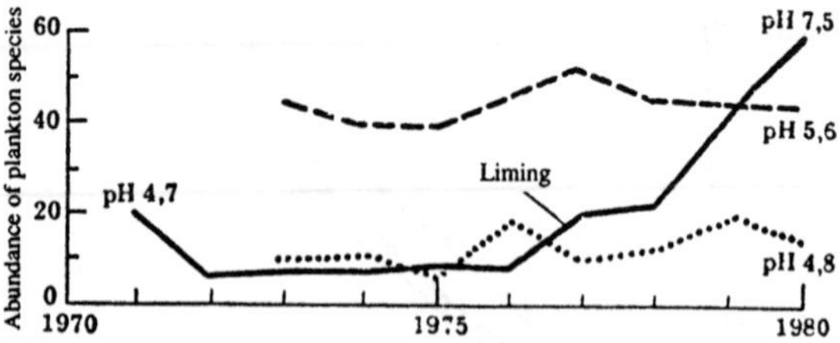

Figure 9.15. Abundance of plankton species in a number of Swedish lakes with various water acidities. One of the lakes was limed in 1977.

lichens in solution also simulate the real natural processes observed when sulfuric-acid-bearing droplets land on leaves and needles. In recent years the damage done to forests by acid rain has become increasingly pronounced [15, 36–38, 44, 52, 57].

Acidification also brings about a significant change in the other elements of terrestrial ecosystems [26, 57]. The eggs of many birds are easily destroyed by acid rain (in some cases the percent of destruction is 80%). Changes in microflora and insects are observed, too [44].

A substantial effect of acid rain on lake and river ecosystems is noted in a number of regions of Europe and North America. Already at pH < 6.0 in some freshwater bodies, the species diversity of organisms decreases and the food chains are disturbed.

In 'acidified' water bodies (caused by soil washing) the concentrations of aluminum ion, which is hazardous to all living organisms, are high.

When water bodies become acidified, the microbiological processes are disturbed; the bacteria and protozoa which were dominant give way to fungi [23]. The species diversity and the phytoplankton biomass both decrease; the simplification of plant communities leads to a reduced abundance of many zooplankton species.

Figure 9.15 shows the plankton species diversity as a function of the pH of the water in some Swedish lakes [29]. The figure shows that the plankton species number in acidified water is lower, while, on the other hand, after liming, when the pH suddenly increased, the plankton species diversity began to recover. Amphibia, particularly salamanders and frogs, which spawn in shallow water and in meltwater, perish first [26]. The structure of the fish communities is changed as well (a number of fish die). Eggs and young fish are particularly susceptible to the environmental change (the increased acidity).

At pH $= 4.0$ to 5.0 fish perish because of the disturbed ionic and acid-alkaline equilibrium in their gills [18, 25]. The reproductivity of many fish species in freshwater bodies drops at pH < 5.5 and practically ceases at pH < 4.5.

The optimum pH (with respect to the acidity of the environment) in freshwater bodies is from 6.5 to 8.5, while the range of relative safety is from 5.0–5.6 to 9.0 [18, 50].

The acid rain falling on the surface of the World Ocean affects the neuston organisms in the surface film [40] (although the 'buffer action' of the ocean water is high, the pH there stays around 8.0).

Sulfuric acid deposited on buildings and other structures converts calcium carbonate into soluble calcium sulfate (gypsum), causing the structures to deteriorate. Great esthetic damage occurs as well, when ancient monuments are damaged. During the last 30 to 50 years, for instance, the Parthenon and the marble sculptures on the Acropolis in Athens, and old houses in Krakow, have been subjected to such effects [58]. The Taj Mahal in India and the Colosseum in Rome are also in danger [34], although for the Taj Mahal the high content of alkaline dust in the inland regions of India keeps the 'buffer action' of the atmosphere there rather high.

The corrosion of metals induced by acid precipitation is a serious problem [21, 34]. Various metals (copper, aluminum, iron, and even steel) quickly absorb sulfur dioxide and acids at their surfaces and thus are gradually destroyed [55].

Figure 9.16 gives the zinc corrosion factor as a function of the falling density of acid rain (results from 33 stations in West Germany and Great Britain) [49]. Inspection of the figure shows that the corrosion rate is quite high.

Figure 9.17 shows how acid precipitation affects building materials: limestone and sandstone (samples of these materials were exposed). Obviously, the corrosion of various materials can result in quite significant economic damage.

As already mentioned, the emission into the atmosphere of a large quantity of sulfur compounds and nitrogen oxides produces an enormous amount of highly dispersed aerosol particles in the troposphere.

A study of the size distribution of atmospheric particles reveals a bimodal distribution (see, for instance, [13]).

The mode of the smallest particles includes dimensions (diameters) from 0.1 to 0.1 μm. These are principally particles produced by fossil fuel combustion, by the conversion of sulfur and nitrogen; they are spherical, they scatter light well, and they are enriched, as a rule, with sulfates and heavy metals (of anthropogenic origin).

The mode of larger particles corresponds to particles with a diameter $> 1\mu$m, produced by wind erosion. These particles are, as a rule, irregular in shape and they do not scatter light well; they consist of elements like silicon and calcium, i.e., the principal elements of the Earth's crust.

The total amount of anthropogenic particles was estimated above as being around 300 million tons per year. These particles can affect signifi-

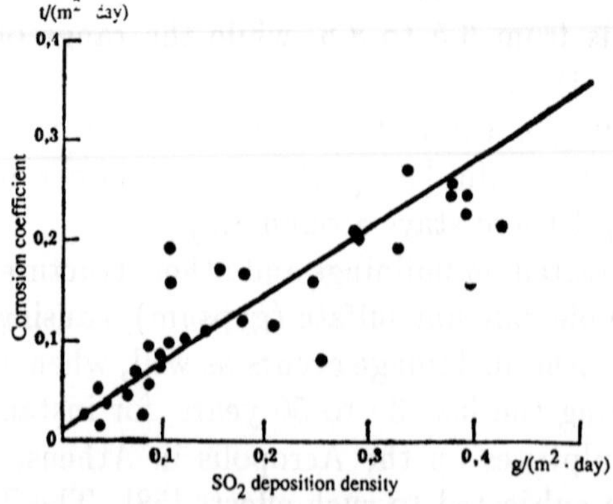

Figure 9.16. **The coefficient of zink corrosion relative to the SO_2 deposition density.**

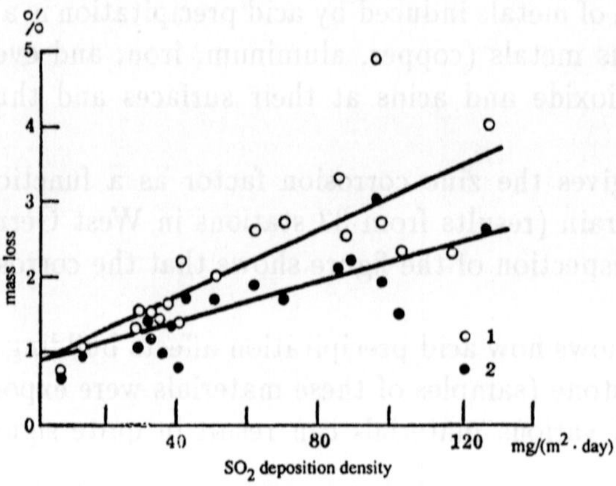

Figure 9.17. **Mass loss (of 60 × 60 × 30 mm plates) of sandstone (1) and limestone (2), as function of SO_2 deposition density.**

cantly the radiation characteristics of the lower atmosphere and its albedo (mainly increasing it). Particles can also enter the stratosphere, producing the stratospheric aerosol.

Particles deposited on a snow or ice surface change the albedo of the underlying surface (mainly decreasing it).

An enormous amount of the highly dispersed particles produced become crystallization and condensation nuclei, i.e., they promote the development of fog, clouds, and precipitation.

Therefore, the emission of sulfur and nitrogen oxides, and the consequent formation of a great amount of aerosol particles, leads to certain changes in the weather and can even affect the Earth's climate as well (this issue was already discussed in the appropriate chapter).

It should be emphasized also that the particles so formed change the transparency of the atmosphere, affecting visibility, and they can make transport movement (including air transport) difficult; they cause great esthetic damage, too (by producing mist that obscures natural landscapes). This effect is being given a thorough study. Approaches are being developed for setting up standards for this impact [13, 14]. In clean areas an increase in dust concentration by only 4.5 $\mu g/m^3$ can result in a 50% decrease in visibility. This phenomenon can be particularly perceptible in a region with a picturesque landscape, where the visibility usually exceeds 100 km. Information on the atmosphere transparency is being collected at nature reserves and national parks in the USA [13].

The concentration of sulfate particles in the western USA is relatively low: up to 3 to 6 $\mu g/m^3$ (excluding vicinities of some local sources). On the other hand, the average concentration in the eastern USA varies from 8 to 13 $\mu g/m^3$ [13], and in a number of cases it is much higher (35–40 μ g/m^3 [13]).

9.4. The Problem of Transboundary Transport

As a whole an enormous amount of SO_2 is emitted into the atmosphere, particularly in the regions of Europe and North America (more than 100 million tons per year). Substantial quantities of sulfur compounds are transported into other countries: for instance, from Western Europe to Scandinavia, and from the United States to Canada and back. Every year about 5 million tons of SO_2 and its transformation products are transported into the USSR across its western boundary, while only 1/5 to 1/10 as much is transported back [8, 34].

The urgent need to reduce transboundary transport was affirmed at the All-Europe Conference on Environmental Protection in 1979, by authorized representatives from 32 European countries, the United States, and Canada. The Convention on Long-Range Transboundary Air Pollution was signed at this conference. The convention came into force in 1983.

The convention defined the following basic commitments for the participating countries.

1. Each country undertakes to develop the best possible complex of measures to control pollutant discharges into the air. These measures are to be consistent with a balanced development of the economy and based on the best available technology. The countries are to exchange information and consider jointly their policy and strategy in the battle against pollutant emission, particularly with regard to limiting and preventing transboundary air pollution.
2. The countries will set up and implement a Joint Program for Observing

Table 9.3. Mean monthly sulfur depositions (in hundreds

Acceptor country	Donor														
	1	2	3	4	5	6	7	8	9	10	11	12	13	14	15
1. Albania	10	0	0	3	0	0	0	0	0	0	2	2	0	0	8
2. Austria	0	52	3	0	35	0	0	20	24	32	0	14	0	0	56
3. Belgium	0	0	67	0	0	0	0	28	2	24	0	0	0	0	0
4. Bulgaria	0	0	0	153	8	0	0	2	6	4	7	12	0	0	8
5. Czechoslovakia	0	22	10	2	483	2	0	45	195	108	0	68	0	0	39
6. Denmark	0	0	0	0	3	39	0	3	12	11	0	0	0	0	0
7. Finland	0	0	2	0	8	3	77	5	19	15	0	3	0	0	2
8. France	0	0	33	0	7	0	0	629	20	98	0	0	0	2	41
9. GDR	0	0	8	0	56	3	0	20	497	85	0	4	0	0	3
10. FRG	0	8	37	0	48	4	0	108	118	561	0	6	0	0	24
11. Greece	3	0	0	34	4	0	0	3	3	3	93	5	0	0	12
12. Hungary	0	9	0	0	45	0	0	7	18	13	0	194	0	0	27
13. Iceland	0	0	0	0	0	0	0	0	0	0	0	0	0	0	0
14. Ireland	0	0	0	0	0	0	0	2	0	0	0	0	0	18	0
15. Italy	0	6	2	0	12	0	0	53	11	22	0	11	0	0	793
16. Luxembourg	0	0	0	0	0	0	0	3	0	2	0	0	0	0	0
17. Netherlands	0	0	16	0	2	0	0	15	6	45	0	0	0	0	0
18. Norway	0	0	4	0	8	6	2	11	22	20	0	2	0	0	0
19. Poland	0	7	7	3	136	8	0	26	213	80	0	43	0	0	18
20. Portugal	0	0	0	0	0	0	0	0	0	0	0	0	0	0	0
21. Romania	0	3	0	30	37	0	0	7	26	16	4	71	0	0	27
22. Spain	0	0	2	0	0	0	0	38	4	20	0	0	0	0	2
23. Sweden	0	0	5	0	18	16	10	15	42	35	0	5	0	0	4
24. Switzerland	0	0	0	0	2	0	0	23	3	13	0	0	0	0	47
25. Turkey	0	0	0	28	5	0	0	3	6	5	20	7	0	0	13
26. USSR (considered area only)	3	14	18	69	189	25	50	67	283	190	22	159	0	2	94
27. Great Britain	0	0	7	0	4	0	0	27	11	21	0	0	0	7	2
28. Yugoslavia	3	14	2	32	38	0	0	20	25	22	6	72	0	0	131
29. Remaining area considered in calculations	10	13	51	85	106	52	30	295	214	252	79	68	0	20	448
Total emission (approximate)	37	165	292	448	1272	171	175	1490	1793	1712	241	760	0	581	812

and Evaluating the Long-Range Spreading of Air Pollution in Europe (EMEP, the European Monitoring Program). Further technological and organizational reinforcement and development of the joint program is also planned.

The convention recognizes two equivalent approaches to the exchange of information regarding the contribution of a state to the transboundary transport:

(1) data can be given on transboundary flows across sections of the national boundary; or
(2) data can be given on pollutant emissions in grid squares of the country

of tons) between October 1978 and September 1980.

country														Depositions from unknown sources	Total deposition
16	17	18	19	20	21	22	23	24	25	26	27	28	29		
0	0	0	0	0	2	0	0	0	0	0	0	13	0	12	67
0	2	0	13	0	0	2	0	3	0	0	8	35	0	30	341
0	4	0	0	0	0	0	0	0	0	0	18	0	0	9	161
0	0	0	7	0	34	0	0	0	3	12	2	45	0	29	346
0	7	0	95	0	14	4	0	2	0	14	32	53	0	91	1301
0	0	0	4	0	0	0	0	0	0	0	11	0	0	11	109
0	2	2	15	0	2	0	14	0	0	43	14	3	0	55	293
4	12	0	4	0	0	66	0	7	0	0	99	6	0	166	1212
0	6	0	23	0	0	0	0	0	0	3	22	6	0	26	778
5	23	0	18	0	0	6	0	6	0	3	72	15	0	80	1158
0	0	0	2	0	8	2	0	0	4	4	0	27	0	31	253
0	0	0	21	0	16	0	0	0	0	4	3	75	0	20	467
0	0	0	0	0	0	0	0	0	0	0	0	0	0	18	24
0	0	0	0	0	0	0	0	0	0	0	12	0	0	26	65
0	0	0	6	0	3	17	0	7	0	2	9	65	5	93	1132
3	0	0	0	0	0	0	0	0	0	0	0	0	0	0	11
0	40	0	0	0	0	0	0	0	0	0	27	0	0	11	173
0	3	20	10	0	0	0	10	0	0	8	40	2	0	74	255
0	7	0	565	0	13	2	4	0	0	34	32	40	0	77	1339
0	0	0	0	20	0	17	0	0	0	0	0	0	0	29	73
0	0	0	38	0	287	0	0	0	4	48	6	115	0	60	797
0	3	0	0	9	0	367	0	0	0	0	15	0	2	111	583
0	5	8	31	0	3	0	83	0	0	24	35	7	0	113	472
0	0	0	0	0	0	2	0	14	0	0	5	3	0	17	141
0	0	0	5	0	12	3	0	0	175	17	0	22	0	79	416
3	23	5	386	0	205	10	38	3	45	3610	103	208	3	1061	6901
0	5	0	2	0	0	4	0	0	0	0	675	0	0	72	847
0	2	0	21	0	26	7	0	0	0	8	7	557	2	85	1093
6	53	11	130	9	76	175	45	10	85	226	611	216	21	1074	4436
38	213	50	1414	44	713	703	206	63	323	4072	1877	1526	44	3571	25300

(the length of a boundary section and the size of a grid square are both assumed to be 150 km).

The equivalence of these two approaches can be demonstrated theoretically from the standpoint of computing the pollutant flows at any point on the continent.

The Joint Program for Observing and Evaluating the Long-Range Spreading of Air Pollution in Europe (EMEP) has already been implemented, and (Eastern and Western) Meteorological Synthesizing Centers have been established in Moscow and Oslo.

The transboundary flows are computed on a regular basis; a direct trajectory analysis is carried out, as well as an objective analysis of the wind field and an exchange of calculated data on the meteorological elements, with a sufficiently long forecasting period.

Table 9.3 gives the mean monthly depositions of sulfur between October 1978 and September 1980 (according to data of the Western Meteorological Synthesizing Center [11]). The contribution of any given country (donor country) to the sulfur deposition onto any other European country (acceptor country) can be determined from this table.

Monitoring of the long-range transport of SO_2 and its transformation products, and in particular of the transboundary flows of sulfur dioxide and sulfates, has been organized in the Soviet Union (both aircraft and ground-based methods are used) [11, 34].

The Convention on Long-Range Transboundary Air Pollution is certain to play a large part in reducing the transboundary transport of pollutants over long distances.

There are a number of organizational and technological measures for controlling the long-range transport of pollutants.

The first link in the chain of such measures is the organization of a system for monitoring long-range pollutant transport, together with an analysis of the resultant changes in the state of the biosphere and the presentation of the results of the observation and analysis.

The second link in the chain is setting up standards for limiting pollutant emissions, and thus for the subsequent pollution as well. The introduction of standards is based on ensuring high quality environmental norms and standards for man and for natural ecosystems. Here it is important to take into account:

(1) the various chains of transformation and transition and the diverse ecological effects;
(2) the massive continuous emission of pollutants.

Major engineering and technological measures should be aimed at the direct reduction of discharges of hazardous pollutants (rather than aimed at the dispersal of these).

It should emphasized that international cooperation, with the wide and efficient participation of different countries and international organizations, is of prime importance in providing a solution to the problem of acid rain.

At an international conference in Munich (June 1984) on establishing the causes of and ways of preventing damage to forests and water by air pollution in Europe, specific measures were proposed for reducing the emissions of sulfur dioxide and nitrogen oxides in different countries, as well as means of reducing lead emissions into the air from vehicles [15].

At the conference delegations from 18 countries (including the USSR) expressed their intention to reduce the emissions of sulfur dioxide or its transboundary flows by 30% during the next 10 years.

CHAPTER 10

Stressing of the Atmosphere and Climate. The Geophysical and Ecological Consequences of Nuclear War

As mentioned above, the intensive economic activity of mankind is causing large-scale (even worldwide) geophysical and ecological changes on the planet; essentially the entire biosphere is affected by man's activities.

This book deals with such global issues as a possible climatic change produced by human activity, a possible change in the Earth's ozone layer, global pollution of the World Ocean, and the long-range (transboundary) transport of various toxic pollutants.

The impacts which lead to long-term large-scale negative geophysical and ecological changes in the biosphere can be considered as stress impacts. Among natural processes these impacts include falls of large meteorites (such as the one in Tunguska in 1908) and powerful volcanic eruptions. If a nuclear war should break out, involving weapons of mass destruction, the impact on the biosphere would be many orders of magnitude greater. In this case the problem would be whether the biosphere would be preserved in its present form, whether genetic degeneration would occur, whether all life on the Earth would survive, and whether destruction of the climate system would occur.

At all times progressive people have felt a particular revulsion toward wars, which lead to a large-scale destruction of population and which cause damage to the environment.

It is remarkable that calamities similar to the aftermath of a nuclear war were described in the ancient Indian epic, *The Mahabharata*, long before our era:

> "An awful thunder was heard in the heavens,
> Bright lightning flashes dazzled the eye.
> The entire Earth, with its mountains, trees, and forests, began to sway."
>
> "It (that terrible weapon) killed even the fetuses inside women..."

"For twelve long years no rain would fall in this land..."
(From the book 'The Attack on Those Who Slept', Chs. XIV–XV).

These words can serve as a serious warning to the people of our time as well. Information on global anthropogenic effects, as well as ideas about determining such impacts, can be used to evaluate the consequences of stressful extreme impacts, including the aftermath of a nuclear war.

The large-scale consequences of a possible nuclear war have been considered in a number of published works [8, 9, 19, 25, 30] and at two conferences: *Scientists for Saving Mankind from Nuclear War* (Moscow, 17–18 May 1983) and *The World after Nuclear War* (Washington D.C., 31 October–1 November 1983).

This problem was also analyzed by the author in a report to the Ninth Congress of the World Meteorological Organization on 9 May 1983, during a discussion of *Meteorology and Society*. The results of this research indicate once again, with new strength and conviction, that the idea of nuclear war is totally unacceptable, since such a war would inevitably disturb global systems, the climate system, for instance, and it would cause irreversible damage to the biosphere as a whole, and thus to all of mankind.

10.1. The Large-Scale Consequences of a Nuclear War

The damaging effects of nuclear explosions have different scales: from strictly local to global.

In order to estimate the effects of a nuclear war, various scenarios for the possible nuclear impact are used (total explosive force, the force of individual explosions, etc.). Most of these scenarios indicate a possible total explosive force of 5000 to 10.000 Mtons and a force of individual explosions ranging from 1 to 3 Mtons, and sometimes up to 10 Mtons [19, 25].

From the ecological standpoint, the character of the explosion (at the surface, in the air, in the upper atmosphere) is of great importance as well.

This chapter deals primarily with estimates of the possible ecological consequences of a Nuclear War (connected with the large-scale natural rearrangements). Direct injury (by thermal radiation, shock wave, and radiative emission) to human beings or other living organisms is practically not considered here.

The primary effects of nuclear explosions (the release of energy in various forms) are damage, including mechanical damage to everything close to ground zero, and warming of the environment.

Radioactive contamination of the environment due to a nuclear explosion is one of the main damaging factors, acting over enormous distances.

Pollution of the atmosphere (both the troposphere and the stratosphere) by aerosol particles and by various gaseous substances produced by the

Table 10.1 Large-scale effects of nuclear explosions

Main large-scale effects (damaging factors)	Possible geophysical and ecological consequences
1. Pollution of biosphere by radioactive products	Radiation damage to ecosystems Variation of electrical properties of atmosphere Variation of properties of ionosphere
2. Pollution of atmosphere by aerosol products	Variation of radiation properties of atmosphere Variation of weather and climate Deterioration of state of ecosystems (especially marine ecosystems) due to reduction of incoming solar radiation
3. Pollution by various gaseous substances (methane, ethylene, tropospheric ozone, etc.) of troposphere of upper atmosphere	Variation of radiation properties of atmosphere, weather, and climate Variation of radiation properties of upper atmosphere, destruction of ozone layer Variation of possibility of transmission of UV radiation, variation of climate
4. Variation of albedo of Earth's surface	Variation of climate

explosion and by secondary factors (forest fires and fires in cities and in oil and gas fields) can have serious ecological consequences. The change in the albedo of the Earth's surface (mainly due to fires) can also have negative consequences.

Possible geophysical and ecological consequences of these damaging factors are listed in Tables 10.1 and 10.4. These include, first of all, massive radiation damage and other damage to the biotic component of the ecosystems (including damage connected with the reduced transmission of solar radiation through the atmosphere), possible weather and climatic changes, destruction of the Earth's ozone layer, changes in the radiation properties and electrical properties of the troposphere, and disturbance of the state of the ionosphere and magnetosphere.

The primary effects (shock wave, hard electromagnetic radiation, etc.) lead to significant disturbances and direct injury to everything around on a local scale, although they are followed by minor secondary ecological effects (including direct warming of the environment). Fires ignited immediately by the nuclear explosion constitute an exception (the ecological consequences of fires will be considered below). The direct warming of the atmosphere by the heat released during the explosion will be insignificant.

The total energy released during a 5000 Mton explosion amounts to 2.1×10^{19} joules. A continual energy release of 1.8×10^{15} W would increase the temperature of the whole atmosphere by 1°C. If the time for complete mixing of the atmosphere in the Northern Hemisphere is assumed to be 10 days, then we find that the mean temperature of the atmosphere (troposphere) during that time will rise by no more than 0.01°C because of direct warming. Radioactive products of a nuclear explosion, being deposited from the radioactive cloud downwind, produce a stable radioactive contamination of the region at distances of many hundreds, and even thousands, of kilometers [5, 6, 13]. The contamination is particularly great after a surface explosion. It can be assumed that a nuclear explosion of the above total power, occurring in the midlatitudes of the Northern Hemisphere (surface explosion), would produce a zone with a total gamma dose of over 400–1000 rem over almost all of Europe and the central part of North America [19, 25].

The contamination of the region by the long-lived radioactive isotopes, strontium-90 and cesium-137, will be up to 4–10 Ci/km^2. The global latitudinal contamination by these isotopes will be 2–3 Ci/km^2.

The above doses and amounts of contamination are very large and are hazardous to man, although they are less dangerous for vegetation (some damage can be caused, mainly to conifers). On the other hand, populations of many insects, particularly those living in the soil, could suffer greatly.

Some radioactive isotopes, particularly iodine-131 and isotopes of strontium, ruthenium, and barium, are highly mobile and readily migrate (the latter isotopes migrate with the groundwater and the surface water). Plants can easily assimilate isotopes of tungsten, yttrium, strontium, and ruthenium, and to a lesser extent cerium and cesium isotopes, through their root systems [2, 7].

This causes secondary damage to the ecosystem components, and the isotopes enter the human food supply via agricultural products.

10.2. The Influence of Aerosol Particles and Gases on the Atmosphere and Climate

Pollution of the atmosphere by aerosol products and harmful gases will result from the explosion itself and from the inevitable massive fires ignited by the nuclear attack.

A surface nuclear explosion throws an immense amount of soil (dust) up into the atmosphere: up to 5000 tons of soil per kiloton of explosive power, and only a small part of the soil becomes vaporized [5, 6]. The vaporized part is then transformed into a highly dispersed aerosol fraction with an average particle size of a fraction of a micrometer. The size of the aerosol particles produced by an atmospheric nuclear explosion will not exceed 10–25 μm [6].

Most of the soil thrown upward falls back in the vicinity of ground zero; highly dispersed aerosols will fall onto the Earth's surface for weeks, months, and even years after the explosion. The higher the atmospheric explosion, the less will be the total mass of soil entrained by the radioactive cloud. For a high-altitude atmospheric burst this mass will be equal to the mass of the explosive device (bomb), and it will apparently range from fractions of a ton to several tons for each explosion.

The possible ratio of surface explosions and atmospheric explosions is hard to predict. It can only be supposed that atmospheric bursts will predominate, since their shock waves are much stronger.

For the large-scale use of atmospheric explosions (when the fireball does not touch the ground [15]), the total quantity of inert material transformed into highly dispersed aerosol particles will apparently be up to tens of thousands of tons. If surface explosions contribute significantly, then this quantity can be 2 to 3 orders of magnitude greater. This aerosol will be injected into the stratosphere at an altitude of 10–40 km, and it will remain there for a long time (up to several years).

However, we should bear in mind that, in order to strike well-shielded targets during a nuclear war, a substantial number of surface explosions may be necessary. In this case a variant is considered in which half the explosions are surface explosions. The aerosol particles will evidently be formed, right during the explosion, by soil fragmentation and by the condensation of vaporized material, as well as during the fires ignited by the explosion.

Only a small part will remain in the stratosphere for a long time in the form of highly dispersed particles (smaller than 1 μm), for example, particles formed by the condensation of vaporized material (soil and the explosive device itself). The tropospheric aerosols will settle out soon (will be washed out).

A surface explosion will melt 180 to 200 tons of soil per kiloton of yield (0.3 to 0.4 of the amount of soil melted by an underground explosion) [6]. In the case of an underground explosion 50 to 70 tons of rock will be vaporized per kiloton of power. Similarly, it can be assumed that 0.3 to 0.4 of this, i.e., 15 to 25 tons of rock per kiloton power, will be vaporized by a surface explosion, although other sources give 1.5 to 25 tons per kiloton [8].

The size distribution of particles (according to activity, number, mass, volume, and surface area) entering a cloud from a surface explosion is described by a lognormal law [6].

For the number of particles

$$N(d_1, d_2) = \frac{1}{\sigma\sqrt{2\pi}} \int_{\log d_1}^{\log d_2} \exp\left[-\frac{(\log d - \log \overline{d})^2}{2\sigma^2} \mathrm{d}(\log d)\right] \qquad (10.1)$$

for an explosion in silicate soil $\log \overline{d} = 2.053$ and $\sigma = 0.732$; for an explosion in coral soil $\log \overline{d} = 2.209$ and $\sigma = 0.424$.

The above distribution is very close in nature to the distribution for the particle activity, mass, and volume [6], and consequently it can be applied as a whole to the material entrained into the nuclear cloud.

In accordance with this distribution it is easy to calculate that for an explosion in silicate soil the mass of particles $< 1\mu$m in size will be $M(r < 1\mu\text{m}) < 0.2\%$.

According to [30], the total mass of material injected into the nuclear cloud (for a total explosive force of 5000 Mt) is 9.6×10^8 tons; 80% of this remains in the stratosphere; 8.4% of the particles are less than 1μm. This gives a total of 80 Mt of highly dispersed particles in the stratosphere.

If the upper mass limit for particles less than 1 μm in size is assumed to be 0.2% (formula (10.1)), then the mass of such particles in the cloud will not exceed 2 Mt. If we proceed from the total amount of rock ejected (5000 tons per kiloton), then the mass (0.2%) is estimated at 50 Mt.

If, on the other hand, we proceed from the amount of vaporized material which, when it condenses, forms highly dispersed aerosol particles, then (according to [14] no more than 20% of the refractory products of the explosion initially vaporized will remain in the cloud for a long time) we find that the amount of stratospheric aerosol produced by the vaporized material for a 5000 Mt bomb (half such explosions are surface explosions) will total 12.5 Mt. For comparison, the amount of stratospheric aerosol $< 1\mu$m six months after the El Chichón volcanic eruption (1982) was 8.3 Mt [12].

The massive fires in cities, forests, and gas and oil fields which would follow a nuclear attack may have very significant consequences. A substantial amount of carbon monoxide and carbon dioxide, various gaseous pollutants, soot, and other aerosol particles will be released during the fires.

The altitude reached and the ability of the ascending convective flows caused by the massive fires to penetrate into the stratosphere can be estimated approximately in accordance with the theory of turbulent jets. This theory has been developed satisfactorily for sources of comparatively low intensity.

Numerical calculations have been carried out for a two-layer atmosphere [4]. It was shown in [4] that for a source power of 10^6 kW (which corresponds to the burning of about 70 tons of kerosene per hour) a jet would rise in a dry standard atmosphere to an altitude of about 2000 m. The altitude reached by the jet will be proportional to the fourth root of the thermal power of the source. In calculations of the high-altitude rise of the jet performed by N.I. Vul'fson, the tropopause height was taken to be 11 000 m, the troposphere was assumed to have the stratification of the standard atmosphere ($\gamma = 0.65°\text{C}/100$ m), while the stratosphere was characterized by isothermy ($\gamma = 0$). An initial jet radius $R_0 = 500$ m, a vertical air velocity $w_0 = 20$ m/s, and a difference between the temperature in the jet and that of the surrounding air $\Delta T_0 = 100, 200$, and 300°C were assumed as boundary conditions near the Earth's surface. These conditions correspond to heat sources $Q = \pi R_0^2 \rho c_p w_0 \Delta T_0$ (where ρ is the air density and c_p is the heat capacity of the air at constant pressure) of 0.9×10^9, 1.85×10^9, and 2.8×10^9 kW, respectively.

According to the computations, the top of the jet (H) depends on the source power as follows:

Q	10^9 kW ·······	0.9	1.85	2.8
H	m	9700	11650	12350

Therefore, for a source power of 10^9 to 3×10^9 kW, the jet can penetrate to the stratosphere. Theoretically, such a power corresponds to a forest fire 10 km^2 in area, with a rate of burning of 3 to 5 $g/(m^2s)$, i.e., the complete burning of a forest (all the timber) in 1 to 3 hours (hardly realizable in practice).

Phase transitions related to the presence of moisture in the atmosphere, the number of which will increase as the forest burns, promote the rising of air in convective flows due to the release of condensation heat. Calculations indicate that the vertical velocities of the air in jets propagating in a moist atmosphere are approximately twice the velocities in a dry atmosphere. The top of the meteotron jet in a standard saturated atmosphere was approximately 6000 m, compared with a 2000 m rise in a dry atmosphere for the same source power [4].

The largest forest fires result in the burning up of 1% of the total forest area in a month: in an area of 50 000 km^2 up to 40 000 or 50 000 hectares will be burned up. On individual days the increment of the fire area can range from 5000 to 11 000 hectares. The number of new centers of burning per day will be up to 100. The usual altitude of the smokeline from such fires is 1.5 to 2.5 km. Local vortical fires (burning time of 1 to 1.5 hours) produce convective jets which can rise as high as 5 or 6 km.

In urban fires the energy release per unit area is much higher (higher concentration of combustible material), and a so-called 'firestorm' can break out. A firestorm is the most dangerous type of large fire. It is formed when many separate fires merge into one gigantic fire. The convective activity is extremely great in this case. High convective columns and strong air flows moving toward the center of burning are observed, and these take on the characteristics of a storm.

The firestorm almost does not spread outside the perimeter of the continuous center of burning, because of the strong inflows of air toward the center of the fire.

The smoke cloud from the firestorm induced by the bombing of Hamburg during World War II reached an altitude 7.2 km. A wind (velocity 15 m/s) directed toward the storm was observed at distances of up to 3.3 km from the firestorm.

On the basis of all the above, the escape into the stratosphere of a significant amount of aerosol particles produced by forest fires seems hardly possible. This can take place only in individual cases. Accordingly, the amount of these aerosols given as 5% in [30] (or 1.1×10^7 tons by mass for a concentrated nuclear strike) seems to be considerably overestimated. Such a discharge is more probable in the case of intensive urban fires.

Up to 2.5×10^9 tons of fossil fuel (oil and gas) in oil and gas fields and in storage will presumably go up in flames, and the total area covered by forest fires (conservative estimate) will be 10^6 km^2 [19].

This area contains approximately 2.2×10^{10} tons of wood (dry weight) or 10^{10} tons of carbon and 10^8 tons of fixed nitrogen. If we assume that around 25% of the available phytomass will burn up, then 2.5×10^9 tons of carbon will be released into the atmosphere, which together with the burned fuel will make about 4 Gt of carbon. This would increase the CO_2 content of the atmosphere by only 2 or 3 ppm (the present total CO_2 content of the atmosphere reaches 700 Gt).

Forest fires will produce great amounts of highly dispersed particles: up to 75 kg of such particles will be formed per ton of burned material. The total mass of highly dispersed particles will reach 2×10^{14} tons, which is comparable to the present annual production of anthropogenic aerosol particles (smaller than 3 μm).

A number of recent papers [15, 19, 30] mention the possibility of a significant cooling on the planet as a result of nuclear war. The onset of a severe 'nuclear winter', with all its catastrophic ecological consequences, is predicted. The most detailed study of this subject so far is [30]; its results will be given briefly and analyzed below.

Scenarios involving nuclear strikes of 100 to 25 000 Mt power are considered in [30]. A 5000 Mt explosion (in the Northern Hemisphere) is taken as the basic variant. It is extremely important to take into account the amount of highly dispersed dust produced during an explosion, since it can remain in the stratosphere for a long time. The amount of dust produced by a 1 Mt surface explosion can be assumed to be 3.3×10^5 tons. For a total yield of 5000 Mt (and the assumed distribution of surface and atmospheric explosions) the total mass of dust (which is only part of the ejected soil) will, as noted above, be 9.6×10^8 tons; 8.4% of the particles will be smaller than 1 μm in size. Therefore, the injected submicron-sized particles will amount to 25 tons per kiloton, or about 0.5% of the total ejected soil mass (i.e., of the 5000 tons per kiloton of power).

Besides the fine dust particles produced during explosions, highly dispersed smoke and soot particles formed by massive fires can affect considerably the transmission of solar radiation through the atmosphere. It is believed that 52% of the aerosol particles will be produced by fires in cities, where the concentration of combustible material is much higher (on the average 3 g/cm^2, and in the city center 10 g/cm^2); 7% will be formed during firestorms, and 34% will be generated by forest fires during the first 10 days (the concentration of combustible material there is on the average 2.2 g/cm^2) and 7% during the next month.

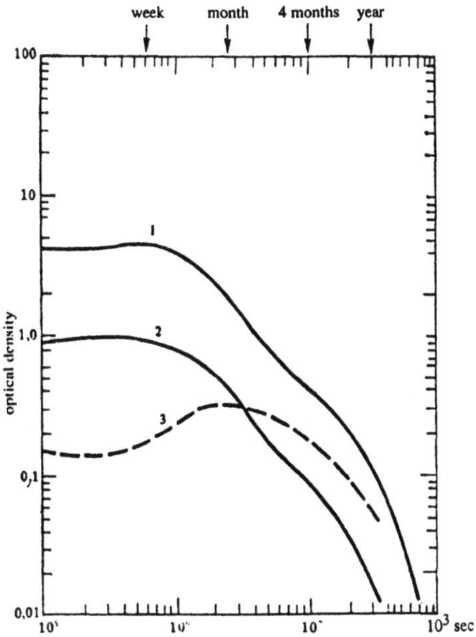

Figure 10.1. Vertical optical density of atmosphere as function of time after nuclear strike (for wavelength of 550 nm): (1) main variant, explosive force of 5000 Mt; (2) explosive force of 100 Mt; (3) El Chichón volcanic eruption.

An amount of 1.9 g/cm^2 of combustible material will be burned in cities, 0.5 g/cm^2 (over an area of 5×10^5 km^2) in forests, giving a total mass of smoke equal to 2.25×10^8 tons, 5% of which will reach the stratosphere.

Highly dispersed aerosols about 0.1 μm in diameter (including soot particles, which make up 25%) produced by fires will absorb very strongly sunlight in the visible and IR ranges.

All this should lead to a sharp increase in the optical density of the atmosphere (Figure 10.1).

For the main variant (5000 Mt), the initial optical density (for a wavelength of 550 nm) will be around 4.0, with 1.0 of this attributable to stratospheric dust and 3.0 to tropospheric smoke. A month later the optical density would still be 2.0. After 2 or 3 months the effect of the dust will prevail in optical effects, since the soot and smoke will already have been washed out by precipitation. This change in the optical density will lead to a drastic temperature drop: according to the baseline scenario the temperature will drop 30° K in 3 or 4 weeks and will reach −23°C; subzero temperatures will then persist for several months (Figure 10.2). In [30] it is emphasized once again that the soot layers in the troposphere will cause a sudden cooling of the Earth's surface for a short time, while the finely dispersed dust particles in the stratosphere will cause long-term (for a year or longer) cooling. It is

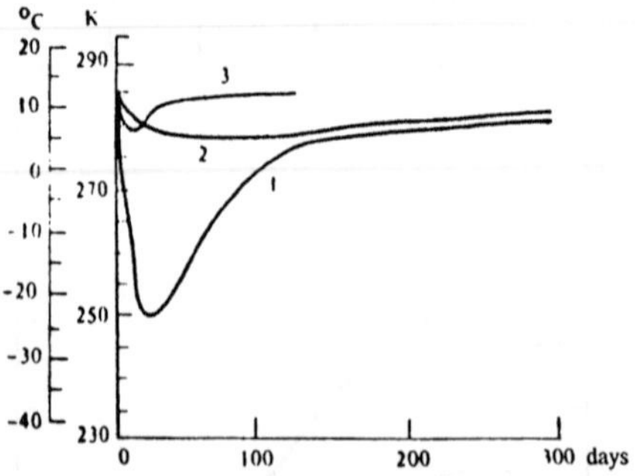

Figure 10.2. Change in mean of temperature of Earth's surface after nuclear strike [30]: (1) main variant, explosive force of 5000 Mt; (2) explosive force of 5000 Mt, only taking 'explosive' dust into account; (3) explosive force of 1000 Mt.

also noted in [30] that, if the amount of submicron-sized particles is no more than 10 t/kt, then the dust optical effect will not exceed the corresponding effect for a large volcanic eruption.

The predicted temperature fluctuations of the World Ocean are relatively low (less than 3°K), because of its heat capacity and the rapid mixing of the surface water.

The amount of solar radiation is reduced so much that for a 10 000 Mt explosion over most of the Northern Hemisphere the mean level of illumination of the Earth's surface will for about 40 days be below the minimum required for photosynthesis.

The effects in the Southern Hemisphere will be significantly weaker than in the Northern Hemisphere, although the exchange of polluted air masses between the hemispheres can be quite rapid (for a week).

Therefore, the absorption and reflection of solar radiation by aerosol particles will change the radiation characteristics of the atmosphere, will cause an overall cooling of the Earth's surface, and will affect the large-scale and mesoscale circulation of the atmosphere. The appearance of numerous condensation nuclei will inevitably affect the development of clouds and precipitation; the albedo of the Earth's surface will change. The evaporation of seawater in the presence of an influx of cold air masses from the continents will cause intensive precipitation in the coastal regions and on the continents; the latent heat of vaporization will be a source of energy, which can cause storms.

However, in [30] the onset of a new ice age was excluded, because of the great inertia of the climate system.

The cooling produced by nuclear explosions was dealt with in [15]. The calculations were made assuming that the troposphere in the Northern Hemisphere is polluted only by soot (up to optical densities of 6.0), and that the

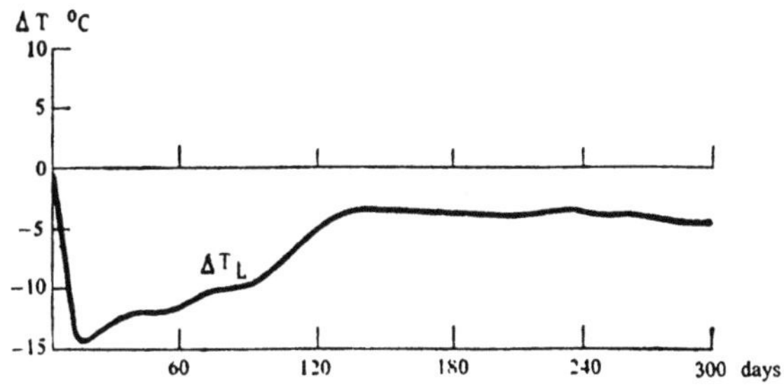

Figure 10.3. Drop in temperature of Earth's surface after nuclear strike [15].

stratosphere is polluted by dust (optical density of 1.2).

This will result in a rapid cooling of the Earth's surface in the Northern Hemisphere (by 15°C in 10 days) (Figure 10.3). The cooling (down to −5°C) will remain for a year.

As a whole the atmosphere will 'warm up', and the heat balance of the climate system will be positive during the first three months. The sea surface will cool less, on the average, by only 1.2°C during ten months. The most severe cooling in the troposphere will be observed in the midlatitudes of the Northern Hemisphere; 40 days after the nuclear explosion the temperature will be −15°C. The subsequent warming of the lower atmosphere will probably occur due to the heat (long-wave radiation) accumulated at first by the upper atmosphere.

The temperature drop in the Southern Hemisphere will be insignificant (less than 5°C).

For comparison, let us consider briefly the geophysical consequences of volcanic eruptions.

During a major volcanic eruption a large amount of crushed rock (the total amount can be 10^9 m^3 or more) is thrown up into the atmosphere; only fractions of a percent by mass of these are highly dispersed particles. Furthermore, during an eruption a huge amount of gaseous pollutants (primarily sulfur dioxide) enters the atmosphere; these are quite rapidly transformed into drops of sulfuric acid, sulfate particles, and other compounds. The particle dimensions are usually < 1 micron, and their residence time in the atmosphere (on condition that they enter the stratosphere) can be 400 days or more [16] (according to some data it can be more than 2 years).

It can be assumed that in the stratosphere the content of sulfate aerosol particles is approximately twice the content of dust (ash). It has been calculated that, in years which were preceded by a number of years without large volcanic eruptions, the background content of aerosol particles in the stratosphere was 0.25 Mt.

During the eruption of Mount St. Helens (18 May 1980) the amount of sulfate aerosol in the stratosphere reached only 0.5 Mt, while during the eruption of Mount Fuego (1974) it was 5 Mt and during the eruption of Mount Agung (1963) it was 30 Mt [12].

As already mentioned, a month after the largest eruption in recent times (El Chichón, 4 April 1982), the amount of stratospheric aerosol was 23 Mt, and six months later it was 8.3 Mt; the optical density of the aerosol cloud in the visible spectrum was 0.3 (for a background value of 0.08) [24].

The flux of direct solar radiation in the vicinity of the Hawaiian Islands was found to be reduced (according to Mauna Loa Observatory) by 20%, and the flux of total radiation by 7%.

Since a relatively small quantity of aerosol particles appeared in the stratosphere after the eruption of Mount St. Helens, its effect on the climate was insignificant (the estimated temperature drop in the Northern Hemisphere was only 0.05 K); the effect on the ozonosphere was insignificant as well (according to assessments, the ozone content did not decrease by more than 0.2%).

The eruption of El Chichón should have had a more significant climatic effect [28]. The most considerable cooling was to have occurred in the northern polar regions in spring and autumn 1984 and in spring 1985. Maximum cooling (by 0.58 K) was to take place in the Northern Hemisphere three years after the eruption.

The settling of a great amount of soot particles onto the surface of glaciers and sea ice (and winter snow) will accelerate their melting considerably and can affect the state of the cryosphere on a large scale.

What will be the role of gases produced by explosions and fires? The amount of carbon monoxide released by massive fires will reach 2×10^8 to 4×10^8 tons (by carbon), the amount of nitrogen oxides will be 0.15×10^8 to 0.3×10^8 tons (by nitrogen) (the same amount as is released by the explosions themselves), which will exceed somewhat the annual anthropogenic production of NO_x [19].

Millions of tons of ethylene and propylene are released into the atmosphere, to become significant components of the photochemical smog.

Exploding gas fields can release, unburnt, considerable amounts of methane (CH_4), ethane (C_2H_6), and propane (C_3H_6). The ethane concentration can increase to 50–100 ppb (by volume), the normal content being 1 or 2 ppb.

If catalysts like nitrogen oxides are present in the troposphere, then a smog situation develops and tropospheric ozone is intensively produced (from methane and oxygen). A dangerous phytotoxicant, peroxyacetyl nitrate, is formed. Its concentration reaches 1 to 10 ppb (for a normal content of 0.1 ppb). Naturally, such a situation will follow the deposition of most of the aerosol particles, during 'clearings' of the atmosphere.

The concentration of tropospheric ozone can be expected to increase to

160 ppb, with formation of smog on a global scale. Such an increased content of tropospheric ozone, ethane, and other minor pollutants would lead to a very substantial change (steady rise) in the mean temperature on the Earth (due to the greenhouse effect) and by at least 4 or 5°C in the Northern Hemisphere [8]. These data were obtained by linear extrapolation of the data of [11, 17].

In [11] it was shown that a rise in just the ozone concentration in the troposhere by 30 ppb (for a background value of 30 ppb) would result in a 0.67°C temperature rise.

All this would lead to a very substantial change in the climate of the entire planet, resulting in extensive droughts, desertification, redistribution of water resources, loss of crop yields, and large-scale drying up of the vegetation.

It should be noted that the effect of the gaseous pollutants produced after explosions, together with the resultant intensive greenhouse effect, were not taken into account and were not considered to be an important factor in possible climatic changes in any work prior to that of the author [8] (and the author's report to the Conference on The World after Nuclear War via the Moscow-Washington satellite channel on 1 November 1983).

The effects of the radioactive products produced by explosions on the properties of the atmosphere have not been considered either. The concentration of radioactive products of nuclear explosions (8×10^{11} Ci a week after the explosion, assuming that 20% of the energy is released due to a fission reaction), evenly distributed over the troposphere (which will account for approximately 10% of the total radioactivity), will be 5×10^{-7} Ci/m^3; this is equivalent to an energy release of approximately 5×10^{-7} MeV (m^3 s). This will cause ionization of the air 500 to 1000 times greater than the ionization needed to change the electrical conductivity of the air by 10% [10]. Such an effect is bound to bring about major changes in the properties of the atmosphere.

If we adopt the scenario with a predominant (or significant) use of surface nuclear explosions, when a vast quantity of aerosol particles (including highly dispersed particles) is released into the atmosphere (including the stratosphere), then in this case the gigantic cloud of stratospheric aerosol particles will cause reflection (absorption) of some of the solar radiation and a consequent decrease in the mean tropospheric temperature at the beginning, after which the 'greenhouse' effect will prevail. The assumed total quantity of aerosol particles injected into the stratosphere from explosions and fires will be the decisive factor in such calculations, since the residence time of these aerosols in the stratosphere is many months, or even years.

10.3. The Effect on the Ozone Layer

The damage to the ozone layer of the atmosphere caused by a nuclear war will be substantial also. This is primarily due to the destruction of the ozone layer resulting from the interaction of stratospheric ozone with the huge amount of nitrogen oxides formed in the high-temperature region of a nuclear explosion.

Assessments show that $(0.9 + 0.6)\times10^{32}$, or about 1×10^{32}, molecules of nitrogen oxides are produced per Mt of explosive power [25]. Taking into account the different heights of ascent of the nuclear cloud for different powers, it has been calculated that a series of 10^4 Mt nuclear explosions would destroy from 30 to 70% of the total ozone in the ozonosphere of the hemisphere where the blasts took place, and that it would take from 2 to 4 years for it to return to normal [9, 19, 25].

Experimental data on ozone measurements in the stratosphere were collected after a series of powerful nuclear explosions in 1961–1962 [1, 22], which produced about 3×10^{34} molecules of NO_2 in the atmosphere. Data were also obtained with the satellite *Nimbus-4*, which measured the ozone content after a French thermonuclear explosion of megaton force (1970). These data raised some doubts, since the measurements did not indicate a significant decrease in the stratospheric ozone content on a global scale [1]. On the average, the ozone content decreased by 2.2% (according to ground-based observations [23]), while according to calculations the decrease in the Northern Hemisphere should have reached 4% [18].

An analysis of the photochemical consequences of the fall of the Tunguska meteorite indicates a considerable decrease in ozone content due to the release of an enormous amount of NO_x into the stratosphere [29]. The amount of NO_x produced at altitudes above 10 km by the shock wave of the falling meteorite is estimated to be 3×10^7 tons, which is equivalent to 6×10^{35} molecules of NO_x. According to the authors of [29], this may have been overestimated by a factor of 10. A comparison of model calculations and observations indicates a possible decrease in the total ozone content in the Northern Hemisphere by as much as 45% during the first year after the meteorite fell.

The short-term effects of powerful atmospheric nuclear explosions in the megaton range (including multiple explosions) on the ozonosphere, taking into account the consequent photochemical processes and atmospheric diffusion, are dealt with in detail in [9].

The results of the calculations depend on the height distribution of the nitrogen-oxide concentration, which depends in turn on the height of ascent of the nuclear cloud.

This cloud height depends on the strength of the explosion and on the meteorological conditions, primarily on the temperature distribution with height. At equatorial latitudes, where the tropopause is higher, the level of cloud stabilization is higher than at polar latitudes. In the midlatitudes, for a 10 to 40 Mt explosion the cloud will stabilize in a height range of about 15–40 km [13], where about 80% of the total amount of radioactive substance will be located.

The volume of the cloud from a megaton explosion is approximately 1×10^{19} cm^3, and it is proportional to the power of the burst. Thus, the average initial concentration of nitrogen oxides NO_x in the stabilized nuclear cloud would be $(0.9 \pm 0.6) \times 10^{13}$ molecules/cm^3, and it would be practically independent of the explosive yield (this value is 20 to 30 times higher than the value in [22]).

The additional number of ozone molecules produced by the UV radiation of the fireball and by the ionization of the air by nuclear radiation was not taken into account in [9]. The role of this (relatively minor) source of ozone requires further study.

The release of nitrogen oxides into the stratosphere will obviously change the equilibrium content of ozone in the layer occupied by the nuclear cloud. Seven principal reactions were used to assess approximately this O_x/NO_x variation:

1. $O_2 + h\nu \rightarrow O + O$,
2. $O + O_2 + M \rightarrow O_3 + M$,
3. $O_3 + h\nu \rightarrow O_2 + O$,
4. $O_3 + O \rightarrow O_2 + O_2$,
5. $NO + O_3 \rightarrow NO_2 + O_2$,
6. $NO_2 + h\nu \rightarrow NO + O$,
7. $NO_2 + O \rightarrow NO + O_2$.

The rates of photodissociation of molecular oxygen (I_2) and ozone (I_3) (because of solar radiation in the wavelength range λ_1, λ_2, reactions 1 and 2, respectively) depend on the altitude and number of O_2 and O_3 molecules in the air column at a level h:

$$I_x = \int_{\lambda_1}^{\lambda_2} F(\lambda)T(\lambda)S_x(\lambda)\,\mathrm{d}\lambda, \tag{10.2}$$

where $F(\lambda)$ is the spectral function of the incoming radiation; $T(\lambda)$ is the atmospheric transparency (governed by the radiation absorption by O_2 and O_3 molecules and by the molecular scattering, and dependent on the content of O_2 and O_3); $S_x(\lambda)$ is the cross section of radiation absorption by an O_x molecule.

Figure 10.4 gives calculated values of $I_2(h, D_3)$ and $I_{3,\mathrm{ex}}(h, D_3)^*$ as functions of the ozone content D_3 (molecule/cm^2) in the air column, obtained for a number of levels of the standard atmosphere at a latitude of 45° and a zenith angle of the sun $\varphi = 45°$ in the wavelength range 183 nm $\leq \lambda \leq$ 242 nm.

It should be noted that the character of the change of $I_2(h, D_3)$ depends

---

* The rate of photodissociation with production of O_2 and O in an excited state.

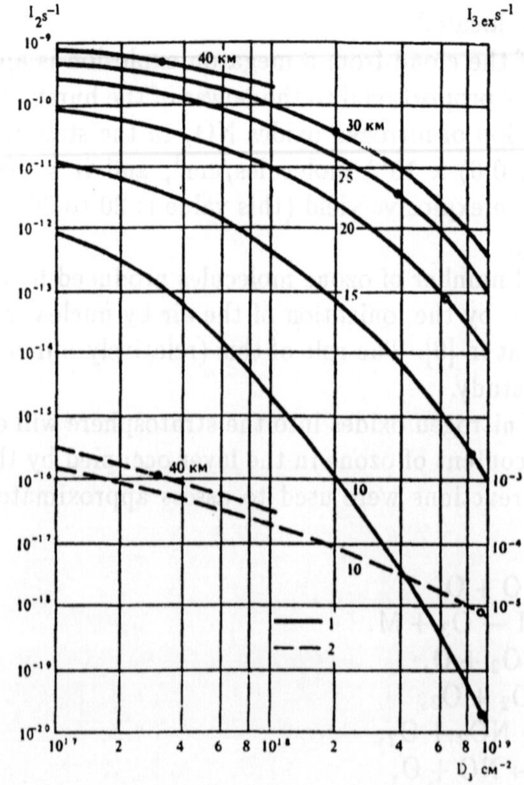

Figure 10.4. Photodissociation rates of oxygen $I_2(1)$ and ozone $I_3(2)$ molecules, as functions of ozone content D_3 in atmosphere at different altitudes.

considerably on the altitude.

The concentration of ozone and other substances in the nuclear cloud will be determined not only by the chemical processes but also by the dynamics of the cloud expansion due to atmospheric diffusion.

In order to simplify the problem, in [9] we considered the average values of the concentrations of material in the cloud and of their change in time, which is connected with both the increase in the volume of the cloud and the entrainment into it of ozone and atmospheric oxygen from the surroundings. The additional income of material (O_3 and O_2), equal to the product of the volume increment times its concentration outside the cloud, was taken as a first approximation. In this case the change in the $[O_3]$, $[O_2]$, $[NO_x]$ concentration in the cloud due to atmospheric diffusion is described by the equation

$$\frac{dc}{dt} = (c_\infty - c)\frac{1}{v}\frac{dv}{dt}, \tag{10.3}$$

where c and c_∞ are the concentrations of material in the cloud and outside the cloud; v is the volume of the cloud.

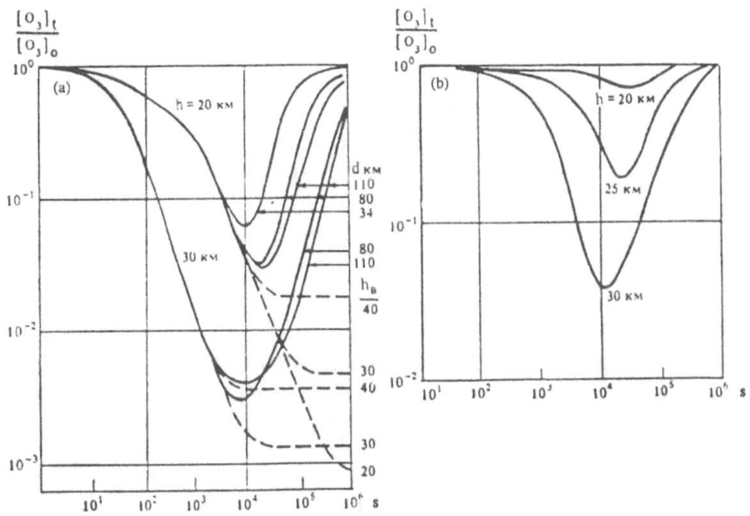

Figure 10.5. Relative change of ozone concentration in time at various altitudes in nuclear cloud: (a) $[NO_x] = 10^{13}$ molecules/cm^3, for different diameters d of cloud (corresponding to different explosive forces); (b) $[NO_x] = 10^{12}$ molecules/cm^3, $d = 80$ km.

If the diffusion time of the cloud is sufficiently long, and it is not very important to take its initial size into account, then the change in cloud volume just due to horizontal scattering will be determined by the expression

$$v = \frac{Q}{c_{\max}} = 2\pi\sigma_h^2(t)\Delta h, \tag{10.4}$$

where $c_{\max}$ is the maximum concentration of the Gaussian distribution of a pollutant with a variance $\sigma_h^2(t)$, Q is the total amount of pollutant, and Δh is the vertical size of the cloud.

Experimental data indicate that the change in the horizontal variance over a wide time interval is defined well by the equation $\sigma_h^2(t) = \alpha^2 u^2 t^2$, where u is the wind velocity and α is a numerical coefficient depending on the cloud altitude $\alpha = \alpha(h)$. For the upper atmosphere $\alpha \approx 0.03$.

The vertical cloud diffusion can be ignored, in comparison with the horizontal diffusion, when considering relatively short-term processes.

A numerical method was used to integrate a system of equations (10.3) for $[O_3]$, $[NO_2]$, and $[O]$. The vertical stratospheric ozone distribution given in [26] was taken as the background concentration in the cloud $[O_3]_0$.

The initial concentration of nitrogen oxides (after the explosion) in the cloud varied from 10^{12} to 10^{13} mol/cm^3. Figure 10.5 shows the calculated relative change in the ozone concentration $[O_3]_t$ / $[O_3]_0$ in time at different altitudes in the clouds produced by explosions of different forces (for $u = 80$ km/h). The explosive force determines the top and bottom of the cloud and

Table 10.2. Relative (%) content of ozone in vertical column at various times.

Explosive force, Mt	Time after explosion, h							
	3.0	10	30	70	3.0	10	30	70
	$[NO_x] = 10^{12}$ mol/cm^3				$[NO_x] = 10^{13}$ mol/cm^3			
35	54	46	54	71	25	25	28	39
10	68	63	76	88	35	34	41	60
1.0	96	98	99	100	75	80	91	97

its initial horizontal diameter d.

As shown by Figure 10.5, the degree of ozone destruction in the nuclear cloud depends primarily on the initial NO_x concentration, and to a lesser extent on the initial size of the cloud. The character of the ozone recovery is mainly determined by the dynamics of cloud expansion and it depends greatly on the initial cloud size. The dashed lines in Figure 10.5a indicate the equilibrium concentrations of ozone for a hypothetical case in which the concentration $[NO_x] = [NO] + [NO_2]$ in the cloud does not change with time.

It is evident from Figure 10.5 (for an altitude of 20 km) that, the greater the initial horizontal size of the cloud, the lower will be the minimum ozone concentration in the cloud. However, with an increase in the explosive force, the vertical size of the cloud will be larger and the thickness of the layer $(h_s - h)$ of NO_x injection will be greater. Consequently, as already noted, the intensity of the UV radiation will increase, due to the reduced ozone content in the injection layer. Here the decrease of the minimum ozone concentration at an altitude h will even be mitigated slightly with increasing explosive force (at very high forces) (see Figure 10.5a, 30 km altitude).

Table 10.2 gives some calculated values of the relative ozone concentration in a vertical column passing through the nuclear cloud at various times and for various explosive forces.

The calculations indicate that, even in the case of single explosions with a force of the order of tens of megatons, the amount of ozone in the vertical column can be significantly reduced; its recovery time is about a week (or several weeks).

The calculations make it possible to estimate the change in the stratospheric ozone content due to the injection of nitrogen oxides from a series of nuclear explosions.

When considering the long-term global effects produced by the reduced ozone concentration resulting from a massive nuclear strike, we assume that an amount of nitrogen oxides corresponding to a total nuclear explosive force of 10^4 Mt is released into the stratosphere. In this case we can presume that after a certain time the nitrogen oxides will be distributed evenly over one of the hemispheres; a steady-state form of the kinetic equations will then be

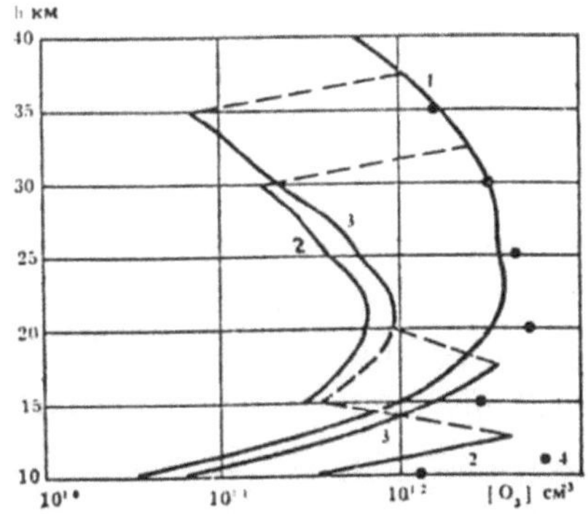

Figure 10.6. Stratospheric ozone distribution: (1) natural distribution; (2) when nitrogen oxides are injected into 15-km layer (15–30 km), $[NO_x] = 2 \times 10^{11}$ molecules/cm^3; (3) the same for 20–35 km layer; (4) ozone distribution at latitude of 45°.

applied for the assessment [9].

A system of equations of type (10.3) is used in [21] to describe the new steady-state distribution of ozone in the stratosphere. This system was solved by us with the help of a numerical method, taking into account the curves for coefficients $I_2(h, D_3)$ and $I_3(h, D_3)$ given in Figure 10.4.

Figure 10.6 (curve 1) shows the calculated steady-state distribution of ozone or the natural height distribution of O_x in the stratosphere [21] and also the natural ozone distribution at the latitude $\varphi = 45°$ N [26]. Despite the simplified character of the model (ignoring vertical diffusion and some traditionally considered chemical reactions), the calculated and actual data on the ozone distribution at altitudes above 20 km show good agreement. Thus the simple model can be used to describe the new disturbed ozone distribution when a large amount of nitrogen oxides has been injected into the stratosphere.

The average concentration of nitrogen oxides in a 15-km-thick layer over the Earth's surface in one hemisphere will be approximately 2×10^{11} mol/cm^3, for a total nuclear explosive force of 10^4 Mt. Figure 10.6 (curves 2 and 3) shows the distribution of ozone in the stratosphere when nitrogen oxides are injected into a 15-km layer with a given NO_x concentration of 2×10^{11} mol/cm^3. The results of calculations for NO_x injection into this layer with different upper and lower boundaries h_1 and h_2 are given for comparison.

The slight increase in the ozone concentration in the lower layers is due to the increased UV radiation caused by the destruction of the ozone layer in the upper stratosphere. The increase in the ozone concentration under the layer of NO_x injection compensates partly for the destruction of ozone in the layer of injection; the higher the injection layer, the more pronounced this effect will be. It is noteworthy that the partial conservation of the

Table 10.3. Ozone content in air column for various altitudes of injection.

Injection layer, km	Ozone content, 10^{18} mol/cm^2	Fraction relative to normal O_3 content, %
15–30	4.1	40
20–35	6.4	62
25–40	7.1	69

ozonosphere under the bottom of the nuclear cloud is vital, for the final estimates of the biological effects on the Earth accompanying the practically complete destruction of the ozonosphere in the case described.

Table 10.3 gives some estimated values of the ozone content in the entire air column of the Northern Hemisphere for a total force of the nuclear explosions of 10^4 Mt (in this same hemisphere).

According to these assessments, a series of nuclear explosions with a total power of 10^4 Mt would destroy from 30 to 60% of the total ozone content of one hemisphere.

The substantial amount of nitrogen dioxide formed can also to some extent shield against the direct solar radiation, although the calculations of A.D. Danilov indicate that this effect would be slight. According to his calculations, the flux of direct solar radiation in the wavelength range $\Delta\lambda = 280$ to 550 nm, which makes up about one fifth of the total flux, namely 250 w/m^2, will for an ozone concentration $q = 2 \times 10^{11}$ mol/cm^3 be reduced by 5% (the solar constant will here decrease by only 1%); for $q = 10^{12}$ mol/cm^3 the flux will be reduced by 27% (the solar constant will decrease by 6%).

In case of a massive nuclear attack the average NO_2 concentration in the atmosphere (everywhere after mixing) will be 2×10^{11} molecules/cm^3 [9]. Consequently, the total decrease in the solar constant will not exceed 1%.

The destruction of the ozone layer affects substantially the passage through the atmosphere of hard ultraviolet radiation, which will have serious biological consequences and will also affect the temperature distribution in the stratosphere, and consequently the Earth's climate as well.

The relative increase with time of the hard ultraviolet radiation (UV-B) after a massive nuclear strike (at various latitudes), and also the possible almost tenfold increase in the UV radiation at midlatitudes 3 to 6 months after a nuclear war (assuming that most of the aerosol particles have been deposited), were given in [19]. Figure 10.7 shows the increase in the flux of UV radiation with a decrease in the ozone content [25].

The ozone layer serves as a shield against the penetration of ultraviolet solar radiation in the wavelength range 240–320 nm. Since UV-B radiation is readily absorbed by the nucleic acids in living cells, it presents a particular danger to all living things. Furthermore, exposure to hard ultraviolet radiation increases the probability (and consequently the frequency of occurrence) of skin cancer (skin melanoma and carcinoma). It has been calculated [25]

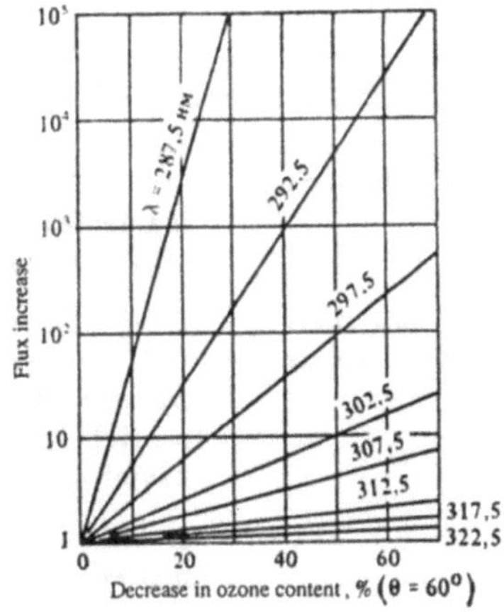

Figure 10.7. Increase in UV radiation flux (for various wavelengths λ) with decrease in total ozone content for height of sun $\theta = 60°$.

that a decrease in the ozone layer by only 5% will increase the number of cases of human skin cancer on the average by 10% (see Section 8.2).

It has been noted that some plants are very sensitive to an increase in UV-B radiation, phytoplankton, in particular [27].

10.4. The Ecological Consequences of Nuclear War

Thus it is obvious that nuclear explosions, particularly in the case of a massive nuclear attack, not only will result in destructive and devastating effects of a local nature, but also will cause serious global disturbances of the biosphere, because of irreversible changes in climate, destruction of the Earth's ozone layer, disturbance of the atmosphere as a whole, and a substantial rearrangement of terrestrial and aquatic ecosystems.

The highly dispersed aerosol particles formed will readily absorb sunlight in all spectral ranges. For at least several weeks after the explosion a substantial drop in the intensity of solar radiation over large areas will be observed (up to 1/2 to 1/150 of the original intensity). This cannot but affect agricultural production and the state and health of the human population, which will have already suffered (in the case of nuclear war) from the direct effects of nuclear explosions. This will influence greatly the state of ecosystems, particularly marine ecosystems, and agricultural products. According to assessments in [19], phytoplankton and plant zooplankton will disappear in more than half of the World Ocean.

Table 10.4 summarizes the most significant ecological short-term (from several weeks to months) and long-term (from months to years) consequences of a massive nuclear attack on terrestrial and marine ecosystems. Data from

Table 10.4. Ecological consequences of nuclear war.

Short-term (several weeks to months)	Long-term (months to years)
	Terrestrial ecosystems
1. Sudden cooling will seriously damage and kill plants, particularly during growth period	1. Subsequent warming (with clearing of atmosphere) will increase aridity, causing decreased bioproductivity and death of plants and animals
2. Atmospheric turbidity and smokiness will suppress photosynthesis and decrease bioproductivity	2. As atmosphere clears, fluxes of hard UV radiation increase. This will have adverse effect on surviving plants, and on sight organs of animals, and on behavior of insects (including pollinators)
	3. Massive destruction of plant and animal species will reduce genetic and species diversity.
	Marine ecosystems
1. Decreased illumination will suppress photosynthesis and phytoplankton production. Many marine animals will lose their food supply, and organisms will die off	1. Subsequent temperature rise will lead to 'blooming' in many water bodies and to an ecological 'explosion', to reproduction of certain aquatic organisms. Dying off of other species will still continue
2. Biosedimentation from ocean surface layer will decrease sharply, reducing removal of toxic substances from this layer	2. Hard UV radiation will suppress photosynthesis
3. Lowering of temperature will cause icingover of fresh-water bodies and subsequent partial destruction of flora and fauna, with disappearance of species	

Notes. 1. Particularly strong effects will be observed at tropical and polar latitudes. 2. The indicated damage will develop against a background of intensive radiation damage to ecosystems.

[8, 20] were used to compile the table.

There is no doubt that in the case of nuclear war the effects on the biosphere as a whole would ultimately create a larger number of victims than would the direct effects of the explosion.

CHAPTER 11

Anthropogenic Effects on the World Ocean

Most of the Earth's surface is covered by oceans, and the oceans are of paramount importance in maintaining life on our planet and in forming its weather and climate.

The oceans participate actively in global processes, which include: the interaction between the ocean and atmosphere, determining the climate of our planet; biogeochemical cycles of chemical elements, closely connected with the circulation of energy and matter in natural ecosystems; the photosynthetic activity of algae, regulating the balance of oxygen and carbon dioxide; etc.

The World Ocean plays a major role in the solution of important problems like the provision of a supply of food and energy and the extraction of minerals.

At the same time, the ecosystems of the World Ocean represent, as a rule, the final stage of various large-scale transformations and flows of various substances. Therefore, the ocean is subjected to a substantial anthropogenic impact, leading to significant negative consequences, including a reduced reproduction of bioresources.

A stressed ecological situation is already observed in a number of regions of the World Ocean, where fields of chronic pollution have been produced. The arrival of anthropogenic pollutants and the irrational extraction of bioresources, including fish (more than 70 million tons per year are caught) are continually acting ecological factors which can transform the marine ecosystems. In recent years the extraction of minerals (oil and gas, iron-manganese concretions, polymetallic sulfides, etc.) has intensified the negative impact on the ocean.

11.1. The Major Sources of Ocean Pollution

Pollutants enter the World Ocean both as a result of natural processes and as a result of man's economic activity. These sources can be:

(1) the direct discharge of pollutants into the ocean (mainly onto its surface), for example, petroleum products during transport, particularly in the

event of tanker accidents;

(2) the direct introduction of pollutants during underwater development and extraction of mineral resources;

(3) river runoff;

(4) direct wastewater discharge (drainage) from land (terrigenous discharge);

(5) atmospheric transport;

(6) underwater discharge of oil and gas.

A detailed review of marine pollution and its consequences is given in [4].

Pollution of the World Ocean by petroleum products during their production and transport is one of the most important problems, since stable fields of oil pollution are observed precisely in the zones of transport (sea lanes) and fishing (mainly in the coastal and offshore areas) [49].

Modern assessments indicate that river runoff and terrigenous discharge, together with atmospheric transport, are the major sources of pollution of the World Ocean. The inputs from these sources are compared in Table 11.1 [7].

The river runoff and the drainage from the land pollute mainly the coastal water of the ocean, inland seas, and bays.

The prevailing anthropogenic pollutants in the terrigenous drainage and river runoff originate in industrial, municipal, and agricultural wastewater. These are primarily heavy metals, biogenic compounds, chlorinated hydrocarbons and petroleum products. It is important to note that in the surface water the anthropogenic runoff entering the World Ocean is often comparable with the natural influxes of chemical elements and their compounds, and it may even exceed these.

As an illustration, let us refer to Figure 9.1, which shows the balance of natural and anthropogenic sulfur, including its sinks and its atmospheric transport into the ocean (and partially onto the land) [42]. It follows from Figure 9.1 that about 90% of the anthropogenic sulfur arrives at the ocean via river runoff, while about 10% is delivered via atmospheric transport. These mechanisms account for 40 and 2.5%, respectively, of the total sulfur received by the ocean.

Chemical toxic substances carried by river runoff and land drainage create local pollution zones, whereas the atmospheric transport and deposition of pollutants onto the underlying ocean surface produce large-scale, and often global, pollution of the World Ocean.

Many investigations indicate that lead, mercury, and other heavy metals, as well as DDT, PCBs, low-molecular petroleum hydrocarbons, and other organic substances, in the gaseous phase or in a suspended state (particles, solid salts) are transported by air currents over thousands of kilometers to the most remote areas of the ocean. This problem is part of the general problem of long-range atmospheric transport. For instance, DDT, PCBs,

Table 11.1. Natural and anthropogenic flows of pollutants into the World Ocean.

Pollutant	Runoff, t/year		Fraction of anthropogenic runoff, %	Flow into ocean, t/year	
	natural	anthropogenic		direct pollution, land runoff	atmospheric deposition
Lead	1.8×10^5	2.1×10^6	92	$(1.0 - 20.0) \times 10^5$	$(2.0 - 20.0) \times 10^5$
Mercury	3.0×10^3	7.0×10^3	70	$(5.0 - 8.0) \times 10^3$	$(2.0 - 3.0) \times 10^3$
Cadmium	1.7×10^4	1.7×10^4	50	$(1.0 - 20.0) \times 10^3$	$(4.0 - 140) \times 10^2$
PCBs	–	7.0×10^3 to 1.3×10^4	100	$(5.0 - 10.0) \times 10^3$	$(2.0 - 30) \times 10^3$
Oil	9.5×10^5	9.6×10^6 *	90	$(5.0 - 10.0) \times 10^6$	6.0×10^5
Sulfur [42]	1.3×10^8	1.2×10^8	47	1.1×10^8	1.1×10^7

* According to [58] 6.1×10^6 to 11×10^6 (total flow into ocean).
Note. Data on pollutant flows were considered earlier in [7].

phthalates, and a substantial amount of vanadium in aerosol particles have been detected in the atmosphere over the North Atlantic [33, 50]. DDT has been found in the atmosphere over Antartica, and high lead concentrations have been detected in Greenland glaciers and in the atmosphere over the Central Pacific [57, 59].

In the case of flows of heavy metals (mercury, lead, vanadium, chromium, etc.), more than half of the total flow of these elements into the ocean is via the atmosphere. This is attributed to the fact that the flow of anthropogenic heavy metals is mainly produced by the combustion of chemical fuel (coal and petroleum products). For instance, around 3000 tons of mercury enter the ocean annually due to atmospheric transport.

The average atmospheric content of anthropogenic lead reached 37 ng/m^3, which is several times higher than the natural lead concentration.

As a whole, annually 2.0×10^5 to 2.0×10^6 tons of lead; 2.0×10^3 to 3.0×10^3 tons of mercury; 5.0×10^2 to 1.4×10^4 tons of cadmium; 2.0×10^3 to 3.0×10^3 tons of PCBs (see Table 11.1); and 10^3 to 3.0×10^4 tons of arsenic are deposited onto the surface of the World Ocean from the atmosphere [62]. Thus it is clear that the atmospheric contribution to the pollution of the World Ocean corresponds approximately to the fraction of the river runoff in the total balance of pollutants in the marine environment.

11.2. Pollution of the World Ocean

Of the many pollutants which enter the World Ocean, a serious danger is presented by those chemical compounds which have a global distribution, which are continually entering the environment, and which have a distinct negative effect on living organisms.

An analysis of available experimental data concerning the effect of ocean pollutants on marine organisms demonstrates that the most hazardous, from the viewpoint of their abundance and influence on marine ecosystems, are the following: hydrocarbons (crude oil, petroleum hydrocarbons, petroleum products); chlorinated hydrocarbons (pesticides, polychlorinated biphenyls); toxic metals; radioactive substances.

The pollution of the World Ocean by oil and petroleum hydrocarbons has the largest scale of all. At present up to 6 million tons of petroleum hydrocarbons from different sources enter the seawater annually [49]. Almost half of the petroleum entering the hydrosphere is connected with transport and offshore production. It has been found that 20 to 30% of the total pollution is from oil spills.

When it enters the sea, the oil forms a surface film, or oil slick. The thickness of the film can vary: from a monomolecular layer to several centimeters. While in a marine environment, oil undergoes the following transformations: evaporation, emulsification, dissolution, oxidation, formation of oil aggregates, sedimentation, and biodegradation.

Under the influence of various physicochemical processes and the subsequent microbial destruction, the oil is transformed into emulsions and then into oil aggregates (tarballs, which consist, as a rule, of heavy fractions of petroleum hydrocarbons); some of the oil passes into the bottom sediments. The most persistent compounds are polyaromatic hydrocarbons (PAH); for example, the halflife of methylnaphthalene is (in the bottom sediments) two years [27, 51].

The petroleum hydrocarbon concentration in the World Ocean varies over a wide range. In the Baltic Sea the content of petroleum hydrocarbons is 0.8 to 8 μg/l, in the Atlantic near Bermuda it is 6 μg/l in the surface layer and 1 μg/l at a depth of 1000 m [40]. In individual regions of the Pacific (the southwestern part, the South China Sea) the content of petroleum products reaches hundreds of μg/l, although the pollution fields are distributed irregularly, and in some vast areas there is practically no oil pollution.

The petroleum concentration in the World Ocean is the highest in the thin surface microlayer [49], which serves as a collector of all the pollutants entering the marine environment: chlorinated hydrocarbons and heavy-metal ions often accumulate in the oil films.

Vast ocean areas, mainly along tanker routes, are permanently covered by oil films (Figure 11.1), while the surface microlayer is polluted by petroleum aggregates, the concentration of which is quite significant in some areas: 100 mg/m^2 (south of Japan), 500 to 600 mg/m^2 (in the Mediterranean Sea) (Figure 11.2) [38]. Such high pollution levels, up to 96 mg/m^2 in the Sargasso Sea, were also cited in [19].

During the MARPOLMON Project (the JOC/WMO Program on Monitoring the Pollution of Seawater by Oil and Petroleum Products) it was found that the average concentration of oil aggregates in the North Atlantic was 4.4 mg/m^2, while their total mass on the water surface reached 15 000 to 20 000 tons. The total amount of petroleum aggregates in the World Ocean is estimated to be 700 000 tons [29, 47].

The oil films or emulsified oil remain on the ocean surface for times from several hours to several days, depending on the environmental conditions and the type of petroleum. The oil aggregates persist for times ranging from several months to a year.

The greatest ocean pollution by heavy metals is observed primarily in coastal waters and inland seas, as a result of river runoff and industrial and municipal waste-water discharge. As already mentioned, atmospheric transport is an important source of ocean pollution by heavy metals.

Heavy metals accumulate readily in suspended solids and, being deposited in the bottom sediments, create a situation of chronic pollution of both the water column and the bottom sediments.

Mercury is the most toxic heavy metal. Upon entering the marine envi-

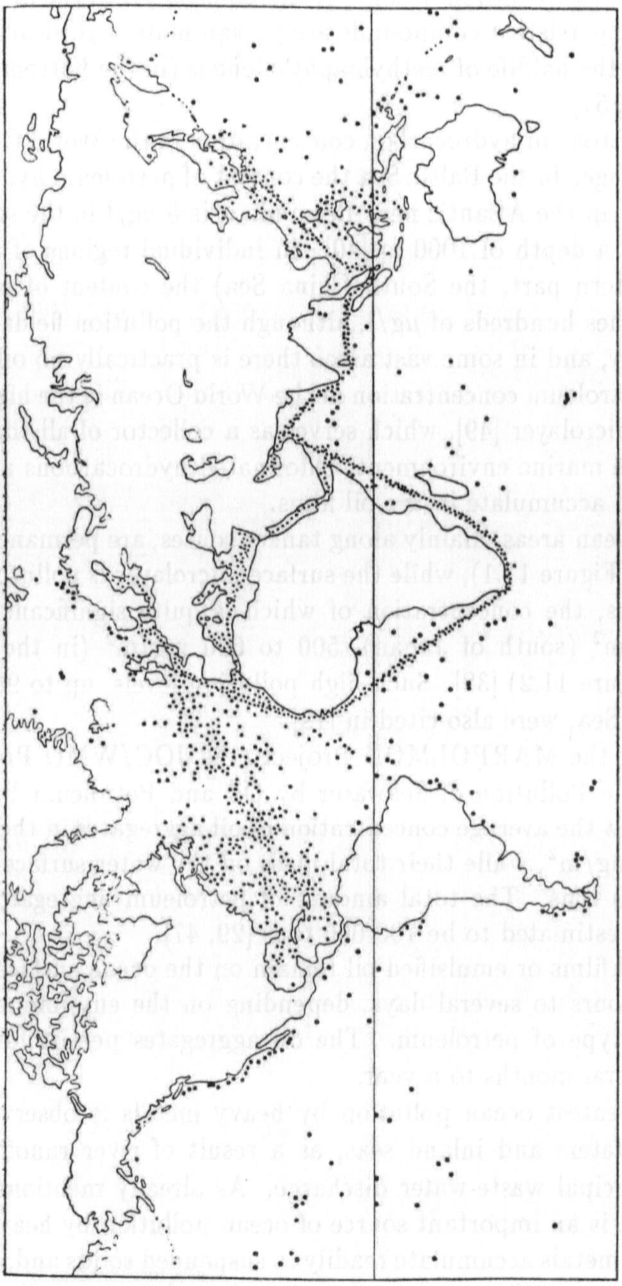

Figure 11.1. Distribution of oil films over surface of World Ocean [38].

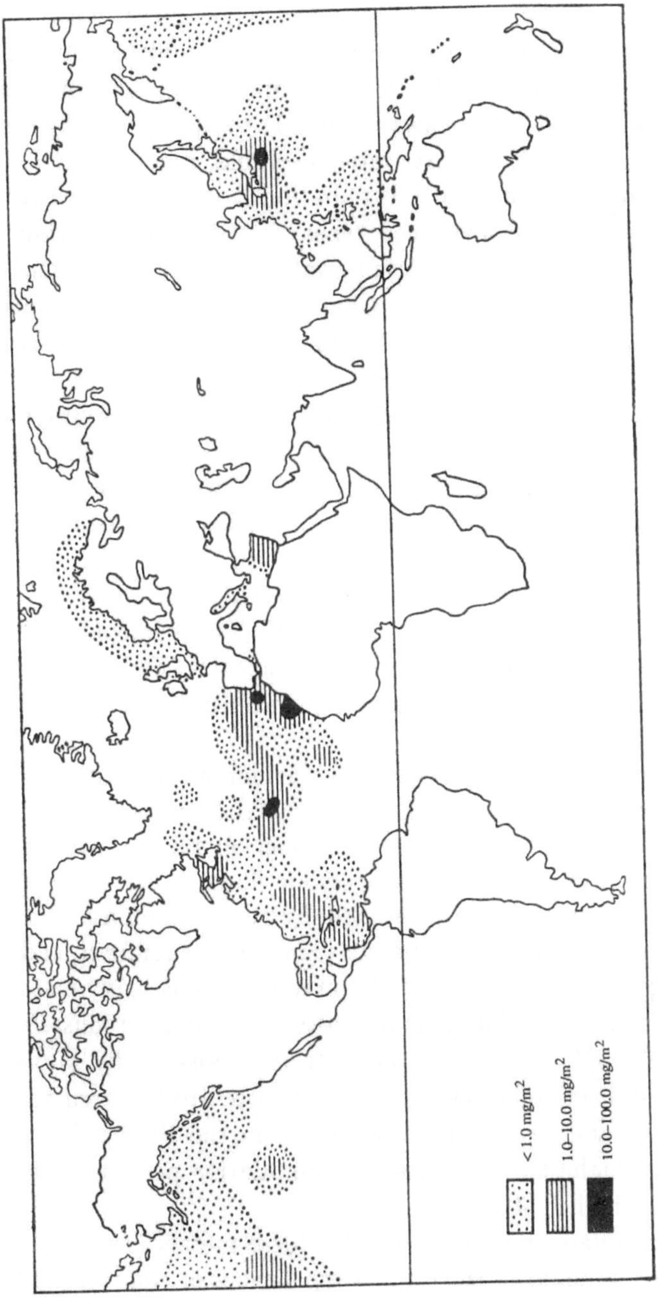

Figure 11.2. Concentration of petroleum aggregates over surface of World Ocean [38].

ronment, it combines with suspended particles and organic aggregates and is deposited on the sea bottom.

In the bottom sediments the mercury is exposed to certain forms of microorganisms [44] and is transformed into highly toxic forms of methylmercury (monomethylmercury and dimethylmercury), with a halflife of up to two years. Monomethylmercury [44] is accumulated intensively by fish (in the muscle tissues) and by benthic animals. Its accumulation factor in the biota (compared with seawater) is 10^3 to 10^4. Dimethylmercury, which is highly volatile and almost water-insoluble, readily enters the atmosphere, where under the effect of UV radiation it is transformed into metallic mercury. This process has been repeatedly described during observations of mercury migration with surface waters and its subsequent transformation in freshwater ecosystems (see Chapter 2). The mercury concentration in the open ocean varies from 5 to 1000 ng/l, being on the average about 30 ng/l [35, 45].

Arsenic, like mercury, when entering the marine environment is deposited on the sea bottom, where as a result of bacterial transformation methylated arsenic is formed. High concentrations of methylated arsenic have been detected in hydrobionta, but this accumulation does not increase substantially along the food chain.

The mean cadmium concentration in Atlantic water is 70 ng/l [30] (with a scatter of values from 40 to 300 ng/l). The lead concentration varies from 2 to 110 ng/l [31]; the mean copper concentration is 1.2 μg/l (it varies from 0.4 to 3.8 μg/l) [30]; the mean zinc concentration is 1.5 μg/l (it varies from 0.9 to 5.2 μg/l [30].

The total influx of heavy metals, including the anthropogenic component, is characterized by the data in Table 11.1 (Section 11.1).

The table shows that at present such metals as lead and mercury enter the ocean primarily via anthropogenic flows.

A serious problem is presented by the accumulation and transport via food chains of such heavy metals as mercury, cadmium, and lead. Some marine organisms are known to accumulate certain heavy metals. For example, oysters can accumulate zinc in quite high amounts; mussels can accumulate cadmium [31]. Mercury is actively incorporated by plankton. The coefficients of accumulation of metals (cobalt, iron, mercury, zinc, chromium, antimony, and scandium) in plankton in the Bering Sea vary from 10^2 to 10^4. The highest coefficients of accumulation pertain to mercury: they can be 10^4 or higher [4].

A grave danger is presented by toxic metals (mercury, lead, cadmium) which accumulate in the organs and tissues of commercial fish.

Unlike petroleum and heavy metals, chlorinated hydrocarbons form a group of xenobiotics, i.e. unnatural components of the environment. These include many pesticides like DDT, dieldrin, aldrin, hexachlorocyclohexane

Table 11.2. Average levels of World Ocean pollution by benzo(a)pyrene.

Type of water area	BP Content, $\mu g/l$	
	surface microlayer	0–100 m
Clean(background area)	0.018	0.003
With heavy navigation	0.040	0.010
With specific sources of PAH (industrial zones, etc.)	0.150	0.030

(HCCH), and polychlorinated biphenyls (PCBs).

These compounds are widely used in agriculture and industry, and their spread has become a global problem. The US production of pesticides alone is about half a million tons per year, about half of which is used within the United States [48].

A substantial amount of pesticides is transported by the wind and by river runoff, and it eventually enters the ocean.

A number of chlorinated hydrocarbons, long-lived DDT and its metabolites (DDE, DDD), and also PCBs, have been detected in various areas of the World Ocean and in many components of marine ecosystems. It is interesting that the ratio between the DDT and PCB concentrations in the surface water is less than 0.05 [40]. This fact indicates that PCBs are more stable than DDT and its metabolites, since the ratio of the world production of DDT and PCBs is much higher than unity (a different sorption ability is also possible).

The concentration of DDT, together with DDE, in the surface microlayer (thickness of about 200 μm) in the North Atlantic reaches 0.09 $\mu g/l$. They have been detected at depths down to 800 m [3]. The average concentration of PCBs in the Western Atlantic is 30 ng/l, while in the Baltic Sea it varies from 30 to 90 ng/l [41].

Chlorinated hydrocarbons are readily soluble in fats and therefore they accumulate actively in lipid fractions of marine organisms, as well as in ocean slicks with high concentration of lipids and petroleum hydrocarbons.

The circulation and accumulation of toxic mutagenic compounds in the ecosystems of the World Ocean have been studied on the example of benzo(*a*)pyrene (BP), a molecular stable compound which is now used as an indicator of environment pollution by carcinogenic polycyclic aromatic hydrocarbons (PAH). However, it should be emphasized that the total PAH content in the marine environment can greatly exceed the detected concentrations of BP.

The distribution and accumulation of BP in marine ecosystems (in the water and the surface microlayer, in the bottom sediments and biota) have a general character in various regions of the World Ocean (Table 11.2). These data indicate a widespread distribution of carcinogenic PAHs in the studied

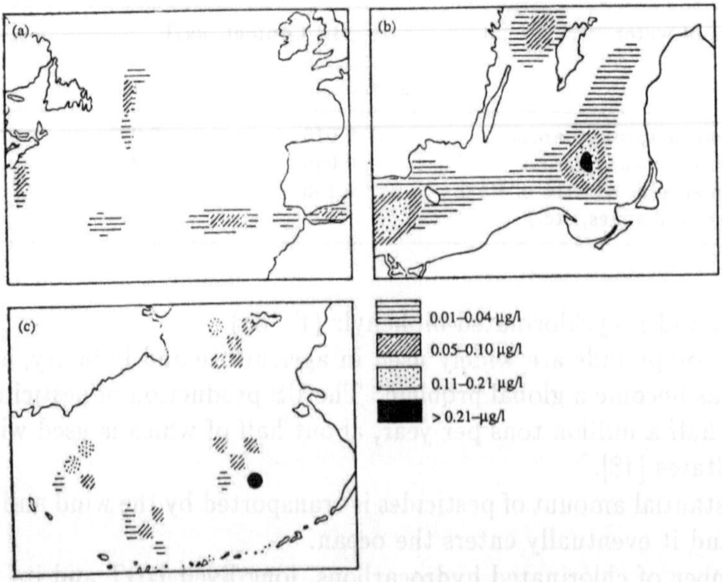

Figure 11.3. Distribution of benzo(a)pyrene in surface microlayer of ocean [4, 11, 43]: (1) North Atlantic, (b) Baltic Sea, (c) Bering Sea.

ecosystems of the World Ocean. The BP content in the water column of these regions mainly varied from 0.003 to 0.02 μg/l, which is close to the PAH background content. Higher concentrations of BP were recorded in the surface microlayer (differences of 1 or 2 orders of magnitude) (Figure 11.3). The tendency of BP to accumulate in the surface microlayer is typical of most zones in the World Ocean, and it is apparently caused by processes of atmospheric PAH transport and deposition [7].

The BP content in Baltic Sea water varied in winter between 0.001 and 0.13 μg/l and in summer between 0.003 and 0.16 μg/l; the maximum values (0.15 to 0.50 μg/l) were observed in regions of intensive navigation, which were also polluted by petroleum products. Pollution of the marine environment with PAH resulted in BP accumulation in bottom sediments (from 10 to 138 μg per kg of dry mass) and also in marine organisms. For example, in winter the BP content in plankton was 4 to 108 μg per kg of dry mass, while in summer, during the period of intensive development of the plankton community, it reached 122 to 326 μg per kg of dry mass (the coefficient of accumulation of BP in plankton is 10^3). In benthic organisms BP was present in smaller amounts: up to 5–30 μg/kg (in winter).

The BP content in the open water column of the World Ocean (North Atlantic between 50° and 36°N) varied from 0.003 to 0.010 μg/l. Higher concentrations of PAH were detected in regions subjected to a significant anthropogenic impact: near the Azores (0.06 μg/l), Gibraltar (0.09 μg/l),

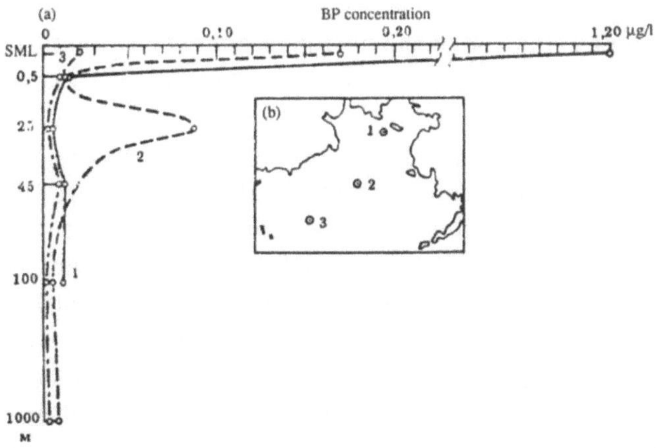

Figure 11.4. Vertical distribution of benzo(a)pyrene in seawater (a) at stations of regions, (b) in Bering Sea, July 1981 [11]: (1) station 12, 64°N, 172°30′W, (2) station 26, 58°30′N, 176°W, (3) station 30, 54°15′N, 175°W; SML = surface microlayer.

Grand Newfoundland Bank (0.015 µg/l), Western branch of Gulf Stream (0.09 µg/l).

PAHs are also widespread in subarctic regions of the Pacific (in the Bering Sea) [11]. In the main water mass of the Bering Sea the BP content varied from 0.002 to 0.007 µg/l, which was 0.02 to 0.1 of the BP pollution level in the Baltic.

The coefficient of BP accumulation in plankton in the Bering Sea was 10^2 to 10^3 (for a content of 112 to 240 µg per kg of dry mass). In a number of regions the BP concentrations in the bottom sediments were 0.51 to 34.6 µg per kg of dry mass (Figure 11.4).

Contamination of the marine environment by radioactive products results from nuclear-weapons tests and from the operation of nuclear-energy installations [32].

Radioactive contamination of the Pacific was produced mainly by the fallout from nuclear-weapons tests in the vicinity of Bikini and Eniwetok atolls and by the global fallout in various parts of the world. The maximum strontium-90 concentrations (3.9 pCi/l) were observed in the Pacific near Bikini and Eniwetok [20]. Radioactive contamination in the northeastern Pacific is apparently due to river runoff of radioactive products from US territory.

Observations carried out from 1962 to 1977 in the Pacific aboard GOSKOMGIDROMET weather vessels showed that in 1964–1966 the concentration of ^{90}Sr (up to 1.0 Ci/l) was a maximum in the northeastern part of the ocean; during subsequent years (1974–1977) in the northern ocean the concentration dropped to 0.15 pCi/l.

It has been noted [18] that the vertical distribution of natural and artificial radionuclides is determined mainly by suspended particles. Of particular importance is contamination of the bottom sediments by long-lived radionuclides, which constitute a source of secondary contamination of the water column by radioactive substances.

Under the conditions of the anthropogenic impact on the World Ocean, important consequences, including some relating to health, are connected with the microbiological pollution of the marine environment. Pathogenic and conditionally pathogenic microorganisms, entering the sea with the runoff from the land, adapt to the new conditions, particularly in brackish and eutrophic regions of the ocean, and then develop intensively in filter-feeding hydrobionts, which are of great ecological importance. Recent investigations [53] also indicate the spread of intestinal microflora in the open water of some seas, and its intensive development in bacterioneuston at the sea-air interface [22].

It should be mentioned, too, that microbial pollution by pathogenic microflora and vibrio cholera (El Tor strain) spreading via coastal waters and mollusks has already created a number of serious situations in coastal cities [28].

Finally, in recent years maricultures, which are of great importance because of the protein problem, have become a source of pollution in coastal areas. Projects for transforming marine farms into wastefree enterprises are being seriously studied, being dictated by considerations of protection of the offshore areas of the ocean.

11.3. The Effect of Pollution on Marine Ecosystems

Anthropogenic pollution causes serious geophysical, geochemical, and biological (and thus ecological, in the broad sense of the word) changes in the ocean.

The geochemical consequences manifest themselves as disturbances of the natural cycles of nitrogen, carbon, sulfur, and other elements in the ocean.

While studying the transport and transformation of pollutants and their effect on geochemical and biological processes, we should pay particular attention to various effects at the sea-air, sea-land, and water-bottomsediment interfaces.

The sea-air interface is impacted significantly by oil spills and also by the atmospheric transport of pollutants. Chemical compounds transported upward with air bubbles during wind-and-wave interactions promote pollution of the surface microlayer. This has been demonstrated using phosphates as an example.

Pollutants change the geophysical properties of the environment at the water-air interface, and they affect substantially the entire 'micro-ecosystem'

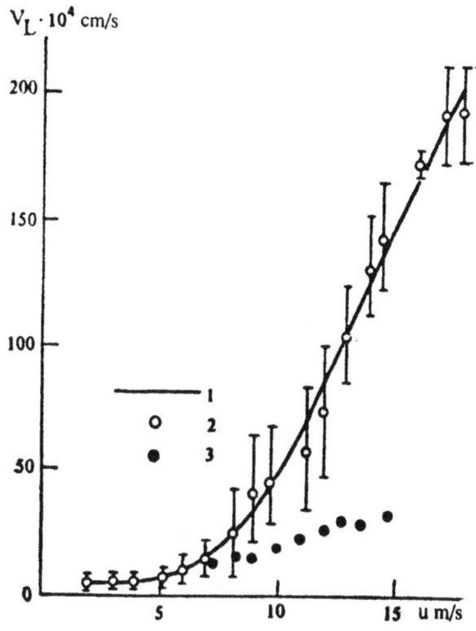

Figure 11.5. Rate of gas exchange between sea and air as function of wind velocity u at altitude of 10 m: (1) averaged curve for polluted surface; (2) the same, experimental data; (3) data for water surface covered by pollution film.

of the interface, which is inhabited by neuston and pleuston. The importance of these changes is further emphasized by the fact that the ocean plays a major role in the global absorption of carbon dioxide (and, consequently, in the problem of the climate): up to 40% of the atmospheric CO_2 is dissolved in the ocean.

Pollution of the surface layer can have the following geophyscial consequences.

1. Disturbance of the mass exchange and energy exchange between ocean and atmosphere; the film formed by petroleum products, detergents, and municipal wastewater presents an additional barrier to gas molecules, damps out capillary waves produced by the wind (thereby changing the gas exchange rate), and affects the radiation properties of the ocean surface [2, 19]. Figure 11.5 shows a substantial change in the rate of gas exchange between ocean and atmosphere in the presence of a pollution film.

The disturbance of the evaporation alters the surface-layer temperature, which in turn hinders the microconvention in the surface layer. The temperature change itself leads to a disturbance of the vital activity of stenothermal organisms.

2. An increase in acidity (due to the absorption of atmospheric SO_2 and H_2SO_4), which affects particularly strongly ecosystems in fresh water and brackish water (coastal water, water of inland seas). The heightened acid reaction of the environment affects directly the bacterioneuston, whose function it is to destroy petroleum hydrocarbons, and it promotes the removal of heavy metals from aerosol particles (this increases the metal concentration

in the surface layer); finally, it also reduces the solubility of CO_2 in water and disturbs the equilibrium for the oxygen exchange.

Neuston organisms are especially sensitive to pollution of the marine environment. Such organisms include many populations of marine organisms at early developmental stages, as well as the abundant microbial population, which is of vital importance for the tranformation of organic substances (including surface-layer pollutants).

The functioning of the near-surface ocean biotope depends significantly on the mixing of the surface layers, which is caused by the wind disturbance and by the heat exchange between sea and air.

The continental area of the ocean, which includes the shelf and the estuaries, is subjected to the greatest anthropogenic impact. At the same time, this is the most productive part of the marine environment.

The zone adjacent to the water-bottom sediment interface is where the coefficients of pollutant accumulation are the highest. Pollutants sorbed onto sinking particles accumulate in the bottom sediments, where they undergo complex transformations, primarily under the influence of the anaerobic microflora.

A possible secondary pollution of the water column by chemical compounds from the bottom sediments should be taken into account. Frequently the near bottom layer of the water mass which has been disturbed spreads over a depth of hundres, or even thousands, of meters.

The pollutants entering the World Ocean become distributed irregularly over it, producing zones of increased pollution in coastal areas, in the euphotic layer, and in hydrofront zones.

Even if pollutants were to fall uniformly onto the surfaces of various ocean areas, hydrological factors (diffusion, vertical advection) would predetermine a 'spottiness' of their distribution (see, for example, [16]).

Thus, the dynamics of the oceanic environment, and the totality of the ongoing most significant physical phenomena, determine the main characteristics of the present ecological situation in the World Ocean.

The transport of industrially-produced pollutants by intensive currents over long distances, and also into the open sea, leads to damage of the most susceptible ocean ecosystems, for example, ecosystems of coral reefs, northern ecosystems, upwelling ecosystems, etc.

The formation of fields of chronic pollution in zones of convergence of diverse water masses, of component currents, in zones of estuaries and quasistationary gyres, and particularly in areas with fragile ecosystems, leads to a stressed ecological situation in these regions of the World Ocean.

The transport of industrial pollutants into the deeper ocean layers, and their accumulation in marine organisms and suspended organic matter, also constitute a hazard.

The phenomena mentioned above lead to increased, and in some cases crit-

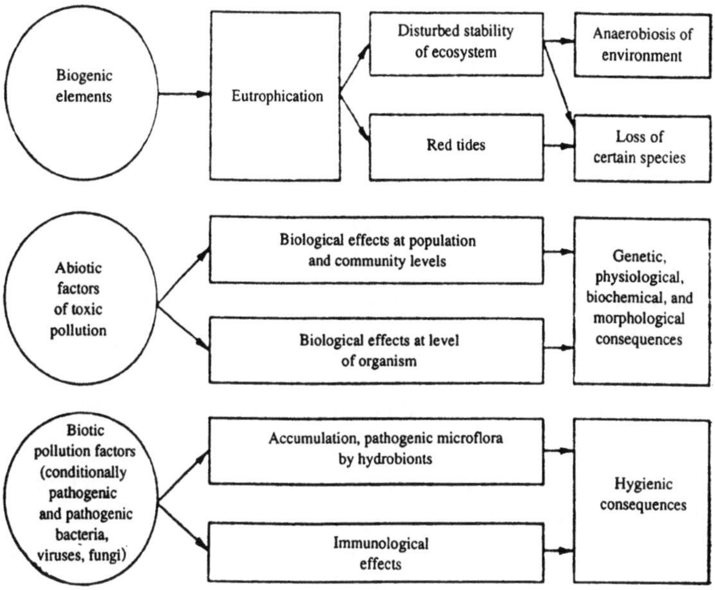

Figure 11.6. Ecological consequences of (1) Biogenic elements; (2) Eutrophication; (3) Disturbed stability of ecosystem; (4) Anaerobiosis of environment; (5) Red tides; (6) Loss of certain species; (7) Abiotic factors of toxic pollution; (8) Biological effects at population and community levels; (9) Biological effects at level of organism; (10) Genetic, physiological, biochemical, and morphological consequences; (11) Biotic pollution factors (conditionally pathogenic and pathogenic bacteria, viruses, fungi); (12) Accumulation of pathogenic microflora by hydrobionts; (13) Immunological effects; (14) Hygienic consequences.

ical, concentrations of chlorinated petroleum hydrocarbons and toxic metals in coastal regions and also in some regions of the open ocean.

In this case there occurs a dysfunction of the biotic component of the marine ecosystems in the areas of chronic pollution. This in turn disturbs the mechanism of the major biological and geochemical processes which bring about the 'self-purification' of the oceanic environment: suppression of the activity of marine microbiocenoses, accumulation of toxic chemicals in marine organisms and the resultant reduction of their ecological value, loss of certain species, disturbance of the production of organic matter, closely related to the biosedimentation removal of pollutants from the active layer of the ocean. Of particular danger is suppression of the photosynthetic activity of algae, which can even alter the balance of oxygen and carbon dioxide, leading to not only biological, but also large-scale geophysical, consequences.

The responses of the marine biota to toxicants can vary: from a gradual reduction in the size of individuals and a rearrangement of enzymatic systems to cessation of reproduction and the death of organisms. Therefore, anthropogenic pollution causes the following phenomena (Figure 11.6):

- accumulation of toxic chemicals in the biota;
- microbiological pollution of coastal areas;
- reduction of biological production;
- progressive eutrophication;
- mutagenesis and carcinogenesis;
- disturbance of the stability of ecosystems.

It should be noted that ecological changes and various rearrangements of ecosystems are also produced by natural factors. Thus the anthropogenic changes take place against a background of very complex natural rearrangements (many effects become especially pronounced during extreme natural fluctuations).

The main objective of integrated ecological monitoring is to detect anthropogenic changes against the background of very complex natural processes. The present situation is analyzed using a systems approach (a comprehensive analysis of the environment, see Chapter 1).

Hazardous ecological consequences depend not only on the pollution level, the lifetimes of the pollutants, and the processes of their transport and transformation, but also on the accumulation of chemical compounds in marine organisms. For instance, the average coefficients of accumulation of various pollutants in planktons, as mentioned above, can be quite high: for lead up to 4.0×10^5, for mercury up to 3.4×10^3, for cadmium 2.1×10^4, for PCBs up to 4.0×10^4, for benzo(a)pyrene 5.0×10^3 [15, 37].

The coefficients of accumulation of mercury in benthic organisms and fish can reach 10^3 to 10^4 [25].

Accumulated pesticides have also been detected in lipid fractions of hydrobionts.

Marine organisms which adsorb and incorporate pollutants in amounts several times greater than their concentrations in seawater are usually called species-accumulators.

Anthropogenic effects manifest themselves both at the individual level and at the population-biocenotic level.

The primary critical disturbances in the functioning of marine organisms under the influence of pollutants take place at the level of biological effects: changes in cell chemistry and in the character of enzymatic systems, respiration processes, osmotic regulation, growth, and reproduction; mutation and carcinogenesis begin; pathological forms appear; cell sizes change; the motion and orientation of marine organisms in space are disturbed. All these are consequences of biochemical, morphological, ethological, physiological, and genetic nature [4].

Pollutants affect significantly the biochemical processes in an organism. In some cases the changes in the biochemical processes result in the appearance of new physiological characteristics of the organism, while in other cases the organism compensates these effects by means of changes in the biochemical system (for instance, rearrangement of the enzymatic apparatus). Thus, an increase in oxygenase activity has been observed in the tissues of fish and invertebrates, in the presence of chlorinated and aromatic hydrocarbons. Changes have been found in organometallic complexes of enzymatic systems, due to the effect of heavy metals (Cd or Hg dislodges Cu or Zn from the complex). Enzymatic activity in marine microorganisms can be induced by the presence of pollutants.

The physiological consequences of pollution can be adaptive in nature. In this case they lead to more effective functioning of the marine organisms in the habitat. It is known that the adaptation of enzymatic systems to organic substances new to microorganisms leads to the appearance of an indicator microflora [43, 61].

Recent investigations of the physiological effects of the anthropogenic impact on marine organisms have led to the development of a system of bioassay, to the use of marine biota as 'monitors' of the state of the marine environment.

The pollutant-induced morphological (or pathobiological) changes in hydrobionta are numerous. They include: pathology of the internal organs, changes in the size of the organism, appearance of malformations, changes in the thickness of the gill membrane in fish, disturbance of the spine, and histopathology of the internal organs.

In most cases the morphological consequences are connected with a disturbance of biochemical and physiological processes in the organism. For instance, in some Baltic areas many fish with tumors have been recorded [52]. This phenomenon is attributed to an increase in the concentrations of chemicals in the Baltic region. It has been proposed to use hydrobionta with tumors as an indicator of the spread of carcinogenic hydrocarbons in the environment.

It is universally accepted that any changes in the composition of populations are caused by a rearrangement of the genotype of the individuals making up the populations. Recent investigations demonstrate that, the greater the environmental pollution, the more mutations dangerous to the organism there will be.

At the level of populations and communities, the genetic consequences of pollution are assumed to manifest themselves as reduced production. At the level of organisms, all signs of viability change, the sizes of individuals are smaller, and the total biomass is reduced.

At present it is known that mutations are often combined with carcinogenesis, possibly at the initial stage [54]. Consequently, an increase in the number of tumors can be used as indirect evidence of an increase in the rate of mutation in the environment. There are also other possible approaches, corresponding to the various objectives of marine biomonitoring. These consist in studying the mutagenic effect of the tissues of marine organisms (of the components of the ecosystem).

It should be noted that monitoring the physiological consequences and monitoring the genetic consequences of the pollution of the marine environment are closely interconnected, since a disturbance of the physiological responses can be a function of those changes which occur at the genetic level. In ocean regions where an unfavorable ecological situation has been produced by a significant anthropogenic impact stable changes in the structure and functioning of marine biocenoses are observed. These include: a change in the average biomass of populations of plankton and benthic organisms; a reduction in the number of higher taxons (genera or families of hydrobionta); the appearance of organisms new to the marine environment (e.g., conditionally pathogenic forms of microorganisms of the *Bacillus coli* group); a change in the ratio between the abundances of individual taxonomic groups of hydrobionta and in the ratio between the indices of production and destruction of organic substances; disturbance of the processes of ecological metabolism; the appearance of antibiotic-resistant forms of microorganisms.

In recent decades, because of eutrophication of some regions of the World Ocean, an increase in the average biomass of microorganisms, phytoplankton, protozoa, zooplankton, meiofauna, and benthos has been observed, particularly in semilandlocked and landlocked seas. For example, microbiological investigations in the Caspian Sea [4] reveal an increase in the average biomass of microflora in this basin by approximately an order of magnitude over the last 20 years. During the last 10 years the average biomass of microorganisms in the northwestern part of the Black Sea also increased by almost an order of magnitude [4, 21]. The confinement of fields of high biomass of microorganisms, phytoplankton, zooplankton, and protozoan organisms to the following zones in the Baltic Sea is noteworthy [4]: the southern Baltic, including the region of the Danish straits, and the central part of the Gotland Basin. It is noteworthy, too, that the fields of high biomass of microorganisms have persisted in the above Baltic regions for the last several years [61] and that they result from the anthropogenic eutrophication of individual regions of the Baltic Sea.

A change in the abundance of higher taxons (genera or families of marine organisms) indicates a disturbed state of the ecosystem. Recent investigations also indicate the appearance of forms of hydrobionta which are new to the marine environment. For example, taxonomic groups of microorganisms inhabiting soil and rivers have been found in the open waters of the Black Sea [24]. The high abundance of these microbial forms in the seawater, macrophytes, and bottom sediments demonstrates their adaptation to new environmental conditions and also indicates the effect of river runoff and terrigenous discharge on the biological regime of the Black Sea. The disappearance of certain species, and the appearance of forms new to the region, are indicative of a change in the abiotic background, which may be associated with rearrangements in an ecosystem which are either natural or anthropogenic.

The changes in a biogeocenosis can also be produced by the joint impact of anthropogenic and natural factors.

A progressive eutrophication of marine water bodies is, as a rule, accompanied by a change in the ratios of production and destruction processes, because of an increase in the rate of destruction of organic substances.

For instance, in the open water of the Baltic Sea [4], in the zone of photosynthesis down to depths of 20 to 25 m, the amount of organic matter destroyed in summer and winter is 2 to 10 times greater than the amount produced by the photosynthetic activity of plankton algae (in this zone).

The impact of toxic substances results in a 'bloom' of populations of organisms which have adapted to new chemical conditions and in a displacement from the system of those which cannot adapt to the impact of foreign substances.

Particular attention should be given to highly sensitive marine organisms ('critical species'), which when exposed to toxic substances manifest negative effects (pathology, mutation, sharply reduced reproduction, death of some individuals).

In chronically polluted, and also eutrophic, sea regions the dominant species of hydrobionta are replaced, and hydrobiont forms adapted to certain chemical compounds undergo intensive development (e.g., nematodes adapted to petroleum hydrocarbons, crustaceans adapted to chlorinated hydrocarbons, marine bacteria adapted to various organic substances, including some unnatural toxic substances). Such hydrobiont forms are widely used as bioindicators of the marine environment, and they have been singled out by the author as well as by several others [4, 14, 43] as indicator organisms. For instance, investigations made during the last decade demonstrated wide spreading in polluted areas and biotopes of the World Ocean of taxonomic groups of marine microorganisms, whose enzymatic system is adapted to polycyclic aromatic hydrocarbons and polychlorinated biphenyls. Such indicator microorganisms have been detected, in particular, in the Azov, Baltic, and Black Seas and in the near-surface biotope of the Pacific and Atlantic Oceans. The abundance and spreading of these microorganisms are found to correlate with the concentrations of the above toxic chemicals in the seawater. It is typical that, in relatively clean ocean areas, particularly in the Pacific, the indicator microflora is scanty and its spread is local in nature, which is in sharp contrast to the fields of spread of these bacteria in more polluted seas and zones of the World Ocean, for example, in the Baltic Sea.

11.4. The Ecological Reserve of the Ocean. The Assimilative Capacity

As noted above, the anthropogenic impact on the World Ocean reduces the natural ability of marine systems to reproduce themselves and to regulate themselves, particularly where the conditions are critical for the normal functioning of marine ecosystems.

The concept of a critical or permissible anthropogenic impact on marine ecosystems, and on the ocean as a whole, is closely connected with the concept of an ecological reserve of the ocean, and it determines the approaches to an ecological standardization of the anthropogenic impact on the entire biosphere, formulated by us in this monograph and in [5]. Such an approach should be based on a complex consideration of the physical, biological, and chemical processes which determine the activity of the natural 'self-purification' and, in particular, the 'flexibility' of marine ecosystems.

This approach differs from that used to set up standards or to limit the discharge of pollutants into coastal ocean waters, taking into account only factors of the spreading and transport of pollutants in the ocean (see, for instance, [39]).

It should be noted that an inherent characteristic of a normally functioning marine ecosystem (in contrast to the air environment) is the predominant importance of its biotic component, with regard to the destruction and removal of chemicals. Therefore, when establishing ecological standards for the anthropogenic impact, we must consider collectively the geophysical, chemical, and biological processes of transport and destruction of toxicants. In this case the concept of the assimilative capacity of the World Ocean, introduced by us in [5] and further developed in [6, 7, 8], is of prime importance.

The assimilative capacity of the marine environment A_{mi} by a given pollutant i (or by all pollutants) is the maximum dynamic capacity of this amount of pollutants (in terms of the whole zone or a unit volume of the environment with the mth marine ecosystem), which can per unit time be destroyed, accumulated, transformed (due to biological or chemical transformations), and removed by sedimentation, diffusion, or any other process of transport beyond the ecosystem, without disrupting its normal functioning. The assimilative capacity thus defined has dimensions of a flow of matter: mass moved per unit time.

The assimilative capacity A_{mi} characterizes the ability of a marine ecosystem to ensure the dynamic accumulation and active removal of pollutants while at the same time retaining its basic properties.

The value of A_{mi} depends on many characteristics of the environment: currents, turbulence, water exchange, water temperature, the structure and functioning of the biotic component, and also the chemical and physical properties of the pollutants (including anthropogenic pollutants) entering the marine ecosystem. The totality of the large number of natural phenomena which promote 'self-purification' of the marine environment can be reduced to a few of the most important processes. These are: hydrodynamic transport, microbiological oxidation, biological transformation, biochemical oxidation, chemical and physical transformations, sedimentation of abiogenic particles, and biosedimentation.

In order to determine the assimilative capacity A_{mi} (for the ith substance), we have to know two factors: the maximum permissible (to provide normal functioning of the ecosystem) concentration of the given ith substance (or totality of substances) C_{mi0} (determined when the substance is not being removed from the given volume); and the multiplicity factor K_{mi} of the amount of substance (in a unit volume), which can be either reprocessed by the ecosystem or else removed from the unit volume by different processes (during the (long) averaging time determined for the permissible concentration C_{mi0}). Then

$$A_{mi} = C_{mi0} K_{mi} V, \tag{11.1}$$

where V is the volume of the considered region of the marine environment.

Let us recall (see Chapter 2) that the concept of the maximum permissible concentration C_{mi0} for ecosystems, which corresponds to the maximum permissible ecological load, is based on the ecological relations

$$\frac{\int_{\mathbf{R}} C_{mi0}(\mathbf{R}) n_m(\mathbf{R})\, \mathrm{d}\mathbf{R}}{\alpha_{mi}(LC_{50})_{mi}} \leq 1 \tag{11.2}$$

(where $\alpha_{mi} = 0.05\ldots0.01$, is the 'reserve' coefficient). For a small portion of the organisms it may be that

$$C_i(\mathbf{R})/LC_{50} \geq 1. \tag{11.3}$$

Here LC_{50} is the concentration (lethal concentration) for which 50% mortality of the organisms of the given population is observed during the averaging time assumed for C_{mi0}.

In general we can write

$$A_{mi} = \alpha C_{mi0} V + \sum_l B_{mil}, \tag{11.4}$$

where B_{mil} is the quantity of substance removed from the ecosystem per unit time by means of the lth process and where α is a coefficient with dimension M^{-1}; the numerical value of $\alpha = 1$.

In this case the multiplicity factor is determined from the equation

$$K_{mi} = \frac{A_{mi}}{C_{mi0}V} = \alpha + \frac{\sum_l B_{mil}}{C_{mi0}V}, \tag{11.5}$$

from which $K_{mi} = 1$ if there are no processes (including biological processes) acting to remove the substance from the given volume of the environment ($B_{mil} = 0$).

It is noteworthy that the value of K_{mi} depends considerably on the averaging time (including the averaging time for which the concept of the permissible concentration is introduced).

If we consider the instantaneous (initial) state of the system, then

$$\sum_l A_{mil} = \alpha C_{mi0} V \quad \text{and} \quad K_{mi0} = 1. \tag{11.5a}$$

This means that it is not allowed for an amount of substance greater than $C_{mi0}V$ to be present in the given volume at any time (without disturbing the ecosystem). However, if we take an averaging time equal to a long time interval, e.g., one day or one month, then due to dilution and removal from the volume caused by various processes, including biological processes, during this time

$$\sum_l B_{mil} \gg 0 \quad \text{and} \quad \int_t K_{mi}(t)\,dt \gg 1.$$

The concentration of the pollutant (harmful substance) in the water (and in the air) is a complex function of the previously mentioned characteristics:

$$C_i(\mathbf{R}, t) = F[g(\mathbf{R}), v_{\mathbf{R}}\omega_{\mathbf{R}}, \mu_{ij}\nu_{ij}, \mathcal{K}_i, \lambda_i], \tag{11.6}$$

where $g(\mathbf{R})$ is the quantity of pollutant emanating from a point source; $v_{\mathbf{R}}$ is the velocity of orderly transport (current velocity, gravitational sedimentation rate); $\omega_{\mathbf{R}}$ is the eddy diffusion; μ_{ij}, ν_{ij} are the rate factors of physical and chemical (biochemical) transformation from the ith substance to the jth substance, respectively; λ_i is a coefficient characterizing the rate of microbial decomposition of the ith substance; $\mathcal{K}_i$ is a coefficient characterizing the rate of bioaccumulation of the ith pollutant by suspended solids; t is the averaging time (used to determine the concentration of the substance when establishing the standard).

The value of $C_i(\mathbf{R}, t_0)$ for a distributed source and the totality of substances can be obtained by integrating over the entire source (and summing up the contributions made by various substances to the pollution).

The principal difference between a water environment and an air environment is that for an air environment practically $\mathcal{K}_i$, $\lambda_i = 0$.

For practical assessments the various processes of natural 'self purification' can be reduced to an even lower number, since microbiological and biochemical oxidation are closely connected, while the biological transformation by plankton and benthic organisms is much less, with regard to its scale and results, compared with the processses of microbiological oxidation of the organic matter.

Even the chemical transformations of organic substances are less significant, compared with microbiological transformations, if the rather low average temperature of the deep waters of the World Ocean (4–5°C) is taken into account. Finally, the sedimentation of abiogenic particles constitutes less than 10% [13] of the total flow of suspended particles leaving the euphotic zone and settling onto the sea bottom.

Thus, we can single out three major processes, which together practically determine the assimilative capacity: hydrodynamics, the microbiological oxidation of organic pollutants, and biosedimentation.

When assessing A_{mi}, it is important to introduce the concept of the critical (permissible) anthropogenic load and to diagnose the critical processes in the ecosystem, as well as the critical (to the impact) species of marine organisms.

Therefore, the value of A_{mi} is determined by three main components, corresponding to the removal of substance by hydrodynamic processes B_{miv}, microbiological oxidation $B_{mi\lambda}$, and biosedimentation $B_{mi\mathcal{K}}$:

$$A_{mi} = C_{mi0}V + B_{miv} + B_{mi\lambda} + B_{mi\mathcal{K}}, \tag{11.7}$$

i.e. the 'assimilative capacity' is composed of a 'geophysical capacity' (determined by the removal of toxicants by geophysical processes) and the 'biological capacity' (determined by the permissible load on the ecological system and the removal of toxicants by biological processes). It should be noted that the rates of biological processes are comparable to the rates of all other processes. A similar approach was used by us in [10] for freshwater ecosystems.

Now, using the concept of the 'assimilative capacity', let us formulate the condition of 'ecological safety' for a given large volume (ensuring the required ecological relations) of the marine environment. If the amount of the ith substance entering volume V (integrated over a quite long time interval) is called ψ_{vi}, and the amount removed (by all processes) is called φ_{vi}, then the condition of ecological safety can be written as follows:

(1) complete safety

$$Q_i \leq A_{mi}, \qquad \varphi_{vl} \geq \psi_{vi}; \tag{11.8}$$

(2) relative safety

$$Q_i \leq A_{mi}, \qquad \varphi_{vi} < \psi_{vi}; \tag{11.9}$$

(3) no safety

$$Q_i > A_{mi}. \tag{11.10}$$

Here Q_i is the total amount of the ith substance contained in volume V.

The condition of relative safety is determined by the dynamics of the process of arrival and removal. It means that, in the event of an intensive arrival of the given substance, it may be disturbed in the very near future.

The data on the pollution of the World Ocean and on the effects of pollutants on marine systems presented in this chapter make it possible to assess the present and future state of the ocean, with the help of the indicated criteria. Although the present level of pollution of the World Ocean causes great concern, it should be noted that in most regions of the ocean the ecological systems can still cope with the incoming pollutants; this is because the assimilative capacity of the World Ocean is quite great.

11.5. The Role of Microbiological Oxidation and Biosedimentation in the Removal of Pollutants from Seawater

As already mentioned, the microbiological oxidation of organic pollutants and the biogenic sedimentation of chemical toxicants are of prime importance with regard to balancing the arrival, destruction, and removal of pollutants in the marine environment. Hydrodynamic processes, on the other hand, tend to redistribute substances throughout the ocean, and thus they have practically no effect on balance assessments for the ocean as a whole.

Marine microorganisms, which have polyfunctional enzyme systems, high biochemical activity, and the ability to proliferate rapidly, play a significant part in the functioning of World Ocean ecosystems. Bacterial populations are the most important component of the marine ecosystems [55].

The biochemical activity of marine microorganisms regulates to a significant extent the biogeochemical cycle of carbon in the World Ocean. For example, the destruction of organic matter in the World Ocean, mainly due to the catabolic activity of aerobic heterotrophs in the upper 100-meter layer, amounts to, on the average, around 30×10^9 t C/year.

The biochemical activity of marine microbiocenoses appears to be closely related to the distribution, size, and concentration of suspended matter. The higher the content of suspended organic matter in the seawater masses, particularly in the upper, intensively aerated layers, the more intensive will be the microbial oxidation of organic pollutants.

The functioning of the microbial population in the zone of interaction between ocean and atmosphere acquires particular importance. The specific complex of microorganisms which develops here (the bacterioneuston) is a very important biological structure, since it destroys many organic pollutants which become localized in the surface film of the ocean [60].

Recent investigations have shown that marine microorganisms are important in the oxidation of paraffin hydrocarbons and even of high-molecular aromatic and chlorinated hydrocarbons resistant to biooxidation.

However, a quantitative assessment of the microbiological oxidation of organic pollutants is needed, in order to estimate the assimilative capacity of the ocean and to study the ecological reserve.

For instance, *in-situ* experiments in the Bering Sea have demonstrated, with the help of isotope methods, that microorganisms can in 24 hours destroy around 2 μg/l of paraffin hydrocarbons of octadecane, and in the Sea of Japan up to 5 μg/l. Model *in situ* experiments with a subsequent chemical analysis have shown that marine microorganisms can destroy around 4 μg/(l 10 days) of benzo(*a*)pyrene in the Bering Sea, around 5.5 μg/(l 10 days) in the Baltic Sea; around 2.3μ g/(l 10 days) in the North Atlantic, and around 1 μg(l 10 days) in the Pacific (Marianas Trench). It has been calculated that the bacterioneuston of the World Ocean inhabiting its nearsurface microbiotope oxidizes around 1.6 million tons of peteroleum hydrocarbons a year, which amounts to about one fourth of the total quantity of oil entering the ocean because of human activity.

It is estimated as well that the bacterioneuston of the World Ocean can destroy around 400 tons of benzo(*a*)pyrene per year, which constitutes about 10% of the total quantity of benzo(*a*)pyrene entering the World Ocean from natural and anthropogenic sources.

The data presented, as well as earlier [23] regional assessments of microbial oxidation of oil at the sea surface, in particular in the Baltic Sea, indicate the great destructive ability of marine microorganisms.

However, the activity of marine microorganisms cannot accomplish the complete destruction of the organic pollutants entering the marine environment. First, many molecularly stable compounds (e.g., aromatic, chlorinated hydrocarbons) are only partially destroyed by the microorganisms. Second, any organic compounds can be oxidized by bacteria only under the appropriate ecological conditions (the presence of certain concentrations of biogenic elements, easily oxidized organic substances, oxygen, etc.), which are not always realized, since they are obtained only during certain seasons and since they are not characteristic of all biotopes of the marine environment.

Furthermore, in arctic and subarctic regions of the World Ocean under low-temperature conditions, as well as in the ocean depths, the activity of microbiological processes is sharply reduced.

Marine organisms not only accumulate, destroy, and transform pollutants, but also transport them to different biotopes, thereby determining the distribution of toxic chemicals in the marine environment.

Of prime importance is the accumulation of chemicals by marine organisms. The primary producers of living substance extract chemical compounds directly from the marine environment, while the consumers get them by sorption and by feeding. As a result of this process the chemical elements (pollutants included) accumulate in the biota, migrate repeatedly along the food chain of the marine organisms, and when the hydrobionta die off they return to the seawater or are removed from the photic ocean layer via biogenic sedimentation and partially reach the ocean bed. Here they are deposited in the bottom sediments, being partially reentrained into the biogeochemical cycles by the action of benthic animals. Due to the continual activity of digging animals, chemical compounds are moved into deeper sediment layers and also into the ocean water column via biological turbidization. In some ocean regions the seawater is also enriched by chemical elements from the sediments under the influence of hydrological factors.

The idea of an active role of 'living substance in the metabolism of the sea' was first introduced by Vernadskii [1].

The hydrosphere and the atmosphere are the regions into which suspended particles of terrigenic, cosmic, and biogenic origin spread. The World Ocean constitutes a huge basin for the accumulation and sedimentation of such particles.

The intensity of the biosedimentation of suspended organic substances, as well as the sedimentation of terrigenic materials, are closely related to the intensity of photosynthetic processes [24]. The more active the primary production of organic material in the ocean, the more intensive will be the bioelimination of chemical toxicants in the euphotic zone of the ocean and their transport to the underlying layers and thence to the bottom of the ocean.

Biogenic particles from several microns to hundreds of microns in size remove many pollutants from the photic layer as they settle out and in this way bring about their redistribution over the depth. The pollutants are thus removed from the water and buried in the bottom sediments. Recent investigations have shown that settling phytoplankton cells actively remove PAHs and copper from the euphotic layer, while fragments of dead zooplankton remove paraffin hydrocarbons, PCBs, BPs, and metals [34, 36].

According to data of direct observations [36], the sedimentation rate of biogenic aggregates is different, and it depends on their sizes. For instance, large biogenic particles several hundred micrometers in size are deposited rather quickly (traveling from 50 to 850 m/day), and they apparently account for the main removal of chemical toxicants from the ocean water column.

Table 11.3 gives data on the biosedimentation removal of some toxic metals and their deposition in the bottom sediments, compared with the total influx of these substances into the World Ocean (see Table 11.1) [8].

In order to determine the average concentrations of pollutants (C) in the suspended organic matter at different depths for shelf and open regions, we used data of various authors (cited in [26]), primarily for the Pacific and Atlantic Oceans. Besides horizontal irregularities, the difference in the qualitative composition of the suspension along the vertical was also taken into account in the calculations, and in particular the fact that at a depth of 100 m the suspension consists mainly of phytoplankton and zooplankton, while at a depth of 400 m mainly of zooplankton and its metabolism products.

In order to estimate the biosedimentation extraction, we proceeded from

Table 11.3. Influx of heavy metals into World Ocean, their removal at depths of 100 and 400 m and accumulation in bottom sediments (t/year).

Substance	Total influx*	Biosedimentation removal		Accumulation in bottom sediments
		at depth of 100 m	at depth of 400 m	
Lead	4.0×10^6	2.5×10^5	8.5×10^4	1.4×10^4
Cadmium	3.4×10^4	1.5×10^5	2.6×10^4	4.3×10^3
Mercury	1.0×10^4	3.8×10^4	3.2×10^3	5.4×10^2

* Upper estimate.

the experimental data and theoretical ideas, according to which about 50% of the synthesized organic matter is removed from the euphotic zone of the ocean, and only 5 to 20% of the particles settling out of the photosynthesis zone (or 2 to 10% of the primary production) reach the bottom [56]. The remaining 80 to 95% of the organic matter removed from the photic layer is destroyed during the course of microbial and chemical oxidation. Heavy metals settling out of the photic zone accumulate at suspended particles and either reach the bottom or else return to the water via the destruction of biogenic aggregates and are in part reentrained into the biogeochemical cycles.

Inspection of the data in Table 11.3 indicates that for the photic layer (0 to 100 m) the sink of metals and their biosedimentation removal are approximately equal. However, in some cases (for instance, for cadmium) the biosedimentation flow appeared to be larger than its influx. The observed difference is evidently caused by the hydrodynamic transport of metals in dissolved form from the subsurface ocean layer (100 to 400 m) to the surface layer.

It is of particular interest to estimate the flow of metals with the biogenic suspension from the 0-400 m layer (see Table 11.3). Since at a depth of 400 m the hydrodynamic transport is on the whole negligibly small compared with the rate of the sedimentation flow, the biosedimentation processes will determine primarily the removal of metals from this layer of the ocean. Despite the approximate nature of this estimate, referred to the entire World Ocean, it becomes evident that the total amount of lead and mercury which is presently being removed from the active layer of the ocean by biosedimentation is much less than the mass of lead and mercury which is entering the ocean from anthropogenic and natural sources. This difference is even more pronounced when we compare the external load with the rate of deposition of heavy metals in the bottom sediments (see Table 11.3). These facts indicate a tendency to accumulate mercury and lead in the marine components of ecosystems. This conclusion agrees well with earlier estimates of the anthropogenic components (see Table 11.1) of the influx of lead, mercury, and cadmium into the ocean: 92, 70, and 50%, respectively.

A further increase in the rates of pollution of the World Ocean with lead and mercury in a number of regions can lead to an increased assimilative capacity and to the attainment of critical concentrations of chemical toxicants in the marine environment, thereby causing the biotic element to malfunction.

According to our classification (see expressions (11.8)–(11.10)), the state of most regions of the ocean can for the present be considered to be only relatively safe ($Q_i < A_{mi}$ but already $\varphi_{vi} < \psi_{vi}$).

11.6. Forecasting the State of the World Ocean

In order to forecast the future state of the ocean, the following elements should be estimated:

(1) the rate of entry of pollutants into the marine ecosystem;
(2) the components of the process of 'self purification' of the marine ecosystem (assessment of microbial and chemical oxidation and of biosedimentation accumulation in bottom sediments, determination of hydrodynamic removal of pollutants from the ecosystem);
(3) the changes in the concentration of pollutants in the marine ecosystem;
(4) the critical concentrations of pollutants which are most hazardous for the given ecosystem;
(5) the ecological consequences of the anthropogenic impact.

An increased influx of various pollutants to the ocean is forecast either via extrapolation with certain assumptions, or else allowing for the development of the respective branches of industry emitting these pollutants. For example, the development of power production and the combustion of fossil fuels lead to increased emissions of Hg and Cd. It is known that by the year 2000 fossil-fuel combustion will have doubled or tripled, and that by the year 2025 it will be 4 or 5 times as great. This may already exceed the critical levels of mercury pollution in some zones of the ocean. An increase in the amount of oil entering the ocean is also predicted, to approximately twice the 1971 level by the year 1990 [19].

On the basis of the above consideration we can suppose that the total average contents of many pollutants in the ocean water will at least double by the year 2000. It should be noted that at present the amount of Pb and Hg being removed from the upper ocean layer (400 m) is less than the amount entering the World Ocean.

An analysis of data (Table 11.4) on the pollution of the World Ocean indicates that at present the average concentration of the most widespread pollutants is 1 to 2 orders of magnitude less than mean critical values (for which the primary phytoplankton production is reduced by 50%). However, if we take into account the fact that the pollution levels in different ocean

Table 11.4. Global pollution of the World Ocean and its impact on primary production of organic matter.

Pollutant	Concentration in seawater, μg/l		Mean critical concentration for phytoplankton, μg/l	Possible change in bioproductivity in 2000 because of given element, %
	by 1979	by 2000		
Lead	0.01–20.0 / 0.03	0.02–400 / 0.06	100	0
Mercury	0.01–10.0 / 0.05	0.02–20.0 / 0.1	5.0	30
Cadmium	0.01–0.5 / 0.1	0.02–1.0 / 0.2	40.0	0
PCBs	0.01–0.1 / 0.035	0.02–0.2 / 0.07	0.1	25

Notes 1. The most probable average value is given in the denominator. 2. The average critical concentration is here the concentration at which the primary production of phytoplankton is reduced by 50%.

regions are different, while the critical concentration values also vary, then it becomes obvious that even now there are vast areas where the situation is nearly critical. This situation (a spread of values and the presence of zones with increased pollution levels) will change for the worse by the year 2000, when the average pollution level will be 2 to 3 times as high. According to the data of [19], the mean level of pollution by petroleum products in the North Atlantic will reach 2 MPC by the year 1990.

Model experiments and calculations indicate that the decrease in the mean primary productivity of some organisms over vast areas may be 25 to 30% by the year 2000 (see Table 11.4), which is very worrisome. This is bound to result in a reduced bioproductivitiy of the ocean; the effects may even be intensified if there is a simultaneous impact of pollutants.

The danger of global pollution of the ocean is becoming real, especially for some 'susceptible' or vulnerable marine ecosystems. For instance, the increased danger is clearly observable in enclosed and semienclosed basins, at active ocean surfaces, and at the sea-land, water-sediment, and sea-air interfaces, as well as in current-convergence zones.

In conclusion, when estimating the actual and forecast state of the ocean, it is especially disturbing to note that in most cases the ocean is the final link in the chain of transformations, migration, and accumulation of various pollutants in the biosphere.

This is apparently why a general estimate of the state of the ocean is of greater concern than estimates of the state of the atmosphere (the impact in the atmosphere does not, as a rule, exceed 10% of the natural values and variations, whereas in the ocean the anthropogenic change of the bioproductivity of individual communities over vast areas is expected to be from 25 to 30%).

CONCLUSION

What To Do About Anthropogenic Changes in the Biosphere

In this book, beginning with the *Introduction* it has been demonstrated convincingly that the anthropogenic loading of the biosphere has intensified during the last decades, and that this has led to significant consequences and changes. As early as 1944, V.I. Vernadskii pointed out the importance of this problem and its large scope, when he wrote that 'mankind, taken as a whole, is becoming a powerful geological force...' [2].

However, only very recently greatly increased attention has been paid to the anthropogenic changes in the biosphere and to the possible depletion of natural resources. Various proposals concerning action at the national and international (even global) levels have been put forth and widely discussed.

During the discussions of recent years, it was often suggested that we are heading toward an ecological crisis, toward an insoluble conflict between human society and nature [1, 4, 15, 17, 20, 21].

For instance, in [17] Forrester maintains that, if the present tendencies in the development of society continue, then a crisis in the relationship between man and the environment is inevitable.

According to the model of Meadows [20], after a certain period characterized by an increase in human population, the wealth of nationals, and the standard of living, a catastrophe should occur in 2030–2070: a drastic increase in the pollution of the environment, depletion of natural resources, and a reduction of human population [4]. The authors of the model in [20] see only one way out: transition to a 'global equilibrium', which will be possible only if strict limitations are introduced to stabilize the population of the Earth and industrial capital (at the level of 1980), to reduce to one eighth the per capita consumption of resources, and to reduce to one quarter the total pollutant emission into the environment.

Soviet scientists do not share this pessimistic opinion about an approaching ecological crisis or about the irreversible processes which have already taken place in the biosphere [14].

However, it is well understood that we must not underestimate the danger associated with the increased anthropogenic impacts and that active measures should be taken to protect man's environment. Experience gained during the last ten years has indicated that environmental problems can be solved and are being solved [18].

The problem can be solved not by retarding the ongoing development of society, but rather by optimizing the interaction between human society and nature (this is most completely attainable only under the conditions of socialism, a social system in which there is no private ownership of the natural resources and the means of production), and by a reasonable transformation and rational use of nature in the interests of the present and future generations.

At the same time, society should also preserve its ability to achieve versatile and controlled development, since according to, for instance, Academician E.K. Fedorov the possibilities for transforming substances and producing energy are growing more rapidly than specific types of raw materials and energy sources are becoming exhausted; since the cultivation of renewable natural resources makes possible a severalfold increase in the productivity of nature as a whole; since the introduction of wastefree technology can result in minimal environmental pollution; since there exist specific ways of controlling or stabilizing climate; and since the population growth rate can be kept within certain limits by either self-regulation or public regulation [15]. From our viewpoint, the introduction of the negative feedback needed to control systems (based on monitoring), depending on the extent of anthropogenic impact and its effects (allowing for achievements in scientific and technological progress), is the necessary condition for ensuring the ecological stability of large ecologoeconomic systems.

Therefore, understanding and interpreting the present and future states of the natural environment on the basis of objective information, estimates (provided by a scientifically established monitoring system, such as the one we have discussed in this monograph), and ecological knowledge constitute the prime condition for achieving ecological stability of the biosphere.

The progress of science and technology provides every new possibility for satisfying the fundamental human requirements, while at the same time conserving, restoring, and improving natural conditions [14]. Thus, mankind faces a single general problem, so that we have to agree upon the measures which should be taken to solve it, and to put at man's disposal the necessary means, while coordinating the efforts being made at the international level. All this will be possible only under conditions of a durable peace and cooperation, involving the joint efforts and firm determination of all the countries in the world [3, 14, 15, 22].

The main points of the general plan of action formulated in [22] are close to the arguments put forth by Fedorov in [15]. The main attention there is focused on the problem of stabilizing the size of human population, halting the 'waste' of renewable resources, using the new technology to raise the 'yield' of renewable resources, using renewable sources of energy, and putting an ecological limit on human activity (this is also discussed by Commoner [9]).

When making use of nature, we must try to maintain the stability of the biosphere, its ecological reserves, and its assimilative capacity (under the anthropogenic impact). Much attention should be focused on a possible 'instability' of the biosphere: its critical elements, the disturbance of which can destroy or cause significant damage to various ecological chains and ecosystems. What is more: "Man should not take over the functions of the biopshere; rather he should facilitate its operation" (S.S. Shvarts, see the *Introduction* of this book).

The main approaches (plans of action) to solving complicated ecological problems of various scales (right up to global), taking the appropriate goals into account, have been discussed at different international conferences, including the Stockholm Conference of 1972, and these were reflected in the resolutions taken (see, e.g. [12, 16, 22]). In the author's opinion, the principal points of the respective plans of action can be summarized as follows (some of the approaches were also discussed by the author earlier [5, 7]):

I. *Organizing a system for monitoring anthropogenic impacts (first and foremost, pollution) and their geophysical (e.g., climatic) and ecological consequences.*

This is precisely the problem which has been dealt with in this book (especially in the first seven chapters).

The development of a system for monitoring human health, as well as studies of human ecology, naturally constitute a separate undertaking.

In addition to the organization of a comprehensive system of observations based on various signs and indicators, the following operations should be carried out within the framework of the monitoring system:

(1) An assessment of the state of the elements of the biosphere and of the anthropogenic loads and their possible ecological consequences (the development of ecologoeconomical approaches to the problem, taking into account the possible esthetic or psychological damage, should also be included here);
(2) Ecological short-term and long-term forecasts of the state of the biosphere on various scales, right up to global;
(3) Development of a system of ecological standards, making it possible to determine the limits of possible anthropogenic loads (including the discharge of pollutants into the environment), taking into consideration the ecological reserve of the biosphere and the assimilative capacity of the environment, while at the same time completely safeguarding human health (setting up standards for the permissible concentrations of pollutants can be included here, too).

Significant work has already been completed with regard to the above: a scientific basis for monitoring has been created; in a number of countries (the USSR included) systems for monitoring anthropogenic pollution at different levels have been established; under the aegis of the UNEP, the Global Environmental Monitoring System (GEMS) is already providing valuable information. This system is designed to provide an 'ecological service', so as to monitor the state of the renewable resources and their natural capacity for reproduction.

The main organization in the Soviet Union responsible for carrying out Part I of the plan is the USSR State Committee for Hydrometeorology and Environmental Monitoring. A substantial number of recent papers have dealt with methods of forecasting the state of the environment on both local and regional scales, analyzing the present state of the biosphere, predicting future states up to the year 2000, and so on (see the earlier mentioned works on modelling the possible development of society and the state of the biosphere, as well as [1, 9, 10 ,13, 15, 19, 23, 24]). Report [24], based on information supplied by experts from various countries, contains an analysis of the state of the environment during the last ten years and gives a forecast for the future.

In this book (see Chapters 8–11) we have presented in detail the contemporary point of view with regard to the present and future states of the biosphere as a whole (atmosphere and World Ocean), including possible catastrophic changes in the event of a nuclear war.

In [19] the author has considered in detail some global ecological problems and ways of solving these; this report was one of six fundamental reports presented at the Interparliamentary Conference on the Environment in Nairobi (26 November – 1 December 1984).

II. *Making decisions and taking appropriate measures with regard to issues of the biosphere, based on the information provided by the monitoring system, taking into account the scientific and economic possibilities.*

This program includes the selection of ways to solve the given problems. These are:

(1) Determination of the priority alternatives (for taking measures) with the aid of ecologoeconomic approaches, as well as ways of regulating the state of the environment (including ecological standards for the impacts);
(2) Determination of the technological (technical) means of solving the problem;
(3) Determination of ways of improving the use of natural resources (in different branches of industry, power production, and agriculture);
(4) Selection of the optimum organizational measures for developing the economy, while at the same time preserving nature; these include plan-

ning the development of different branches, implementing regional programs, organizing territorial-industrial complexes, establishing reserves, etc.

A complete solution of the problem of environmental pollution will obviously require the wide-scale introduction of industrial and agricultural processes which are wastefree (or low-waste, that is, ecologically compatible with the biosphere).

The search for ways of improving the use of natural resources is one of the main tasks of the ecologists working in the Soviet ecological program. The wide use of ecological knowledge and information makes it possible to produce higher yields and greater production without damaging the biosphere, while consuming the same amount of natural resources. This includes, for example, saving biological and water resources, reusing these, and increasing the soil fertility. In the Soviet Union this problem is being solved via an integrated reclamation project, directed at the comprehensive improvement of the soil quality [11]. Information (hydrometeorological information included) about the state of the environment will do much to increase the effectiveness of the reclamation measures being taken.

When implementing the measures under discussion, particularly organizational measures, it is of prime importance to meet the legal norms and to develop state management in the field of nature protection, which ensures first of all the practical realization of the goals reflected in the legislation [8].

For instance, setting up standards to limit the pollution of the environment is an important legal lever in the battle against pollution. The sanctions provided for by legislation with regard to nature protection can obviously be used effectively only if there is monitoring of the state of the environment. In many countries legal issues involving the environment are given serious attention.

There is a reliable legal basis for the rational use of nature and for nature protection in the Soviet Union. These have been accorded the status of a fundamental law, in the Constitution of the USSR.

III. *Determining ways of solving large-scale and global problems within the framework of international cooperation.*

Various approaches can be used to solve global problems of the biosphere: preventing the occurrence of irreversible consequences (e.g., when protecting the ozone layer); weakening the impact (e.g., by the transboundary transport of pollutants and acid rain); adapting human activity to the new conditions (e.g., to slight climatic changes) [19]. Any international efforts will, obviously, be based on specific national actions.

A special problem is presented by the threat of a nuclear war, which would be a major catastrophe for man and the environment, and which would lead to irreversible changes in the biosphere [6, 22]. Obviously, the only possible solution in this case is prevention; the attainment of a durable peace is an indispensable condition for nature protection.

Any global problems regarding the natural environment will undoubtedly be solved only with the help of cooperative international efforts. For instance, such issues are constantly being proposed for consideration at the UN General Assembly. On the initiative of the USSR, the UN General Assembly at its 36th Session adopted a resolution: *On the historical responsibility of states to protect the environment for present and future generations.* A positive contribution has been the signing of different conventions, designed to prevent or reduce anthropogenic impacts on the environment [16].

Obviously, protection of the natural environment is not only an ecological issue: rather, it is also social and political in nature, being closely connected with the fight to limit the arms race and to achieve social progress.

In conclusion, let us emphasize once again that, not by limiting the development of human society, but rather by optimizing the interaction between man and the environment on the basis of scientific and technological progress, by means of cooperative international efforts and on condition of a durable peace on earth, will we be able to guarantee a successful solution of the global problems of the biosphere.

References

Introduction

1. Вернадский В. И. Живое вещество.— М.: Наука, 1978.
2. Всемирная конференция «Климат и человечество». Женева, февраль 1979.— Публикация ВМО, 1979.— 406 с.
3. Всесторонний анализ окружающей природной среды. Труды I советско-американского симпозиума.— Л.: Гидрометеоиздат, 1975.— 326 с.
4. Всесторонний анализ окружающей природной среды. Труды II советско-американского симпозиума.— Л.: Гидрометеоиздат, 1976.— 308 с.
5. Всестороннний анализ окружающей природной среды. Труды III советско-американского симпозиума.— Л.: Гидрометеоиздат, 1978.— 276 с.
6. Всесторонний анализ окружающей природной среды. Труды IV советско-американского симпозиума.— Л.: Гидрометеоиздат, 1980.
7. Герасимов И. П. Научные основы современного мониторинга окружающей среды.— Изв. АН СССР. Сер. геофиз., 1975, № 3, с. 13—25.
8. Израэль Ю. А. Глобальная система наблюдений. Прогноз и оценка изменений состояния окружающей природной среды. Основы мониторинга.— Метеорология и гидрология, 1974, № 7, с. 3—8.
9. Израэль Ю. А. Гидрометеорология и контроль состояния природной среды.— В кн.: Проблемы современной гидрометеорологии. Л.: Гидрометеоиздат, 1977, с. 230—254.
10. Израэль Ю. А. Мониторинг климата и служба сбора климатических данных, необходимых для определения климатических изменений и колебаний.— В кн.: Всемирная конференция «Климат и человечество», Женева, февраль, 1979. Публикация ВМО, 1979, с. 111—130.
11. Ребане К. К. К вопросу охраны окружающей среды и обнаружения внеземных цивилизаций.— Труды Тартуской астрофизической обсерватории им. В. Струве, 1972, т. 40, с. 107—111.
12. Федоров Е. К. Экологический кризис и социальный прогресс.— М.: Гидрометеоиздат, 1977.
13. Форрестер Дж. Мировая динамика.— М.: Наука, 1978.
14. Фратрич И., Халупа К., Кралик Ю. Троянский конь цивилизации.— М.: Мир, 1977.

15. Хильми Г. Ф. Основы физики биосферы.— Л.: Гидрометеоиздат, 1966.
16. Шварц С. С. Теоретические основы глобального экологического прогнозирования.— В кн.: Всесторонний анализ окружающей природной среды. Труды II советско-американского симпозиума. Л.: Гидрометеоиздат, 1976, с. 181—191.
17. Meadows D. H., Meadows D. L. et al. The limits to growth.— New York: Universe Books, 1972.
18. Mesarovic M., Pestel E. Mankind at the turning point.— In: The Second Report to the Club of Rome. New York, 1974.
19. Munn R. E. Global environmental monitoring system. SCOPE, rep. 3. Toronto, 1973.— 130 p.
20. Principles of ecotoxicology/Ed. G. C. Butler. SCOPE, rep. 12. Chichester: John Wiley and Sons, 1978.
21. Global Possible: Resources, development, and the new century. The statement and action agenda of an International Conference sponsored by the World Resources Institute, May 1984.

Chapter 1

1. Алехин А. Б. и др. Проблема оптимизации структуры экономико-экологических систем и особенности описания экологических процессов.— В кн.: Проблемы экологического мониторинга и моделирования экосистем. Л.: Гидрометеоиздат, 1980, т. 3, с. 248—262.
2. Анохин Ю. А., Израэль Ю. А. Системный анализ и имитационное математическое моделирование как методологическая основа определения допустимых нагрузок антропогенных загрязнений окружающей среды — региональный подход.— В кн.: Всесторонний анализ окружающей природной среды. Труды I советско-американского симпозиума. Л.: Гидрометеоиздат, 1975, с. 68—82.
3. Анохин Ю. А. и др. Атмосферный перенос загрязнений в региональном масштабе.— Труды ИПГ, 1978, с. 60—75.
4. Антоновский М. Я. и др. Методология построения балансовых эколого-экономических моделей.— В кн.: Проблемы экологического мониторинга и моделирования экосистем. Л.: Гидрометеоиздат, 1980, т. 3, с. 229—247.
5. Ащепкова Л. Я. Исследования динамики экосистемы озера Байкал методом математического моделирования. Автореф. дисс.— Иркутск, 1979.— 25 с.
6. Воинов А. А. и др. Моделирование озерных экосистем.— В кн.: Проблемы экологического мониторинга и моделирования экосистем. Л.: Гидрометеоиздат, 1981, т. 4, с. 204—234.
7. Всесторонний анализ окружающей природной среды. Труды I советско-американского симпозиума.— Л.: Гидрометеоиздат, 1975.— 326 с.
8. Всесторонний анализ окружающей природной среды. Труды II советско-американского симпозиума.— Л.: Гидрометеоиздат, 1976.— 308 с.
9. Всесторонний анализ окружающей природной среды. Труды III советско-американского симпозиума.— Л.: Гидрометеоиздат. 1978.— 276 с.

10. Всесторонний анализ окружающей природной среды. Труды IV советско-американского симпозиума. Л.: Гидрометеоиздат, 1980.— 317 с.
11. Горстко А. Б. и др. Метод дискретного оптимального управления в решении задачи оптимизации мониторинга экосистемы.— В кн.: Проблемы экологического мониторинга и моделирования экосистем. Л.: Гидрометеоиздат, 1982, т. 5, с. 219—229.
12. Гринфилд С. М. Всесторонний подход к управлению окружающей средой.— В кн.: Всесторонний анализ окружающей природной среды. Труды I советско-американского симпозиума. Л.: Гидрометеоиздат, 1975, с. 26—34.
13. Израэль Ю. А. Комплексный анализ окружающей среды. Подходы к определению допустимых нагрузок на окружающую природную среду и обоснованию мониторинга.— В кн.: Всесторонний анализ окружающей природной среды. Труды I советско-американского симпозиума. Л.: Гидрометеоиздат, 1975, с. 17—25.
14. Израэль Ю. А. Допустимая антропогенная нагрузка на окружающую природную среду.— В кн.: Всесторонний анализ окружающей природной среды. Труды II советско-американского симпозиума. Л.: Гидрометеоиздат, 1976, с. 12—19.
15. Израэль Ю. А. Роль всестороннего анализа окружающей природной среды в организации оптимального взаимодействия человека с природой.— В кн.: Всесторонний анализ окружающей природной среды. Труды III советско-американского симпозиума. Л.: Гидрометеоиздат, 1978, с. 14—22.
16. Израэль Ю. А., Зенин А. А. Об охране озера Байкал.— Метеорология и гидрология, 1973, № 1, с. 15—19.
17. Израэль Ю. А., Анохин Ю. А., Кожова О. М. и др. Комплексный анализ окружающей природной среды и обоснование мониторинга в регионе оз. Байкал.— В кн.: Всесторонний анализ окружающей природной среды. Труды IV советско-американского симпозиума. Л.: Гидрометеоиздат, 1981, с. 43—59.
18. Имитационно-балансовая математическая модель региона оз. Байкал как инструмент всестороннего анализа, долгосрочного прогнозирования и определения допустимых нагрузок воздействия народнохозяйственной деятельности на качество окружающей среды и состояние экологических систем/Ю. А. Израэль, Ю. А. Анохин, А. Х. Остромогильский и др.— В кн.: Всесторонний анализ окружающей природной среды. Труды II советско-американского симпозиума. Л.: Гидрометеоиздат, 1976, с. 246—265.
19. Канторович Е. Г., Семенов С. М. Моделирование сукцессии в многовидовой двухуровенной экосистеме.— В кн.: Проблемы экологического мониторинга и моделирования экосистем. Л.: Гидрометеоиздат, 1983, т. 6, с. 146—159.
20. Кожова О. М. Задачи НИИБиологии в организации исследований по экосистемному анализу и гидробиологическому мониторингу.— В кн.: Гидробиологические и ихтиологические исследования в Восточной Сибири. Чтения памяти проф. М. М. Кожова, Иркутск, 1977, с. 19—27.
21. Кожова О. М. Влияние антропогенных факторов на состояние экологической системы Байкала.— В кн.: Международный симпозиум по комплексному глобальному мониторингу загрязнения окружающей среды. Тезисы докладов. Л.: Гидрометеоиздат, 1978, с. 19—20.

22. Меморандум I сессии советско-американской смешанной комиссии по сотрудничеству в области охраны окружающей среды. Москва, сентябрь, 1972.
23. Парк Р. Экологическое моделирование и оценка экологического стресса.— В кн.: Всесторонний анализ окружающей природной среды. Труды II советско-американского симпозиума. Л.: Гидрометеоиздат, 1976, с. 230—245.
24. Протокол I советско-американского симпозиума по всестороннему анализу окружающей среды.— В кн.: Всесторонний анализ окружающей природной среды. Труды I советско-американского симпозиума. Л.: Гидрометеоиздат, 1975, с. 10—13.
25. Семевский Ф. Н. Прогноз в защите леса.— М.: Лесная промышленность, 1971.
26. Семевский Ф. Н., Семенов С. М. Схема моделирования экзогенной сукцессии.— В кн.: Проблемы экологического мониторинга и моделирования экосистем. Л.: Гидрометеоиздат, 1978, т. 1, с. 95—106.
27. Семевский Ф. Н., Семенов С. М. Математическое моделирование экологических процессов.— Л.: Гидрометеоиздат, 1982.— 280 с.
28. Ульведал Ф. Правильно ли мы оцениваем воздействие окружающей среды на человека? — В кн.: Всесторонний анализ окружающей природной среды. Труды II советско-американского симпозиума. Л.: Гидрометеоиздат, 1976, с. 49—56.
29. Филиппова Л. М., Израэль Ю. А. и др. Комплексный фоновый мониторинг окружающей природной среды на Байкале: цели и задачи, история и перспективы.— В кн.: Комплексный глобальный мониторинг загрязнения окружающей природной среды. Труды II Международного симпозиума. Л.: Гидрометеоиздат, 1982, с. 78—85.
30. Форрестер Дж. Мировая динамика/Пер. с англ.— М.: Наука, 1978.— 167 с.
31. Шварц С. С. Теоретические основы глобального экологического прогнозирования.— В кн.: Всесторонний анализ окружающей природной среды. Труды II советско-американского симпозиума. Л.: Гидрометеоиздат, 1976, с. 181—191.
32. Цветова Е. А. Численное моделирование динамики и термического режима озера Байкал. Автореф. дисс. Новосибирск, 1979.— 28 с.
33. Яворский Н. Я. Всесторонний анализ верхнего эстуария реки Потомак. Основные задачи борьбы с эвтрофикацией.— В кн.: Всесторонний анализ окружающей природной среды. Труды I советско-американского симпозиума. Л.: Гидрометеоиздат, 1975, с. 189—206.
34. Ecological modelling in an environmental management framework/Ed. C. S. Russel.— Resources for the future, 1975.
35. Logofet D. O., Svirezhev Yu. M. Ecological modelling, catastrophy theory and some new concepts of stability.— In: EURO IFIP. Amsterdam, 1979, p. 287—293.
36. Secondary impacts of urbanization of ecosystems methodology/Ed. D. L. Jameson.— Social-economic Environmental Studies Series. U. S. EPA, 1975.

Chapter 2

1. Акоев И. Г., Максимов Г. К., Малышев В. М. Лучевое поражение млекопитающих и статистическое моделирование.— М., 1972.
2. Базилевич Н. И., Родин Л. Е., Розов Н. Н. Сколько весит живое вещество планеты? — Природа, 1971, № 1, с. 46.
3. Базыкин А. Д., Березовская Ф. С. Эффект Олли, нижняя критическая численность популяции и динамика системы хищник—жертва.— В кн.: Проблемы экологического мониторинга и моделирования экосистем. Л.: Гидрометеоиздат, 1979, т. 2, с. 161—175.
4. Базыкин А. Д., Апонина Е. А. Модель экосистемы трех трофических уровней с учетом существования нижней критической плотности популяции продуцента.— В кн.: Проблемы экологического мониторинга и моделирования экосистем, 1981, т. 4, с. 186—203.
5. Гершкович И. Генетика.— М.: Наука, 1968.— 201 с.
6. Глобальный баланс и предельно допустимые выбросы ртути в атмосферу/Б. П. Абрамовский, Ю. А. Анохин, В. А. Ионов и др.— В кн.: Всесторонний анализ окружающей природной среды. Труды II Советско-американского симпозиума. Л.: Гидрометеоиздат, 1976, с. 35—48.
7. Глобальный баланс ртути в биосфере/Ю. А. Анохин, Г. Н. Воронская, И. Я. Николишин и др.— Обнинск: Изд-во информ. центра ВНИИГМИ, 1978.
8. Гречушкина М. П., Израэль Ю. А., Петров В. Н. Возможные технические подходы для выработки норм безопасного проведения ядерных взрывов в мирных целях.— В кн.: Peaceful Nuclear Explosions IV, IAEA, Vienna, 1975, p. 377—388.
9. Джонсон Х. Е. Влияние загрязнения на виды и популяции рыб и птиц.— В кн.: Всесторонний анализ окружающей природной среды. Труды I Советско-американского симпозиума. Л.: Гидрометеоиздат, 1975, с. 158—176.
10. Израэль Ю. А., Петров В. Н. Распространение радиоактивных продуктов подземных ядерных взрывов на большие расстояния.— В кн.: Атомные взрывы в мирных целях. М.: Атомиздат, 1970, с. 87.
11. Израэль Ю. А., Петров В. Н. Диффузия и осаждение радиоактивных продуктов из облака подземного ядерного взрыва.— В кн.: Peaceful Nuclear Explosions IV, IAEA, Vienna, 1975.
12. Израэль Ю. А., Теверовский Е. Н. Возможность широкого мирного использования атомной энергии при обеспечении радиационной безопасности населения.— В кн.: Peaceful Uses of Atomic Energy. Vol. 11. UN, New York. IAEA, Vienna, 1972, p. 703—709.
13. Израэль Ю. А. Мирные ядерные взрывы и окружающая среда.— Л.: Гидрометеоиздат, 1974.— 135 с.
14. Израэль Ю. А. Допустимая антропогенная нагрузка на окружающую природную среду.— В кн.: Всесторонний анализ окружающей природной среды. Труды II Советско-американского симпозиума. Л.: Гидрометеоиздат, 1976, с. 12—19.
15. Израэль Ю. А., Филиппова Л. М. Оценка и пути достижения желаемого качества окружающей природной среды. В кн.:

Мониторинг состояния окружающей природной среды. Труды I Советско-английского симпозиума. Л.: Гидрометеоиздат. 1977, с. 34—40.

16. Израэль Ю. А. и др. Экологический подход к оценке состояния и регулированию качества окружающей природной среды.— ДАН СССР, 1978, т. 241, № 3, с. 723—726.

17. Израэль Ю. А. и др. Экологический подход к оценке состояния и регулирования качества окружающей среды.— ДАН СССР, 1978, т. 241, № 3, с. 723—726.

18. Израэль Ю. А., Цыбань А. В. Об ассимиляционной емкости Мирового океана.— ДАН СССР, 1983, т. 272, № 3, с. 702—704.

19. Имитационно-балансовая математическая модель региона оз. Байкал как инструмент всестороннего анализа, долгосрочного прогнозирования и определения допустимых нагрузок воздействия народнохозяйственной деятельности на качество окружающей среды и состояние экологических систем/Ю. А. Израэль, Ю. А. Анохин, А. Х. Остромогильский и др.— В кн.: Всесторонний анализ окружающей природной среды. Труды II Советско-американского симпозиума. Л.: Гидрометеоиздат, 1976, с. 246—265.

20. Кислотные дожди/Ю. А. Израэль, И. М. Назаров и др.— Л.: Гидрометеоиздат, 1983.— 206 с.

21. Кист А. А. О граничных химических условиях существования жизни.— В кн.: Проблемы экологического мониторинга и моделирования экосистем. Л.: Гидрометеоиздат, 1980, т. 3, с. 64—76.

22. Кнэльсон Дж. Загрязнители со множественными путями воздействия на человека: ионизирующая радиация, асбест и свинец.— В кн.: Всесторонний анализ окружающей природной среды. Труды I Советско-американского симпозиума. Л.: Гидрометеоиздат, 1975, с. 83—98.

23. Левич А. П. Понятие устойчивости в биологии. Математические аспекты.— В кн.: Человек и биосфера, вып. 1. М.: Изд-во МГУ, 1976, с. 138—174.

24. Логофет Д. О., Свирежев Ю. М. Концепция устойчивости биологических систем.— В кн.: Проблемы экологического мониторинга и моделирования экосистем. Л.: Гидрометеоиздат, 1983, т. 6, с. 159—171.

25. Ньюилл В. А. Максимально допустимые нагрузки на человека.— В кн.: Всесторонний анализ окружающей природной среды. Труды I Советско-американского симпозиума. Л.: Гидрометеоиздат, 1975, с. 121—137.

26. Одум Ю. П. Основы экологии.— М.: Мир, 1975.— 740 с.

27. Олла Б. Л., Стадхолм А. Л. Воздействие окружающей среды и поведение. Способности к ответным реакциям у морских рыб.— В кн.: Всесторонний анализ окружающей природной среды. Труды II Советско-американского симпозиума. Л.: Гидрометеоиздат, 1976, с. 64—75.

28. Петросян А. М.— Атомная энергия, 1967, т. 23, вып. 1.

29. Предельно допустимые концентрации вредных веществ в воздухе и воде/Г. М. Беспамятнов, К. К. Богушевская, А. В. Беспамятнова и др.— М.: Химия, 1975.— 456 с.

30. Рабинович Э. З. Переходные процессы в живых системах; нагрузка, устойчивость, адаптация — важный аспект биологиче-

ского мониторинга.— В кн.: Проблемы экологического мониторинга и моделирования экосистем. Л.: Гидрометеоиздат, 1978, т. 1, с. 153—163.

31. Развитие атомной энергии в СССР/Под ред. И. Д. Морохова, А. П. Александрова.— М.: Атомиздат, 1977.

32. Рапопорт И. А., Филиппова Л. М. Дифференциация мутагенного действия препаратов, синтезируемых для химиотерапии.— Бюл. МОИП, 1965, № 4, с. 117—124.

33. Салиев А. В. Задачи экологического районирования региона ЕТС с использованием критерия допустимой экологической нагрузки.— В кн.: Проблемы экологического мониторинга и моделирования экосистем. Л.: Гидрометеоиздат, 1979, т. 2, с. 258—263.

34. Саноцкий И. В. Концепция пороговости реакции живых систем на внешние воздействия и ее следствия в проблеме противохимической защиты биосферы.— В кн.: Всесторонний анализ окружающей природной среды. Труды I Советско-американского симпозиума. Л.: Гидрометеоиздат. 1975, с. 112—120.

35. Семенов С. М. Об устойчивости стационарных состояний динамических систем.— В кн.: Проблемы экологического мониторинга и моделирования экосистем. Л.: Гидрометеоиздат, 1979, т. 2, с. 197—205.

36. Свирежев Ю. М., Логофет Д. О. Устойчивость биологических сообществ.— М.: Наука, 1978.

37. Сивинцев Ю. В. Радиационная безопасность на ядерных реакторах.— М.: Атомиздат, 1967.

38. Федоров В. Д., Соколова С. А. Об устойчивости планктонного сообщества и некоторых характеристик внешней среды.— Океанология, 1972. вып. 6, с .1057—1065.

39. Федоров В. Д. Новый показатель неоднородности структуры сообщества.— Вестник МГУ. Сер. биол. 1973, № 2.

40. Федоров В. Д. Концепция устойчивости экологических систем.— В кн.: Всесторонний анализ окружающей природной среды. Труды I Советско-американского симпозиума. Л.: Гидрометеоиздат, 1975, с. 207—217.

41. Федоров В. Д. Оценки приоритета в ряду загрязнителей.— В кн.: Всесторонний анализ окружающей природной среды. Труды III Советско-американского симпозиума. Л.: Гидрометеоиздат, 1978, с. 138—145.

42. Федоров В. Д. и др. Количественные подходы к проблеме оценки нормы и патологии экосистем.— В кн.: Человек и биосфера, 1982, вып. 6, с. 3—42.

43. Филиппова Л. М. К вопросу о генетической опасности загрязнителей внешней среды.— В кн.: Всесторонний анализ окружающей природной среды. Труды I Советско-американского симпозиума. Л.: Гидрометеоиздат, 1975, с. 145—151.

44. Филиппова Л. М. Генетические аспекты определения допустимых нагрузок.— В кн.: Всесторонний анализ окружающей природной среды. Труды II Советско-американского симпозиума. Л.: Гидрометеоиздат, 1976, с. 136—143.

45. Филиппова Л. М., Израэль Ю. А. Проблема пороговости действия загрязняющих веществ и ее значение в регулировании качества окружающей природной среды.— В кн.: Теоретические и практические подходы к проблеме мутагенеза и канцерогенеза окружающей среды. Материалы советских участников III Совет-

ско-американского симпозиума. М.: Гидрометеоиздат, 1976.
46. Шварц С. С. Теоретические основы глобального экологического прогнозирования.— В кн: Всесторонний анализ окружающей природной среды. Труды II Советско-американского симпозиума. Л.: Гидрометеоиздат, 1976, с. 181—191.
47. Экологический подход к оценке состояния и регулированию качества окружающей природной среды/Ю. А. Израэль, И. М. Назаров, Л. М. Филиппова и др.— В кн.: Всесторонний анализ окружающей природной среды. Труды III Советско-американского симпозиума. Л.: Гидрометеоиздат, 1978, с. 101—130.
48. Acid precipitation and its effects in Norway.— Oslo: Ministry of Environment, 1974.— 16 p.
49. Air pollution across national boundaries. The impact on the environment of sulfur in air and precipitation.— Sweden's case study for the UN conference on the human environment. Stockholm, 1971.— 48 p.
59. Akagi H., Takabatane E. Photochemical formation of methylmercuric compounds from mercuric acetate.— Chemosphere, 1973, vol. 2, p. 131—133.
51. An approach to environmental monitoring via the dynamics of environmental processes particulary the wider applicability of proved models.— MARC/WS (1)/Doc. 3. 3 Sept. 1976.
52. Bond V. P. Evaluation of potential hazards from tritiated water.— In: Environmental aspects of Nuclear Power Stations (Proc. Symp., New York, 1970). IAEA, Vienna, 1971, p. 287.
53. Bourdeau P., Treshow M. Ecosystem response to pollution.— In: Principles of Ecotoxicology/Ed. G. C. Butler.— SCOPE, rep. 12, Chichester: John Wiley and Sons, 1978, p. 313—330.
54. Butler G. C. Estimation of doses and integrated doses.— In: Principles of Ecotoxicology/Ed. G. C. Butler.— SCOPE, rep. 12, Chichester: John Wiley and Sons, 1978, p. 91—114.
55. Brollonin L. Science and information theory.— New York, 1962.
56. Edwards C. A. Environment toxicology of pesticides.— London: Plenum Press., 1972.— 432 p.
57. Energy in a finite world, Al. Mc Donald, IIASA, Laxenburg, Austria, 1981. p. 68.
58. Environmental mutagenic hazards.— Science, 1975, vol. 187, p. 503—514.
59. Farmer F. R. Siting criteria — a new approach.— In: Containment and Sitting of Nuclear Power Plants (Proc. Symp., Vienna, 1967) IAEA, Vienna, 1967, p. 303.
60. Fimreite N. Mercury uses in Canada and their possible hazards as sources of mercury contamination.— Environ. Pollut., 1970, vol. 1, p. 1.
61. French J. G. Effects of suspended sulfates on human health.— Environ. Health Perspectives, 1975, vol. 10, p. 35—37.
62. Garland J. E. e. a. Deposition of gaseous sulfur dioxide to the ground.— Atm. Environ., 1974, vol. 8, p. 75—79.
63. Hulbert S. H. Some dark conception of the species diversity.— Ecology, 1971, vol. 52, N 4.
64. Iernelöv A., Iensen S. Biocidinformation, 1967, vol. 10, N 4.
65. Jensen A. L. The effect of increased mortality on the young in

a population of brook trout, a theoretical analysis.— Trans. Amer. Fish. Soc., 1971, vol. 100, N 3, p. 456—459.

66. Johnels A. G., Westermark T. e. a. Pike (*Esox encins* L.) and some other aquatic organisms in Sweden as indicators of mercury contamination in the environment.— Oikos, 1967, vol. 18, p. 323—333.
67. Junge C. The cycle of atmospheric gases — natural and man-made.— Quart. J. Roy. Met. Soc., 1972, vol. 98, N 418, p. 711—729.
68. Kellog W. W. The sulfur cycle.— Science, 1972, vol. 175, N 4022, p. 587—596.
69. Kitamure S. Determination of mercury content in bodies of inhibitants, cats, fishes and shell fishes in Minamata district and the mud in Minamata Bay.— In: Minamata Disease. Kumamoto Univ., Japan, 1968. 257 p.
70. Knabe W. Effects of sulfur dioxide on terrestrial vegetation.— Ambio, 1976, vol. 5, N 5—6, p. 213—218.
71. Krenkel P. A. Mercuryi Environmental considerations. P. 1.— Critical Reviews in Environmental Control. May 1973.
72. Kurland L. e. a. World Neurology, 1960, vol. 1, N 370.
73. Martin I. E. e. a. Radioactivity from fossile full and nuclear power plants.— In: Environmental Aspects of Nuclear Power Stations (Proc. Symp., New York, 1970). IAEA, Vienna, 1971, p. 325.
74. McArthur R. M. Fluctuations of animal populations and measure of community stability.— Ecology, 1955, vol. 36, N 3.
75. McKim J. M., Beneir D. A. Effects of long-term exposure to copper on survival, growth and reproduction of brook trout (Salvelinus Fontinalis).— J. Fish. Res. Bd. Canada, 1971, vol. 28, N 5, p. 655—662.
76. Miller D. R. Models for total transport.— In: Principles of Ecotoxicology/Ed. G. C. Butler.— SCOPE, rep. 12, Chichester: John Wiley and Sons, 1978, p. 71—90.
77. Mount D. I., Stephan C. E. A method for establishing acceptable toxicant limits for fish-mamathion and the butoxyethanol ester of 2, 4-D.— Trans. Amer. Fish. Soc., 1967, vol. 96, p. 185—193.
78. Nisbet I. C., Sarofim A. F. Rates and routes of transport of PCB's in the environment.— Environ. Health Perspectives, 1972, vol. 1, p. 21—38.
79. Odum E. P. Diversity as a function of energy flow.— In: Unifying concepts in Ecology: Eds. W. H. van Dobben and R. H. Lowe — McConnell, The Hague, 1975, p. 11—14.
80. Patric R. I. Stability and diversity.— Proc. Nat. Acad. Sci. Phyladelphia, 1949, vol. 101, N 2.
81. Peterson H. T. e. a. Environmental tritium contamination from increasing utilization of nuclear energy sources.— In: Environmental Contamination by Radioactive Materials (Proc. Seminar, Vienna, 1969). IAEA, 1969, p. 35.
82. Position paper on regulation of atmospheric sulfates.— US EPA 450/2-75-007, 1975.— 78 p.
83. Stalling D. L., Mayer F. L. Jr. Toxicities of PCB's to fish and environmental residues.— Environ. Health Perspectives, 1972, vol. 1, p. 159—164.
84. U. S. Reactor Containment Technology, 1969, vol. 1, N 1.

85. P. Van Voris P. et al. Functional complexity and ecosystem stability.— Ecology, 1980, vol. 61, N 6, p. 1352—1360.
86. Veith G. D. Recent fluctuations of chlorobiphenyls (PCB's) in the Green Bay, Wisconsin region.— Environ. Health Perspectives, 1972, vol. 1, p. 51—54.
87. Westermark T. Mercury monitoring in Sweden international problems in environmental monitoring.— In: A Bellagio Conf., 16—18 Feb. 1977. Rockefeller Found., Aug. 1977, p. 89—104.
88. Wood I. E. e. a. Synthesis of methyl-mercury compounds by extracts of a methnogenic bacterium.— Nature, 1968, vol. 220, p. 173.

Chapter 3

1. Алехин А. Б. и др. Проблема оптимизации структуры экономико-экологических систем и особенности описания экологических процессов.— В кн.: Проблемы экологического мониторинга и моделирования экосистем. Л.: Гидрометеоиздат, 1980, т. 3, с. 248—262.
2. Антоновский М. Я. Выбор переменных в некоторых эколого-экономических моделях.— В кн.: Проблемы экологического мониторинга и моделирования экосистем. Л.: Гидрометеоиздат, 1980, т. 3, с. 122—140.
3. Антоновский М. Я. и др. Методология построения балансовых эколого-экономических моделей.— В кн.: Проблемы экологического мониторинга и моделирования экосистем. Л.: Гидрометеоиздат, 1980, т. 3, с. 229—247.
4. Антоновский М. Я., Корзухин М. Д. Математическое моделирование экономических и эколого-экономических процессов.— В кн.: Комплексный глобальный мониторинг загрязнения окружающей природной среды. Труды II Международного симпозиума. Л.: Гидрометеоиздат, 1982, с. 373—378.
5. Балацкий О. Ф. Экономика чистого воздуха.— Киев: Наукова думка, 1979.
6. Гофман К. Г. Экономическая оценка природных ресурсов в условиях социалистической экономики.— М.: Наука, 1977.
7. Гофман К. Г., Лемешев М. Я. Социально-экономическая эффективность мероприятий по охране окружающей среды.— В кн.: Актуальные проблемы охраны окружающей природной среды в Советском Союзе и Федеративной Республике Германии. Труды научного симпозиума, Мюнхен, 9—14 ноября 1981, с. 199—211.
8. Гофман К. Г., Гусев А. А. Экологические издержки и концепция экономического оптимума качества окружающей природной среды.— Экономика и математические методы, 1981, т. 17, вып. 3.
9. Гусев А. А., Досчанов Т. Д. Определение сравнительной эффективности применения ядохимикатов с учетом последствий загрязнения окружающей среды.— В кн.: Первая школа «Методы планирования и управления природными ресурсами». М., ЦЭМИ АН СССР, 1975, с. 210—212.
10. Гусев А. А. Модели прогноза согласования развития экономики и охраны атмосферы.— В кн.: Проблемы экологического мониторинга и моделирования экосистем. Л.: Гидрометеоиздат, 1979, т. 2, с. 230—248.

11. Израэль Ю. А. Гидрометеорология и контроль природной среды.— В кн.: Проблемы современной гидрометеорологии. Л.: Гидрометеоиздат, 1977, с. 230—254.
12. Израэль Ю. А. Роль всестороннего анализа окружающей природной среды. Труды III советско-американского симпозиума. Л.: Гидрометеоиздат, 1978, с. 14—22.
13. Израэль Ю. А. Мониторинг климата и служба сбора климатических данных, необходимых для определения климатических изменений и колебаний. Мониторинг данных, связанных с климатом.— Всемирная конференция по климату. Женева, февраль 1979 г. Публикация ВМО, с. 111—130.
14. Израэль Ю. А. Охрана природной среды и рациональное природопользование в СССР.— Вопросы философии, 1979, № 6, с. 119—127.
15. Израэль Ю. А., Филиппова Л. М. Экономические и внеэкономические аспекты проблемы регулирования качества окружающей природной среды.— В кн.: Охрана окружающей природной среды. Экономические аспекты проблемы и вопросы стандартизации. Материалы Советско-американского симпозиума. М., 1978, с. 5—10.
16. Израэль Ю. А., Назаров И. М., Рябошапко А. Г. Подходы к нормированию атмосферных загрязнений при дальнем и трансграничном переносе.— В кн.: Комплексный глобальный мониторинг загрязнения окружающей природной среды. Труды II Международного симпозиума. Л.: Гидрометеоиздат, 1982, с. 351—358.
17. Израэль Ю. А. и др. Экологический подход к оценке состояния и регулирования качества окружающей среды.— ДАН СССР, 1978, т. 241, № 3, с. 723—726.
18. Кислотные дожди/Ю. А. Израэль, И. М. Назаров, Л. М. Филиппова и др.— Л.: Гидрометеоиздат, 1983.— 206 с.
19. Лемешев М. Я. Эколого-экономическая модель природопользования.— В кн.: Всесторонний анализ окружающей природной среды. Труды II советско-американского симпозиума. Л.: Гидрометеоиздат, 1976, с. 266—276.
20. Лемешев М. Я., Гофман К. Г., Гусев А. А. Социально-экономическое обоснование развития производства с учетом охраны окружающей среды. Труды III советско-американского симпозиума. Л.: Гидрометеоиздат, 1978, с. 240—254.
21. Литвин В. А. и др. Об определении экономической оценки ущерба, связанного с изменением состояния природной среды развивающегося территориально-производственного комплекса.— В кн.: Проблемы экологического мониторинга и моделирования экосистем. Л.: Гидрометеоиздат, 1982, т. 5, с. 248—256.
22. Охрана окружающей среды: Модели управления чистотой природной среды.— М.: Экономика, 1977.
23. Синельщиков В. С., Ушаков Е. П. Совместный учет гидрологических свойств водного объекта и экономических показателей ущерба от загрязнения вод при планировании очередности водоохранных мероприятий.— В кн.: Проблемы экологического мониторинга и моделирования экосистем. Л.: Гидрометеоиздат, 1980, т. 3, с. 268—279.
24. Федоров Е. К. Экологический кризис и социальный прогресс.— Л.: Гидрометеоиздат, 1977.
25. Федоренко Н. П. и др. Социально-экономическая эффективность охраны природы.— Природа, 1980, № 10.

26. Хачатуров Т. С. Эффективность капитальных вложений.— М.: Экономика, 1979.— 335 с.
27. Шварц С. С. Теоретические основы глобального экологического прогнозирования.— В кн.: Всесторонний анализ окружающей природной среды. Труды II советско-американского симпозиума. Л.: Гидрометеоиздат, 1976, с. 181—191.
28. Экономические проблемы оптимизации природопользования/Под ред. Н. П. Федоренко.— М.: Наука, 1973.
29. Nevers N. Air pollution control philosophies.— J. Air Poll. Control Assoc., 1977, vol. 27, N 3, p. 197—205.

Chapter 4

1. Берлянд М. Е. и др. Вопросы взаимосвязи глобального и локального мониторинга загрязнения атмосферы.— См. [19], с. 135—151.
2. Виерсма Д. Б. Комплексный мониторинг загрязняющих веществ в национальном парке Олимпик — биосферном заповеднике США.— См. [19], с. 94—106.
3. Гвинн М. Д., Селла Ф., Вален К. К. Глобальная система мониторинга окружающей среды: принципы и прогресс.— См. [18], с. 35—42.
4. Герасимов И. П. Научные основы современного мониторинга окружающей среды.— Изв. АН СССР. Сер. геогр., 1975, № 3, с. 13—25.
5. Герасимов И. П., Израэль Ю. А., Соколов В. Е. Об организации биосферных заповедников (станций) в СССР.— В кн.: Всесторонний анализ окружающей природной среды. Труды II советско-американского симпозиума. Л.: Гидрометеоиздат, 1976, с. 29—34.
6. Герасимов И. П. и др. Слежение за биогеохимическими циклами миграции веществ в биосферном (глобальном) и геосистемном (региональном) мониторинге.— См. [19], с. 26—36.
7. Джонсон Х. Е. Влияние загрязнения на виды и популяции рыб и птиц.— В кн.: Всесторонний анализ окружающей природной среды. Труды I советско-американского симпозиума. Л.: Гидрометеоиздат, 1975, с. 158—176.
8. Израэль Ю. А. Глобальная система наблюдений. Прогноз и оценка изменений состояния окружающей природной среды. Основы мониторинга.— Метеорология и гидрология, 1974, № 7, с. 3—8.
9. Израэль Ю. А. Комплексный анализ окружающей среды. Подходы к определению допустимых нагрузок на окружающую природную среду и обоснованию мониторинга.— В кн.: Всесторонний анализ окружающей природной среды. Труды I советско-американского симпозиума. Л.: Гидрометеоиздат, 1975, с. 17—25.
10. Израэль Ю. А. Основные принципы мониторинга окружающей природной среды и климата.— См. [18], с. 5—14.

11. Израэль Ю. А. Об оценке состояния биосферы и обосновании мониторинга.— ДАН СССР, 1976, т. 226, № 4, с. 955—957.

12. Израэль Ю. А. Допустимая антропогенная нагрузка на окружающую природную среду.— В кн.: Всесторонний анализ окружающей природной среды. Труды II советско-американского симпозиума. Л.: Гидрометеоиздат, 1976, с. 12—19.

13. Израэль Ю. А. Концепция мониторинга состояния биосферы.— В кн.: Мониторинг состояния окружающей природной среды. Труды I советско-английского симпозиума. Л.: Гидрометеоиздат, 1977, с. 10—25.

14. Израэль Ю. А. Фоновый мониторинг и его роль в оценке и прогнозе глобального состояния биосферы.— См. [19], с. 15—26.

15. Израэль Ю. А., Гасилина Н. К., Ровинский Ф. Я. Система наблюдений и контроля загрязнений природной среды в СССР.— Метеорология и гидрология, 1978, № 10, с. 5—12.

16. Израэль Ю. А., Цыбань А. В. Экологические основы мониторинга состояния Мирового океана.— См. [19], с. 44—62.

17. Ковда В. А., Керженцев А. С. Комплексный мониторинг экологического региона.— См. [19], с. 121—129.

18. Комплексный глобальный мониторинг загрязнения окружающей природной среды. Труды Международного симпозиума.— Л.: Гидрометеоиздат, 1980.

19. Комплексный глобальный мониторинг загрязнения окружающей природной среды. Труды II Международного симпозиума.— Л.: Гидрометеоиздат, 1982.— 392 с.

20. Манн Р. Е. Основные принципы и критерии системы комплексного мониторинга.— См. [18], с. 15—18.

21. Осуществление в СССР системы мониторинга загрязнения природной среды/Ю. А. Израэль, Н. К. Гасилина, Ф. Я. Ровинский, Л. М. Филиппова.— Л.: Гидрометеоиздат, 1978.— 115 с.

22. Престон А., Порт Дж. Комплексный глобальный мониторинг окружающей среды: политика и принципы.— См. [18], с. 19—28.

23. Рекомендации II Международного симпозиума по комплексному глобальному мониторингу загрязнения природной среды.— См. [19], с. 391—392.

24. Ровинский Ф. Я. и др. Фоновое загрязнение природных сред на Евразийском континенте.— См. [19], с. 106—114.

25. Симмс Д. Л., Томас Дж. Ф. А., Бишоп Дж. Ф. Комплексный мониторинг, его основы и практическое применение.— См. [19], с. 36—44.

26. Соколов В. Е., Смирнов Н. Н. Мониторинг биологической составляющей биосферы.— См. [18], с. 29—34.

27. Федоров В. Д. К стратегии биологического мониторинга.— Докл. Высшей школы. Сер. биол., 1974, № 10.

28. Федоров В. Д. Концепция устойчивости экологических систем.— В кн.: Всесторонний анализ окружающей природной

среды. Труды I советско-американского симпозиума. Л.: Гидрометеоиздат, 1975, с. 207—217.

29. Ягусевич А., Пжибыльска Г. Концепция организации фонового мониторинга окружающей среды на территории Польши.— См. [19], с. 90—94.

30. Adams R. L. A. Weather information and outdoor recreation decisions: A case study of the New England beach trip.— Worcester (Mass.): Clark Univ. Deport of Geography, 1971.

31. Altouney E. C. The role of uncertainties in the economic evaluation of water resourse projects.— In: Engineering-economic Systems, rep. EEP-7, Palo Alto, Calif., Stanford Univ., 1963.

32. Bulloch B. C. Reliability engineering in the chemical industry.— Risley, England: Systems Reliability Service, UK Atom., Energy Authority, 1975.

33. Burton I., Wilson I., Munn R. E. Environmental impact assessment: National approaches and international needs.— ibid, vol. 3, N 2, p. 133—150.

34. Dion K. L. e. a. Why do groups make risker decision than individuals.— In: Advances in Experimental Social Psychology. Vol. 5. New York: Academic Press, 1970, p. 305—377.

35. Environmental Monitoring and Assessment/Ed. G. B. Wiersma and A. I. Sors, 1982, vol. 2, N 4.

36. Grun C. H. Concepts of safety: A critical bibliography.— B. Dundu, Scotland, Res. Unit., School of Architecture, Duncan of Iordanstone College of Art, 1975.

37. Izrael Yu. A. The problem of air pollution and other aspects of environmental pollution. The concept of monitoring and regulating the quality of the environment.— In: Meteorology and the human environment, WMO N 517. Geneva, 1979, p. 1—9.

38. Izrael Yu. A. et al. Atmosphere; Background Report, prepared for the United Nations Environment Programme, 1980. См. [55].

39. Izrael Yu. A., Munn R. E. Environmental and Renewable resource monitoring. Paper presented at the meeting "Sustainable Development of the Biosphere". Laxenburg, Austria, 27—31 aug. 1984.

40. Kates R. W. The pursuit of beauy in the environment.— Landscape, 1966, vol. 16, p. 21—25.

41. Kates R. W. Risk assessment of environmental hazard.— SCOPE, rep. 8, Chichester—New York—Brisbane—Toronto: John Wiley and Sons, 1978.

42. Lawrence W. W. Of Acceptable risk science and the determination of safety.— Los Altos, California: William Kaufmann Inc, 1976.

43. Munn R. E. Global Environmental Monitoring System (GEMS). Action Plan for Phase 1. SCOPE, rep. 3, Toronto, 1973.— 130 p.

44. Martin B., Sella F. The development and implementation of

the global Environmental Monitoring System.— Doc. UNEP, Nairobi, 1977.

45. Otway H. J., Erdman R. C. Reactor siting and design from a risk standpoint.— Nucl. Eng. and Design, 1970, vol. 13, p. 365—376.

46. Otway H. J. A method for estmating the risk from a plowshare detonation, peaceful nuclear explosions. II.— IAEA, Vienna, 1971, p. 305—318.

47. Otway H. J. Risk estimations and evaluation.— In: Proc. IIASA Planning Conference on Energy Systems. Schloss Laxenberg, Austria: IIASA, 1973.

48. Otway H. J., Cohen. Revealed preferences: Comments on the starr benefit-risk relationship.— Schloss Laxenberg, Austria: IIASA RM-75-7, 1975.

49. Pollution monitoring and research in the framework of the MAB Programme, Moscow, 23—26 Apr. 1974.— MAB, rep. ser. N 20, Paris: UNESCO, 1974, p. 58—63.

50. Raiffa H. Decision analysis reading.— Mass.: Addison-Wesley, 1968.

51. Report of the international Meeting on Monitoring Held at Nairobi 11—20 Feb. 1974.— 60 p.

52. Schweitzer G. E. Risk assessment near uncontrolled hazardous waste sites: role of monitoring data.— Environmental Monitoring and Assessment, 1982, vol. 2, N 1—2, p. 15—32.

53. Starr C. Social benefit versus technological risk.— Science, 1968, vol. 165, p. 1332—1338.

54. Summary Report on the status of the WMO Background air pollution monitoring Network (BAPMoN), WMO, May 1982, Geneva.

55. The World Environment 1972—1982. A report by the United Nations Environment Programm. Tycooly Intern. Publ. Lim., Dublin, 1982 (Chapter 2, p. 65).

56. Westcott M. R. Toward a contemporary psychology of intuition.— New York: Holt, Rinehart and Winston, 1968.

57. Wiggins I. H. Earthquake safety in the City of Long Beach based on the concept of balanced risk.— In: Perspectives on Benefit-Risk Decision Making. Wash.: Nat. Acad. Eng., 1972, p. 87—95.

Chapter 5

1. Биосферные заповедники. Труды I советско-американского симпозиума.— Л.: Гидрометеоиздат, 1977.— 272 с.
2. Биосферные заповедники. Труды II советско-американского симпозиума.— Л.: Гидрометеоиздат, 1981.
3. Бурдин К. С. Место биотестирования в системе биологического мониторинга.— В кн.: Комплексный глобальный монито-

ринг загрязнения окружающей природной среды. Труды II Международного симпозиума. Л., 1982, с. 315—321.

4. Виерсма Д. Б. Комплексный мониторинг загрязняющих веществ в национальном парке Олимпик — биосферном заповеднике США.— В кн.: Комплексный глобальный мониторинг загрязнения окружающей природной среды. Труды II Международного симпозиума. Л., 1982.
5. Виноградов Б. В. Космические методы изучения природной среды.— М.: Мысль, 1976.
6. Виноградов Б. В. Аэрокосмические исследования природоохранных территорий в СССР.— В кн.: Концепция биосферных заповедников. М., 1983, с. 146—162.
7. Гасилина Н. К., Ровинский Ф. Я., Ь лтнева Л. И. Программа и методика комплексного мониторинга загрязнений в биосферных заповедниках.— В кн.: Биосферные заповедники. Труды I советско-американского симпозиума. Л.: Гидрометеоиздат, 1977, с. 146—151.
8. Герасимов И. П. и др. Внутренний оборот веществ в главных типах экосистем на территории СССР.— Изв. АН СССР. Сер. геогр., 1972, № 2.
9. Герасимов И. П., Израэль Ю. А., Соколов В. Е. Об организации биосферных заповедников (станций) в СССР.— В кн.: Всесторонний анализ окружающей природной среды. Труды II советско-американского симпозиума. Л., 1976.
10. Герасимов И. П. Биосферные станции-заповедники, их задачи и программа деятельности.— Изв. АН СССР. Сер. геогр., 1978, № 2.
11. Герасимов И. П. Геосистемный мониторинг и его реализация в биосферных заповедниках.— В кн.: Концепция биосферных заповедников. М., 1983, с. 106—116.
12. Гиляров М. С., Покаржевский А. Д. Почвенные беспозвоночные как объект экологического мониторинга.— В кн.: Концепция биосферных заповедников. М., 1983.
13. Григорьев А. А. Космическая индикация ландшафтов.— Л., 1975.
14. Доклад об установлении причин и предотвращении нанесения ущерба лесам, водоемам и строительным сооружениям в результате загрязнения воздуха в Федеративной Республике Германии. Многосторонняя конференция по окружающей среде. Мюнхен, 24—27 июня 1984 г.— 140 с.
15. Израэль Ю. А.— Вестник АН СССР, 1978, № 6, с. 35—36.
16. Израэль Ю. А. О направленности первоочередных исследований и практических мер при организации комплексного экологического мониторинга состояния биосферы и задачи проекта № 14 программы ЮНЕСКО «Человек и биосфера».— В кн.: Изучение загрязнения окружающей природной среды и его влияния на биосферу. Материалы совещания по проекту № 14 программы ЮНЕСКО «Человек и биосфера». Л.: Гидрометеоиздат, 1979.
17. Израэль Ю. А., Симонов А. И., Цыбань А. В. Научные аспекты комплексного исследования Тихого океана с целью предотвращения его загрязнения.— Тезисы докладов на XIV Тихоокеанском конгрессе. Хабаровск, 1979.
18. Израэль Ю. А., Симонов А. И., Цыбань А. В. Научные аспекты комплексного исследования Тихого океана с целью

предотвращения его загрязнения.— В кн.: Труды Тихоокеанского конгресса. М.: Наука, 1981.
19. Израэль Ю. А., Филиппова Л. М., Ровинский Ф. Я. Влияние загрязнений на биосферу и их мониторинг на базе биосферных заповедников.— В кн.: Биосферные заповедники. Труды I советско-американского симпозиума. Л., 1977.
20. Израэль Ю. А., Филиппова Л. М., Семевский Ф. Н. и др. О некоторых принципах экологического мониторинга в условиях фонового загрязнения окружающей природной среды.— ДАН СССР, 1978, т. 241, № 1.
21. Израэль Ю. А., Цыбань А. В. Проблемы мониторинга экологических последствий загрязнения океана.— Л.: Гидрометеоиздат, 1981.
22. Израэль Ю. А., Филиппова Л. М. и др. Экологический мониторинг и регулирование состояния природной среды.— В кн.: Проблемы экологического мониторинга и моделирования экосистем. Л., 1981. т. 4.
23. Израэль Ю. А., Ровинский Ф. Я., Филиппова Л. М. Комплексный мониторинг фонового загрязнения и его экологических эффектов в биосферных заповедниках. Первый Международный конгресс по биосферным заповедникам 26 сентября — 2 октября 1983.— В кн.: Концепция биосферных заповедников СССР. М., 1983, с. 117—132.
24. Израэль Ю. А., Филиппова Л. М. и др. Фоновый мониторинг и анализ причин глобальных изменений в состоянии биоты.— В кн.: Проблемы экологического мониторинга и моделирования экосистем. Л.: Гидрометеоиздат, 1983, т. 6, с. 68—75.
25. Израэль Ю. А., Ровинский Ф. Я. Мониторинг фонового загрязнения природных сред.— ДАН СССР, 1982, т. 264, № 5.
26. Инсаров Г. Э. Об учете лишайников-эпифитов на стволах деревьев.— В кн.: Проблемы экологического мониторинга и моделирования экосистем. Л., 1982, т. 5.
27. Ковач М., Опаускi И., Клинчек П., Подани Я. Листья деревьев как индикаторы загрязнения тяжелыми металлами крупных городов.— В кн.: Комплексный глобальный мониторинг загрязнения окружающей природной среды. Труды II Международного симпозиума. Л., 1982, с. 307—314.
28. Ковда В. А., Керженцев А. С. Комплексный мониторинг экологического региона.— В кн.: Комплексный глобальный мониторинг загрязнения окружающей природной среды. Труды II Международного симпозиума. Л., 1982, с. 121—129.
29. Кондратьев К. Я. и др. Метеорологическое зондирование подстилающей поверхности из космоса.— Л.: Гидрометеоиздат, 1979.
30. Кондратьев К. Я. Современное состояние и перспективы применения спутников для мониторинга климата.— Метеорология и гидрология, 1982, № 7, с. 118—123.
31. Комплексный глобальный мониторинг Мирового океана. I Международный симпозиум. Тезисы докладов.— М., 1983.
32. Криницкий В. В. Государственные заповедники СССР и их роль в наблюдениях за изменениями в биосфере.— В кн.: Биосферные заповедники. Труды I советско-американского симпозиума. Л.: Гидрометеоиздат, 1977.
33. Мониторинг атмосферы — обоснование приоритетности за-

грязнителей, оценка фоновой региональной и глобальной составляющих загрязнения/Ю. А. Израэль, Г. Н. Воронская, В. Н. Колесникова и др.— В кн.: Проблемы экологического мониторинга и моделирования экосистем. Л.: Гидрометеоиздат, 1978, т. 1, с. 7—18.

34. Морган Дж. Б., Уиерсма Дж. Б., Барт Д. С. Мониторинг в биосферных заповедниках в целях выявления регионального фона загрязнений.— В кн.: Биосферные заповедники. Труды I советско-американского симпозиума. Л.: Гидрометеоиздат, 1977, с. 234—248.

35. Налимов В. В. Теория эксперимента.— М.: Наука, 1971.

36. Ниомаркаи М. К., Фридвальски Л., Сас Я. Использование хвои сосны обыкновенной (Pinus silvestris L.) для определения содержания двуокиси серы в воздухе.— В кн.: Комплексный глобальный мониторинг загрязнения окружающей природной среды. Труды II Международного симпозиума. Л., 1982.

37. О программе комплексного фонового мониторинга состояния загрязнения окружающей природной среды/Ю. А. Израэль, Л. М. Филиппова, Ф. Я. Ровинский и др.— Метеорология и гидрология, 1978, № 9.

38. Орлов Д. С. и др. Количественные закономерности отражения света почвами. О возможности крупномасштабного картирования почв по спектральной отражательной способности.— Научн. докл. Высшей школы. Сер. биол. науки, 1982, № 1, с. 92—96.

39. Ровинский Ф. Я., Филиппова Л. М., Израэль Ю. А. Фоновый мониторинг: региональные и базовые станции, биосферные заповедники.— В кн.: Мониторинг состояния окружающей природной среды. Труды I советско-английского симпозиума. Л.: Гидрометеоиздат, 1977.

40. Секи Х. Мониторинг эвтрофикации методом исследования динамики микробного поглощения растворенной в воде органики.— В кн.: Комплексный глобальный мониторинг загрязнения окружающей природной среды. Труды II Международного симпозиума. Л., 1982.

41. Семенов С. М., Филиппова Л. М. Прогнозирование состояния биоты в системе экологического мониторинга.— В кн.: Комплексный глобальный мониторинг загрязнения окружающей природной среды. Труды II Международного симпозиума. Л., 1982, с. 364—370.

42. Слепян Э. И. Пороки развития растений как тесты при мониторинге загрязнения окружающей среды.— В кн.: Комплексный глобальный мониторинг загрязнения окружающей природной среды. Труды Международного симпозиума. Л.: Гидрометеоиздат, 1980, с. 367—369.

43. Соколов В. Е. и др. Сукцессии наземных сообществ и биологический мониторинг.— В кн.: Комплексный глобальный мониторинг загрязнения окружающей природной среды. Труды II Международного симпозиума. Л., 1982.

44. Соколов В. Е. и др. Критерии размещения биосферных заповедников.— В кн.: Концепция биосферных заповедников. М., 1983, с. 77—89.

45. Федоров В. Д. Особенности организации биологических систем и гипотеза «вспышки» вида в сообществе.— Вестник МГУ. Сер. биолог., 1970, № 2, с. 71—81.

46. Федоров В. Д. К стратегии биологического мониторинга.— Биологические науки, 1974, № 6.
47. Федоров В. Д. Концепция устойчивости экологических систем.— В кн.: Всесторонний анализ окружающей природной среды. Труды I советско-американского симпозиума. Л.: Гидрометеоиздат, 1977, с. 207—217.
48. Федоров В. Д. Проблема предельно допустимых воздействий антропогенного фактора с позиций эколога.— В кн.: Всесторонний анализ окружающей природной среды. Труды II советско-американского симпозиума. Л.: Гидрометеоиздат, 1976.
49. Федоров В. Д. Оценки приоритета в ряду загрязнителей.— В кн.: Всесторонний анализ окружающей природной среды. Труды III советско-американского симпозиума. Л.: Гидрометеоиздат, 1978, с. 138—145.
50. Федоров В. Д. Принципы организации биологического мониторинга.— В кн.: Изучение загрязнения окружающей природной среды и его влияния на биосферу. Материалы совещания по проекту № 14 программы ЮНЕСКО «Человек и биосфера», Ташкент, 3—7 мая 1978 г. Л.: Гидрометеоиздат, 1979, с. 8—14.
51. Федоров В. Д. Роль эксперимента для обоснования и осуществления биологического мониторинга. Докл. раб. группе по Проекту № 14 программы ЮНЕСКО «Человек и биосфера». М., 1975.
52. Физические основы теории климата и его моделирования.— Труды международной научной конференции. Стокгольм, 29 июля — 10 августа 1974 г. Л.: Гидрометеоиздат, 1977.
53. Шварц С. С. Теоретические основы глобального экологического прогнозирования.— В кн.: Всесторонний анализ окружающей природной среды. Труды II советско-американского симпозиума. Л.: Гидрометеоиздат, 1976, с. 181—191.
54. Экологический мониторинг в связи с задачей регулирования качества окружающей природной среды/Л. М. Филиппова, Ф. М. Семевский, С. М. Семенов и др.— В кн.: Всесторонний анализ окружающей природной среды. Труды III советско-американского симпозиума. Л.: Гидрометеоиздат. 1978, с. 172—190.
55. Badgley P. C. e. a. Oceans from space.— Houston, 1969.
56. Breymeyer A. I. Monitoring of the functioning of ecosystem.— Environmental Monitoring and Assessment, 1981, vol. 1, N 2, p. 175—183.
57. Colwell I. E. Use of remote sensing in the inventory of agriculture crops.— Michigan, 1972.
58. Early action on the Global Environmental Monitoring System.— Wash.: Nat. Acad. Sci., 1976.— 23 p.
59. Estes I. E. Geoforum, 1972, vol. 5, N 9.
60. Farrar J. F., Thompson J. R. Monitoring requirements for biological systems MARC.— London: Chelsea College, 1977.
61. Final Report of the Expert Meeting on the Operation of Integrated Monitoring Programmes (Geneva, 2—5 Sept. 1980) Annex V/Yu. A. Izrael et al.
62. Goodman G. T. Outline proposal for a collaborative (MAB/WMO) Project Pollutant Dynamics and Biotic Response in Ecosystem.— London: Chelsea College, 1977.
63. Holter M. Remote sensing with special reference to forestry and agriculture.— Wash., 1970, p. 73—163.

64. Izrael Yu. e. a. Observation, evaluation and prediction of environmental changes: Principles of monitoring.— MAB, rep. ser. N 20. Paris: Unesco, 1974.
65. Izrael Yu. A., Tsiban A. V. Problems of monitoring the ecological consequences of ocean pollution.— J. Oceanogr. Japan, 1981, vol. 36, N 6, p. 293—314.
66. Izrael Yu. A., Filippova L. M., Rovinsky F. Ya. The programm of ecological monitoring in biosphere reserves.— In: Proc. of Second US—USSR Symp. on Biosphere Reserves. March 10—15, 1980, Florida, pp. 207—220.
67. Izrael Yu. A., Novikov Y. V., Afanasiev Yu. A. Application of space imagery for the study of anthropogenic pollution and its impaction the environment.— IAF-83-134, 34th Congress of the International Astronautical Federation, Oct. 10—15, 1983, Budapest.
68. Martin B., Sella T. The Global Environmental Monitoring System.— In: A Bellagio Conf., 16—18 Feb., 1977. Rockefeller Found., Aug. 1977, p. 16—24.
69. Programme on Man and the Biosphere (MAB). Task Force on: Pollution monitoring and research in the framework of the MAB Programme, organized jointly by Unesco and UNEP.— MAB, rep. ser. N 20. Paris: Unesco, 1974.
70. Randerson D. e. a. J. Appl. Met., 1971, vol. 10, N 6.
71. Rao P. K. Bull. Amer. Met. Soc., 1972, vol, 53, N 7.
72. Report of the Intergovernmental Meeting on Monitoring held at Nairobi 11—20 Feb. 1974.— 60 p.
73. Short N. M. et al. Mission to Earth: Landsat views the World., NASA, Wash., D. C., 1976.
74. Suomi V. E. The need for climate monitoring energy and climate.— Wash.: Nat. Acad. Sci., 1977, p. 128—132.
75. Tobler W. R. Area, 1969, vol. 1, N 3.
76. Ghilarov M. S. Bodenwirbellose als Indikatoren des Bodenhaushaltes und von bodenbildenden Prozessen.— Podobiologia, 1978, Bd 18, S. 300—309.
77. Izrael Yu. A., Novikov Y. V., Afanasiev Yu. A. A study of anthropogenic pollution and its impact on water bodies using computer analysis of satellite imagery.— IAF-84-125, 35th Congress of the International Astronautical Federation, Oct. 7—13, 1984, Lausanne.

Chapter 6

1. Будыко М. И. Атмосферная углекислота и климат.— Л.: Гидрометеоиздат, 1978.— 32 с.
2. Будыко М. И. Метод воздействия на глобальный климат.— Метеорология и гидрология, 1974, № 2.
3. Будыко М. И. Изменения климата.— Л., 1974.
4. Ветлов И. П. Результаты исследований в области спутниковой метеорологии.— В кн.: Проблемы современной гидрометеорологии. Л.: Гидрометеоиздат, 1977.
5. Всесторонний анализ окружающей природной среды. Труды I Советско-американского симпозиума. Тбилиси, 25—29 марта 1974 г.— Л.: Гидрометеоиздат, 1975.— 326 с.

6. Годовой отчет ВМО за 1977 г. Публикация ВМО, 1978, № 502.
7. Десятый отчет об осуществлении Всемирной службы погоды.— Публикация ВМО, 1978, № 518.
8. Израэль Ю. А. Развитие гидрометеорологической службы в Советском Союзе.— Бюл. ВМО, 1978, т 27, № 2, с. 103—112.
9. Израэль Ю. А. Мониторинг состояния биосферы и климата. Доклад на советско-американском симпозиуме по изменениям климата. Ленинград, июнь 1977.
10. Израэль Ю. А. Мониторинг климата и служба сбора климатических данных, необходимых для определения климатических изменений и колебаний. Мониторинг данных, связанных с климатом.— Всемирная конференция «Климат и человечество». Женева, февраль 1979 г. Публикация ВМО, с. 111—130.
11. Кароль И. Л. Системное моделирование «тепличного эффекта», переноса и фотохимии газовых примесей в атмосфере.— Изв. АН СССР. Физика атмосферы и океана, 1980, № 1.
12. Келлог У. У. Влияние деятельности человека на климат.— Бюл. ВМО, 1977, т. 26, № 4, с. 285—298; 1978, т. 27, № 4.
13. Кондратьев К. Я. Современное состояние и перспективы применения спутников для мониторинга климата.— Метеорология и гидрология, 1982, № 7, с. 118—123.
14. Марчук Г. И. Моделирование изменений климата и проблема долгосрочного прогноза погоды.— Всемирная конференция «Климат и человечество», Женева, февраль 1979 г. Публикация ВМО.
15. Материалы Межправительственного совещания по мониторингу. Найроби, 1974.
16. Сокращенные отчеты XXIX и XXX сессий Исполнительного Комитета Всемирной Метеорологической Организации с резолюциями, Женева, 1977, 1978.
17. Федоров Е. К. Изменение климата и стратегия человечества.— Всемирная конференция по климату. Женева, февраль 1979 г. Публикация ВМО, с. 13—28.
18. Annual Report of the World Meteorological Organization, WMO, N 628. Geneva, 1983.
19. Background paper on climate-related monitoring.— UNEP/WMO Doc. for UNEP/WMO Gov. Expert Meet., Geneva, Apr. 1978.
20. Elements of the research strategy for the United States Climate Programme.— Wash.: Nat. Acad. Sci., 1978.
21. Eleventh Status Report on Implementation World Weather watch, WMO, N 601, Geneva, 1982.
22. Johnson D. S., Vetlov I. P. The role of satellites in WMO programmes in the 1980-s.— World Weather Watch Planning Rep., N 36.
23. Koshlyakov M. N., Monin A. S. Strategy of ocean monitoring for climate research.— In: The Global Climate. Cambridge Univ. Press, 1984, p. 205—212.
24. Marchuk G. I. Formulation of the theory of perturbations for complicated models.— Geofisica International, 1975, vol. 15, N 2, p. 103—156; vol. 15, N 3, p. 169—182.
25. Munn R. E. Global Atmospheric Monitoring System. The physical basis of climate and climate modelling.— GARP Publ., ser. N 16, 1975, p. 258—265.
26. Raschke E. Satellite capability in monitoring climate parameters.— The physical basis of climate and climate modelling.— GARP Publ., ser. N 16, 1975, p. 252—257.

27. Report of the Meeting of a Goverment Expert Group on Climate-related Monitoring. Geneva, 1982, WCP-25.
28. Suomi V. E. The need for climate monitoring. Energy and Climate.— Wash.: Nat. Acad. Sci., 1977, p. 128—132.
29. The physical basis of climate and climate modelling.— GARP Publ., ser. N 16, 1975.

Chapter 7

1. Безуглая Э. Ю., Горичев А. А., Разбегаева Е. А. Годовой и суточный ход содержания атмосферных примесей в городских условиях.— Труды ГГО, 1971, вып. 254, с. 152—161.
2. Безуглая Э. Ю. Особенности загрязнения воздуха городов и роль метеорологических факторов.— В кн.: Проблемы контроля и обеспечения чистоты атмосферы. Л.: Гидрометеоиздат, 1975.
3. Берлянд М. Е. Современные проблемы атмосферной диффузии и загрязнения атмосферы.— Л.: Гидрометеоиздат, 1975.
4. Болтнева Л. И., Израэль Ю. А. и др. Глобальное загрязнение ^{137}Cs и ^{90}Sr и дозы внешнего облучения на территории СССР.— Атомная энергия, 1977, т. 42, вып. 5, с. 355—360.
5. Гвинн М. Д., Селла Ф., Валлен К. К. Глобальная система мониторинга окружающей среды: принципы и прогресс.— В кн.: Комплексный глобальный мониторинг загрязнения окружающей природной среды. Труды Международного симпозиума. Л., 1980.
6. Годовой отчет ВМО за 1977 г.— Публикация ВМО, 1978, № 502.
7. Головастиков Н. А. Аэросиноптические условия загрязнения воздуха в г. Дзержинске Горьковской области и оперативное обслуживание народнохозяйственных организаций.— В кн.: Проблемы контроля и обеспечения чистоты атмосферы. Л.: Гидрометеоиздат, 1975, с. 66—68.
8. Гудман А. Х. Стандарты качества воды.— В кн.: Мониторинг состояния окружающей природной среды. Труды I Советско-английского симпозиума. Л.: Гидрометеоиздат, 1977, с. 203—215.
9. Израэль Ю. А. Комплексный анализ окружающей среды. Подходы к определению допустимых нагрузок на окружающую природную среду и обоснованию мониторинга.— В кн.: Всесторонний анализ окружающей природной среды. Труды I Советско-американского симпозиума. Л.: Гидрометеоиздат, 1975, с. 17—25.
10. Израэль Ю. А. Гидрометеорология и контроль состояния природной среды.— В кн.: Проблемы современной гидрометеорологии. Л.: Гидрометеоиздат, 1977, с. 230—254.
11. Израэль Ю. А. Развитие гидрометеорологической службы в Советском Союзе.— Бюлл. ВМО, 1978, т. 27, № 2.
12. Израэль Ю. А. К стратегии сохранения и регулирования качества природной среды.— Водные ресурсы, 1977, № 6, с. 5—10.
13. Израэль Ю. А. и др. О программе комплексного фонового мониторинга состояния загрязнения окружающей природной среды/Ю. А. Израэль, Л. М. Филиппова, Ф. Я. Ровинский и др.— Метеорология и гидрология, 1978, № 9, с. 5—11.
14. Израэль Ю. А., Гасилина Н. К., Ровинский Ф. Я. Система наблюдений и контроля загрязнений природной среды в СССР.— Метеорология и гидрология, 1978, № 10. с 5—12.
15. Израэль Ю. А. Роль всестороннего анализа окружающей при-

родной среды в организации оптимального взаимодействия человека с природой.— В кн.: Всесторонний анализ окружающей природной среды. Труды III Советско-американского симпозиума. Л.: Гидрометеоиздат, 1978, с. 14—22.

16. Израэль Ю. А. О направленности первоочередных исследований и практических мер при организации комплексного экологического мониторинга состояния биосферы и задачи проекта № 14 программы ЮНЕСКО «Человек и биосфера». Изучение загрязнения окружающей природной среды и его влияния на биосферу.— Материалы совещания по проекту № 14 МАБ ЮНЕСКО, Ташкент, 3—7 мая 1978 г. Л.: Гидрометеоиздат, 1979.
17. Израэль Ю. А., Филиппова Л. М., Ровинский Ф. Я. Глобальные и региональные аспекты мониторинга загрязнения природной среды.— Изв. АН СССР. Серия геогр., 1983, № 1, с. 5—11.
18. Израэль Ю. А., Антоновский М. Я. и др. К статистическому обоснованию компонент регионального загрязнения атмосферы в фоновом районе.— ДАН СССР, 1984, с. 334—337.
19. Израэль Ю. А. Проблемы охраны природной среды и пути их решения.— Л.: Гидрометеоиздат, 1984. 48 с.
20. Кедди А. В. Прогностическое моделирование загрязнения воздуха: исследование в Форт-Вэлли.— В кн.: Мониторинг состояния окружающей природной среды. Труды I Советско-английского симпозиума. Л.: Гидрометеоиздат, 1977.
21. Научные основы контроля качества поверхностных вод по гидробиологическим показателям. Труды советско-английского семинара. Валдай, СССР, 12—14 июля 1976.— Л., 1977.
22. Некоторые методологические аспекты научного обоснования мониторинга загрязнения атмосферы на примере Среднеазиатского региона/Ю. А. Израэль, Г. Н. Воронская, В. Н. Колесникова и др.— В кн.: Актуальные проблемы охраны окружающей среды и рационального использования природных ресурсов. Ташкент, 1977.— 258 с.
23. Осуществление в СССР системы мониторинга загрязнения природной среды/Ю. А. Израэль, Н. К. Гасилина, Ф. Я. Ровинский, Л. М. Филиппова.— Л.: Гидрометеоиздат, 1978.— 118 с.
24. Пичи Дж. Е., Кингслен П., Порт Г. Н. Дж. Стратегия мониторинга и оценка загрязнения окружающей среды.— В кн.: Мониторинг состояния окружающей природной среды. Труды I Советско-английского симпозиума. Л.: Гидрометеоиздат, 1977.
25. Пичи Дж. Е. Экологический анализ как средство контроля.— В кн.: Мониторинг состояния окружающей природной среды. Труды I Советско-английского симпозиума. Л., 1977.
26. Порт Дж. Комплексный глобальный мониторинг окружающей среды. Практический подход.— В кн.: Комплексный глобальный мониторинг загрязнения окружающей природной среды. Труды Международного симпозиума. Л.: Гидрометеоиздат, 1980, с. 134—147.
27. Ровинский Ф. Я. и др. Фоновое содержание озона, пыли, соединений азота и серы.— В кн.: Мониторинг фонового загрязнения природной среды. Л.: Гидрометеоиздат, 1982, вып. 1, с. 3.
28. Ровинский Ф. Я. и др. Содержание полициклических ароматических углеводородов и хлорорганических пестицидов в природных средах (по мировым данным).— См. [27], с. 36.
29. Симпсон Е. А. Система согласованного мониторинга качества

внутренних пресных вод.— В кн.: Мониторинг состояния окружающей природной среды. Труды I Советско-английского симпозиума. Л.: Гидрометеоиздат, 1977, с. 140—153.
30. Сокращенный отчет XXIX сессии Исполнительного Комитета ВМО с резолюциями. Женева, 1977.
31. Стырикович М. А. и др. Пути снижения загрязнения атмосферы городов продуктами сжигания топлива.— Теплоэнергетика, 1976, № 9.
32. Фейерклаф А. Дж. Подход к мониторингу окружающей среды в Великобритании.— В кн.: Мониторинг состояния окружающей природной среды. Труды I Советско-английского симпозиума. Л.: Гидрометеоиздат, 1977, с. 26—33.
33. Шабад Л. М. О циркуляции канцерогенов в окружающей среде.— М.: Медицина, 1973.— 261 с.
34. Яворский Н. Я. Всесторонний анализ верхнего эстуария реки Потомак. Основные задачи борьбы с эвтрофикацией.— В кн.: Всесторонний анализ окружающей природной среды. Труды I Советско-американского симпозиума. Л., 1975.
35. Berry B. J. L., Bruzewicz A. J., Cargo D. B. Land use. Urban formand environmental quality.— Chicago, 1974.
36. Cocoros G., Cahn P. H. Mercury concentrations in fish, plankton and water from three Western's Atlantic estuaries.— J. Fish Biol., 1973, vol. 5, N 6, p. 641—647.
37. Early action on the Global Environmental Monitoring System.— Wash.: Nat. Acad. Sci., 1976.— 23 p.
38. Environmental factors in respiratory disease. Ed. D. H. Lee.— New York: Academic Press, 1972.
39. Environmental quality. The fourth annual report of the council on environmental quality.— Wash.: U. S. Government printing office, 1973.— 499 p.
40. Environmental Protection Agency reports. National air quality improvements (news focus).— J. Air Poll. Control Assoc., 1975, vol. 25, N 3, p. 292—294.
41. Fujiwara W. Environmental and food contamination with PSB's in Japan.— Science Total Environ., 1975, vol. 4, N 3, p. 219—247.
42. Final Report of the Expert Meeting on the Operation of Integrated Monitoring Programmes. Geneva, 2—5 Sept. 1980, WMO.
43. Gardiner J. The chemistry of cadmium in natural water. 1. A Study of cadmium complex formation using the cadmium specific-ion electrode.— Water Research, 1974, vol. 8, N 1, p. 23—30.
44. Giebel J. Globale Aspekte der Luftver-Schnitzung.— Umschau in Wissenschaftlung Technik, 1975, Bd. 75, N 22, S. 85—90.
45. Goldberg E. D. Health of the Oceans.— La Jolla: Scripps Inst. Oceanogr., 1975.
46. GEMS — The Global Environmental Monitoring System UNEP, 1982.
47. Hans-Walter Georgii. Review of the chemical composition of precipitation as measured by the WMO BAPMoN, WMO. Geneva, 1982.
48. Higginson I. Environment and Cancer.— Practitioner, 1974, vol. 198, p. 621—630.
49. Hundson J. Z., Stunel J. J., Solomon R. L. Measurement of the ambient lead concentration in the vicinity of urbana-champaign, Illinois.— Atm. Environ., 1975, vol. 9, N 11, p. 1000—1006.

50. Information sources on water pollution control (WPCF publication).— J. Water Poll. Control Feder., 1974, vol. 46, N 3.
51. International Symposium, USSR, Tallinn, Oct. 2—10, 1983, (Abstracts), 1983.
52. Leithe W. Die Analyse der Luft und ihrer Verunreinigungen.— Stuttgart: Wiss. Verlageges, 1974.
53. Martin B., Sella T. International monitoring activiting. International problems in environmental monitoring.— In: A Bellagio Conf., 16—18 Feb. 1977. Rockefeller Found., Aug. 1977.
54. Martin B., Sella T. The Global Environmental Monitoring System.— In: A Bellagio Conf., 16—18 Feb, 1977, Rockefeller Found., Aug. 1977, p. 16—24.
55. Maugh T. H. Polychlorinated biphenyls: Still prevalent, but less of a problem.— Science, 1972, vol. 178, N 4059, p. 388—480.
56. Munn R. E. Global Environmental Monitoring System (GEMS).— SCOPE, rep. 3, Toronto, 1973.— 130 p.
57. Noller B. N., Bloom H. The determination of atmospheric particulate lead using low volume sampling and non-flame atomic absorption.— Atm. Environ., 1975, vol. 2, N 5, p. 505—511.
58. Oya F., Ochiai S. Konagaku Summogu us Chiiki tokusei to taisaku oijobi kenkyn taisei nitsuite.— Kogai Taisaku to Gyosei, 1971, vol. 3, N 11, p. 58—77.
59. Programme on Man and the Biosphere (MAB). Task forse on: Pollution monitoring and research in the framework of the MAB Programme, organized jointly by UNESCO and UNEP.— MAB, rep. ser. N 20. Paris: Unesco, 1974.
60. Report of the International Meeting on Monitoring held at Nairobi 11—20 Feb. 1974.— 60 p.
61. Report of the Interagency Working Group on Monitoring on the Development of a Global Environmental Monitoring System.— Nairobi, UNEP, IG. 1/2 Nov. 15, 1975.
62. Robertson L. W., Chynoweth D. P. Another halogenated hydrocarbon.— Environment, 1975, vol. 17, N 6, p. 25—27.
63. Report and studies N 15. The Review of the Health of the Oceans. GESAMP, 1982.
64. Summary Report on the status of the WMO Background air pollution monitoring network as at May 1982. WMO. Geneva, 1982.
65. The Global Environment Monitoring System. Selected Works on Intergrated Monitoring. GEMS Int. Series, N 2, apr. 1980, Nairobi.
66. The monitoring of the environment in the United Kingdom.— London, 1974, p. 66.
67. Westermask T. e. a. Mercury monitoring in Sweden, international problems in environmental monitoring.— In: A. Bellagio Conf., 16—18 Feb. 1977. Rockefeller Found., Aug. 1977.
68. Wilson J. G. Environment and birth defects.— New York: Academic Press, 1973.— 305 p.
69. Woodwell G. M. e a. DDT residues in an east coast estuary: A case of biological consentration of a persistent insecticide.— Science, 1967, vol. 156, p. 821—824.

Chapter 8

1. Атмосферный озон и изменение глобального климата/ Э. Л. Александров, И. Л. Кароль, Ю. С. Седунов и др.— Л.: Гидрометеоиздат, 1982.
2. Бломек Дж. Обращение с радиоактивными отходами.— В кн.: Безопасность ядерной энергетики. М.: Атомиздат, 1980, с. 97—106.
3. Будыко М. И. Эволюция биосферы.— Л.: Гидрометеоиздат, 1984.
4. Винников К. Я. и др. Современные изменения климата северного полушария.— Метеорология и гидрология, 1980, № 6, с. 5—17.
5. Влияние увеличения количества углекислоты газа в атмосфере на климат. Материалы советско-американского совещания. 15—20 июня 1981 г.— Л.: Гидрометеоиздат, 1982.
6. Гудков А. Н., Иванов В. И., Колобашкин В. М. и др. Широтное распределение криптона-85 в приземном слое атмосферы.— Морские гидрофизические исследования, 1978, № 2, с. 83—92.
7. Девишева М. Н., Смирнов Ю. В., Круглов А. К. Прогнозирование потребности в энергии и роль ядерной энергетики.— Атомная техника за рубежом, 1979, № 10, с. 3—8.
8. Израэль Ю. А. Развитие энергетики и ее влияние на окружающую среду.— В кн.: Актуальные проблемы охраны окружающей природной среды. Труды советско-западногерманского симпозиума. Мюнхен, 1981.
9. Израэль Ю. А., Назаров И. М., Рябошапко А. Г. Проблема антропогенных выбросов криптона-85 в атмосферу.— Метеорология и гидрология, 1982, № 6, с. 5—15.
10. Израэль Ю. А., Ровинский Ф. Я. Мониторинг фонового загрязнения природных сред.— ДАН СССР, 1982, т. 264, № 5, с. 1109—1112.
11. Источники и действие ионизирующей радиации. Научный комитет ООН по действию атомной радиации. Доклад за 1977 год Генеральной Ассамблеи с приложениями в 3-х томах. Том I. ООН. Нью-Йорк, 1978, 382 с.
12. Кароль И. Л. Системное моделирование «тепличного эффекта», переноса и фотохимии газовых примесей в атмосфере.— Изв. АН СССР. Физика атмосферы и океана, 1980, т. 16, № 1, с. 20—31.
13. Кароль И. Л. и др. Атмосферный озон и изменение глобального климата.— Л.: Гидрометеоиздат, 1982.
14. Кароль И. Л. Газовые примеси в атмосфере и изменения глобального климата.— Метеорология и гидрология, 1983, № 8, с. 108—116.
15. Келлог У. У. Влияние деятельности человека на климат.— Бюллетень ВМО, 1977, т 26, № 4, с. 285—299.
16. Кондратьев К. Я. Извержение вулкана Эль-Чичон: наблюдавшиеся и возможные воздействия на атмосферу.— М., 1983.
17. Криптон-85 в атмосфере. Накопление, биологическое значение и способы контроля. Рекомендации национального комитета США по радиационной защите и измерениям, 1975.— М.: Атомиздат, 1978.— 64 с.
18. Легасов В. А., Кузьмин И. И. Проблемы энергетики.— Природа, 1981, № 2, с. 8—23.

19. Мастерс Р. XI Мировая энергетическая конференция.— Атомная техника за рубежом, 1981, № 4, с. 37—40.
20. Отчет пятой сессии Координационного комитета по озонному слою, ЮНЕП. Копенгаген, 1981.
21. Протоколы 33-й сессии Исполкома Всемирной метеорологической организации.— Женева, 1981.
21 а. Раунер Ю. Л. Климат и урожайность зерновых культур.— М.: Наука, 1981.— 163 с.
22. Славин Г. Б. Мировой энергетический баланс и производство электроэнергии к 2000 г.— Атомная техника за рубежом, 1979, № 5, с. 35—38.
23. Смирнов Б. М. Экологические проблемы атмосферы Земли. Успехи физических наук, 1975, т. 117, вып. 2, с. 313—332.
24. Стырикович М. А., Внуков А. К. Экологические проблемы энергетики. В кн.: Всесторонний анализ окружающей природной среды. Труды IV советско-американского симпозиума. Л.: Гидрометеоиздат. 1981, с. 70—77.
25. Стыро Б. И., Буткус Д. В. Некоторые проблемы инертных радиоактивных газов — продуктов атомных электростанций.— Физика атмосферы. 1977, № 3, с. 15—26.
26. Тальрозе В. Л., Ларин И. К. Ионизация атмосферы и озон.— В кн.: Комплексный глобальный мониторинг загрязнения окружающей природной среды. Труды II Международного симпозиума. Л.: Гидрометеоиздат, 1982, с. 62—69.
27. Тертышник Е. Г., Сиверин А. А., Баранов В. Г. Концентрация криптона-85 в атмосфере над территорией СССР в 1971—75 гг.— Атомная энергия, 1977, т. 42, вып. 2, с. 145.
28. Уланова Е. С. Агрометеорологические условия и урожайность озимой пшеницы.— Л.: Гидрометеоиздат, 1975.
29. Уланова Е. С. Агрометеорологические условия и продуктивность зерновых культур.— Метеорология и гидрология, 1984, № 5, с. 95—100.
30. Barret E. W.— Sol. Energy, 1971, vol. 13, p. 323.
31. Boek W. L. Meteorological consequences of atmospheric krypton-85. Science, 1976, vol. 193, N 4249, p. 195—198.
32. Bolin B., Kaling C. D. Large-scale atmospheric mixing as deduced from the seasonal and meridional variations at carbon dioxide.— J. Geophys. Res., 1963, vol. 68, N 13, p. 3899—3920.
33. Chlorofluorocarbons and their effect on stratospheric ozone. Pollution Paper N 5, London, 1976.
34. Chlorofluorocarbons and their effect on stratospheric ozone. Second Report. Pollution Paper N 15. London, 1979.
35. Climate monitor, 1981, vol. 10, N 5.
36. Hafele W. A global and long-range picture of energy developments,— Science, 1979, vol. 209, p. 174—182.
37. Halocarbons: Effects on Stratospheric Ozone. Panel on Atmospheric Chemistry.— Washington, D. C., 1976.
38. Manabe S., Wetherald R. T. The effect of doubling the CO_2 concentration on the climate of a general circulation model.— J. Atm. Sci., 1975, vol. 32, N 1, p. 3—15.
39. Manabe S., Wetherald R. T. On the distribution of climate change resulting from an increase in CO_2 content of the atmosphere.— J. Atm. Sci., 1980, vol. 37, N 1, p. 99—118.
40. Report of the WMO (CAS) Meeting of experts on the CO_2

concentrations from pre-industrial times to I. G. Y. Report N 10. WMO—ICSU, 1983.
41. Rotty R. M. Present and future production of CO_2 from fossel fuels. A global appraisal. Workshop on the global effects of carbon dioxide from fossil fuels, p. 36—43 (1979).
42. Rozanski K. Krypton-85 in the atmosphere 1950—1977: a data review.— Environment Intern., 1979, vol. 2, N 3, p. 139—143.
43. Stuiver M., Quay P. D. Earth and planetary.— Sci. Letters, 1983.
44. Turco R. P. et al. Stratospheric aerosols and climate.— Nature, 1980, vol. 283, p. 283—286.
45. Coleman J. R., Liberace R. Radiol. Health Data Rep., 1966, vol. 7.
46. Holdgate M. W. et al. World environment trends between 1972 and 1982.— Environ. Conservation, 1982, vol. 9, N 1, p. 11—29.
47. Joint WMO/ICSU/UNEP Meeting of experts on the assessment of the role of CO_2 on climate variations and their impact (Villach, Austria, November 1980), World Climate Programme Series, N 3, 1981.
48. McDonald A. Energy in a finite world, IIASA, Ex. rep. 4, Laxenburg, Austria, 1981.
49. Robock A. El Chichon cloud cause considerable cooling.— Preprint Univ. of Maryland, 1982.
50. Wardszko T., Nidecka. Krypton-85 in atmospheric air: measurement method and results for 1975—77.— Nukleonika, 1978, vol. 23, N 8.

Chapter 9

1. Гречушкина М. П., Израэль Ю. А., Петров В. Н. Возможные технические подходы для выработки норм безопасного проведения ядерных взрывов в мирных целях.— В кн.: Peaceful Nuclear Explosions, IV, IAEA, Vienna, 1975.
2. Гудериан Р. Загрязнение воздушной среды.— М.: Мир, 1979.
3. Израэль Ю. А. Изотопный состав радиоактивных выпадений.— Л.: Гидрометеоиздат, 1973.
4. Израэль Ю. А., Михайлова И. Е., Прессман А. Я. Модель оперативной оценки трансграничных потоков загрязняющих веществ.— ДАН СССР, 1980, т. 253, № 4, с. 848—852.
5. Израэль Ю. А., Петров В. Н. Распространение радиоактивных продуктов подземных ядерных взрывов на большие расстояния— В кн.: Атомные взрывы в мирных целях. М.: Атомиздат, 1970.— 87 с.
6. Израэль Ю. А., Петров В. Н. Диффузия и осаждение радиоактивных продуктов из облака подземного ядерного взрыва.— В кн.: Peaceful Nuclear Explosions, IV, IAEA, Vienna, 1975.
7. Израэль Ю. А., Филиппова Л. М. и др. Влияние фонового загрязнения природной среды на биоту: проблемы оценки и прогноза.— В кн.: Проблемы экологического мониторинга и моделирования экосистем. Л., 1982, т. 5, с. 6—18.
8. Инсаров Г. Э., Филиппова Л. М. и др. Модель влияния фоновых концентраций сернистого ангидрида на растения.—

В кн.: Проблемы экологического мониторинга и моделирования экосистем. Л., 1981, т. 4, с. 235—250.

9. Инсарова И. Д. Влияние сернистого газа на лишайники.— В кн.: Проблемы экологического мониторинга и моделирования экосистем. Л., 1982, т. 5, с. 33—48.
10. Кароль И. Л. Радиоактивные изотопы и глобальный перенос в атмосфере.— Л.: Гидрометеоиздат, 1972.
11. Кислотные дожди/Ю. А. Израэль, И. М. Назаров и др.— Л: Гидрометеоиздат, 1983.
12. Кунина И. М. и др. Действие сернистого ангидрида на метаболизм растительной клетки.— В кн.: Проблемы экологического мониторинга и моделирования экосистем. Л.: Гидрометеоиздат, 1979, т .2, с 87—124.
13. Макнелис Д. Оценка качества окружающей среды западных энергетических районов.— В кн.: Всесторонний анализ окружающей природной среды. Труды IV советско-американского симпозиума. Л.: Гидрометеоиздат, 1981, с. 78—85.
14. Мальм У. К. Методология интерпретации и изменения видимости.— В кн.: Всесторонний анализ окружающей природной среды. Труды IV советско-американского симпозиума. Л.: Гидрометеоиздат, 1981, с. 241—253.
15. Материалы многосторонней конференции по установлению причин и предотвращению ущерба лесам и водоемам в результате загрязнения воздуха в Европе. Мюнхен, 24—27.VI.1984; резолюция М 11/4 от 27 июня 1984 г.
16. Проблемы биосферы, 1981, № 2.
17. Смит Л. Ф., Миллз М. Т. Региональный анализ качества воздуха и его применение в бассейне р. Огайо.— В кн.: Всесторонний анализ окружающей природной среды. Труды IV советско-американского симпозиума. Л.: Гидрометеоиздат, 1981, с. 196—214.
18. Соколов В. А., Виноградов Г. А. Изучение адаптации рыб к различным значениям pH наружной среды.— Биология внутренних вод. Инф. бюлл. АН СССР, 1977. вып. 33.
19. Филиппова Л. М., Ниомаркаи К. и др. Влияние O_2 на растения.— Будапешт, 1982.
20. Acid Rain. Report of the European Conf. Göteborg, 1981.
21. Acidification today and tomorrow. Swedish Ministry of Agriculture, Environment'82 Committee. Stockholm, 1982.
22. Beamish R. I. Loss of fish population's from unexploited remote lakes in Ontario, Canada as a consequence of atmospheric fallout of acid.— Water Res., 1974, vol. 8, N 1, p. 85—95.
23. Cherry et al. Recovery of aquatic bacterial populations in a stream after cessation of chemical pollution.— Water, Air and Soil Pollution, 1977, vol. 1, p. 95—101.
24. Dickson W. The acidification of Swedish lakes.— Rep. Inst. Freshw. Res. Drottningholm, 1975, vol. 54, p. 8—20.
25. Ecological impact of acid precipitation. Proc. of Intern. Conf. Oslo, 1980.— 383 p.
26. Effect of acid precipitation of terrestrial ecosystems, Yuild Inn, Scerborough. May 22—26, 1978.
27. Environment in Europe, air and water pollution, including transboundary flows. Symp. of the Inter-Parlamentary Union, IPU, Geneva, 1980.

28. Environmental migration of long-lived redionuclides. Proc. of Symp. IAEA. Vienna, 1982.
29. Eriksson et al. Ecological effects at lime treatment of acidified lakes and rivers in Sweden.— Hydrologia, 1982.
30. Falconer R. E. et al. Determination of cloud water acidity at a mountain observatory in the Adirondack mountains of New York state.— J. Geophys. Res., 1980, vol. 85, N C12, p. 7465—7470.
31. Grenfelt P. Sources and distribution of acidifying compounds.— In: Acid Rain. Report of the European Conf. Göteborg, 1981, p. 31—47.
32. Haut H. Van. Die Analyse von Schwefeldioxidwirkungen auf pflanzen im Laboratoriumsversuch.— Staub, 1961, Bd 21, S. 52—56.
33. Ivanov M. V. The global biogeochemical sulphur cycle.— In: Some Perspectives of the Major Biogeochemical Cycles. SCOPE, 1981.
34. Izrael Yu. A Acid rain and its environmental effects.— In: Long-range transport of sulphur in the atmosphere and acid rain. Lectures presented at the thirty-third session of the WMO Executive Committee. WMO-N 603, 1983, p. 1—15.
35. Izrael Yu. A., Nazarov I. M., Ryaboshapko A. G. Metodology of investigations of long-range transport of sulphur compounds. Report on Seminar on environmental impact assessment. Villach (Austria), 24—29 Sept. 1979 (Env/sem 10/Com. 2).
36. Jacobson J. S. et al. Ecological effects of acid rain.— JAPCA, 1983, vol. 33, N 11, p. 1031—1054.
37. Jacobson J. S., Troiano J. J. Dose-reponse functions for effects of acidic precipitation on vegetation.— Water Quality Bull., 1983, vol. 8, N 2, p. 67—71.
38. Lesinski I. A. Effects of acid precipitation on forest in Poland: Selected topics.— Water Quality Bull., 1983, vol. 8, N 3, p. 156—159.
39. Leonard R. L. et al. Some measurements of the pH and chemistry of precipitation at Davis and lake Tahoe, California.— Water, Air and Soil Pollution, 1981, vol. 15, p. 153—167.
40. Likens G. E., Borman F. H. Acid rain: a serious regional environmental problem.— Science, 1974, vol. 184, N 4142, p. 1176—1179.
41. Luckat S. Quantitative Unterschung des Einflusses von Luftverunreinigungen bei der Lerstörnag von Naturstein.— Staub—Reinhalt-Luft, 1982, Bd 41, S. 440—442.
42. Memorandum of intent on transboundary air pollution. US—Canada, Fin. Rep., Jan. 1983.
43. Nieboer E. et al. The phytotoxicity of sulfur dioxide in relation to measurable responses in lichen. Effect of air pollutants on plants. Cambridge Univ. Press, 1976, p. 61—65.
44. Nicholson I. A., Paterson I. S. Aspects of acid precipitation in relation to vegetation in the United Kingdom.— Water Quality Bull., 1983, vol. 8, N 2, p. 59—66.
45. Niemann B. L. An analysis of regional acid deposition episodes the feasibility of their deterministic simulation.— Water Quality Bull., 1983, vol. 8, N 3, p. 143—149.
46. Okita T. Acid precipitation and related phenomena in Japan.— Water Quality Bull., 1983, vol. 8, N 2, p. 101—108.

47. Renberg I. et al. The pH history of lakes in Sweden as calculated from the subfossil diatom flore of the sediments.— Ambio, 1982, vol. 1, p. 30—33.
48. Report of Norges Naturvernforbund.— In: Acid Rain. Report of the European Conf. Göteborg, 1981, p. 85—87.
49. Schikorr Y. Die Bedeutung des Scwefeldioxyls für die atmospherische Korrosion der Mettalle.— Werkstoffe und Korrosion, 1964, Bd. 15, S. 457—463.
50. Schofield C. I. Acid precipitation: Effect on fish.— Ambio, 1976, vol. 5, N 5—6, p. 228—230.
51. Sekueria R. Acid rain: an assessment based on acid-base considerations.— J. Air Poll. Control Assoc., vol. 32, N 3, p. 241—245.
52. Stern H. et al. Rettet den Wald.— München, 1983.
53. Summers P. W. A global perspective on acid deposition, its sources and atmospheric transport.— Water Quality Bull., 1983, vol. 8, N 2, p. 81—88.
54. United States — Canada: Memorandum of Intent on Transboundary Air Pollution: Strategies, Development and Implementation. Interim Rep., Febr. 1981, in [16].
55. Vannerberg N. G. Corrosion induced by atmospheric Sulphur contaminants.— In: Acid Rain. Report of the European Conf. Göteborg, 1981, p. 59—62.
56. Whelpdate D. M. Acid deposition: distribution and impact.— In: Long-range transport of sulphur in the atmosphere and acid rain. Lectures presented at thirty-third session of the WMO Executive Committee. WMO-N 603, 1983, p. 39—53.
57. Whelpdale D. M. Acid deposition: Distribution and impact.— Water Quality Bull., 1983, vol. 8, N 2, p. 72—80.
58. Wröbel S. Cracow, its environments and air pollution.— In: Acid Rain. Report of the European Conf. Göteborg, 1981, p. 89—92.

Chapter 10

1. Александров Э. Л. Человек и стратосферный озон.— Л.: Гидрометеоиздат. 1979.
2. Алексахин Р. М. Радиоактивное загрязнение почвы и растений.— М.: Изд-во АН СССР, 1963.
3. Будыко М. И. Глобальная экология.— М.: Мысль, 1979.
4. Вульфсон Н. И., Левин Л. М. Исследование распространения струи метеотрона в облачной среде применительно к активным воздействиям.— Труды ИПГ, 1981, вып. 46, с. 50—68.
5. Действие ядерного оружия/Под ред. Э. Глесстона.— М.: Воениздат, 1966.
6. Израэль Ю. А. Изотопный состав радиоактивных выпадений.— Л.: Гидрометеоиздат, 1973.
7. Израэль Ю. А. Мирные ядерные взрывы и окружающая среда.— Л.: Гидрометеоиздат, 1974.
8. Израэль Ю. А. Экологические последствия возможной ядерной войны.— Метеорология и гидрология, 1983, № 10, с. 5—10.
9. Израэль Ю. А., Петров В. Н., Северов Д. А. О влиянии атмосферных ядерных взрывов на содержание озона в стратосфере.— Метеорология и гидрология, 1983, № 9, с. 5—13.

10. Израэль Ю. А. и др. Проблема антропогенных выбросов криптона-85 в атмосферу.— Метеорология и гидрология, 1982, № 6, с. 5—15.
11. Кароль И. Л. и др. Газовые примеси в атмосфере.— Л.: Гидрометеоиздат, 1983.
12. Кондратьев К. Я. Извержение вулкана Эль-Чичон: наблюдавшиеся и возможные воздействия на атмосферу.— Москва, 1983.
13. Радиоактивные выпадения от ядерных взрывов/Пер. с англ. под ред. Ю. А. Израэля.— М.: Мир, 1968.
14. Фрейлинг Э. и др. Образование частиц-осколков ядерного взрыва.— В кн.: Радиоактивные выпадения от ядерных взрывов. М.: Мир, 1968.
15. Aleksandrov A. A., Stenchikov G. G. On the modelling of the climatic consequences of the nuclear war. The Comp. Centre of AS USSR. Moscow, 1983.
16. Bryson R. A., Goodman B. M. The climatic effect of explosive volcanic activity: analysis of the historical data.— In: Atm. Effects and Potential Climatic. Impact of the 1980 Eruption of Mount St. Helens. Proc. of the Symp. NASA Conf. Publ. N 2240, p. 191—202 (1982).
17. Carbon dioxide and climate: A second assessment. Nat.— Washington D. C.: Academic Press, 1982.
18. Chang J. S. et al.— J. Geophys. Res., 1979, vol. 84, p. 1755.
19. Crutzen P. J., Briks J. W. The atmosphere after a nuclear war: twilight at noon.— Ambio, 1982, vol. 11, N 2—3, p. 114—125.
20. Ehrlich P. R. Long-term biological consequences of nuclear war.— Science, 1983, vol. 222, N 4630, p. 1283—1300.
21. Hidalgo H., Crutzen P. The tropospheric and stratospheric composition perturbed by NO_x emissions of highlatitude aircraft.— J. Geophys. Res., 1977, vol. 82, N 37.
22. Johnston H. S. Expected short-term local effect of nuclear bombs on stratospheric ozone.— J. Geophys. Res., 1977, vol. 82, N 20.
23. Johnston H. S. et al.— J. Geophys. Res., 1973, vol. 78, p. 6107.
24. Lockwood G. W. Spectrally resolved measurements of the El Chichon cloud from Flagstaff Arisona EOS. Trans. Amer. Geophys. Union, 1982, vol. 63, p. 897.
25. Long-term world-wide effects of multiple nuclear-weapons detonation. Washington, D. C., 1975.
26. Nicolet M. Stratospheric ozone: An introduction of its study.— Rev. Geophys. Space Phys., 1975, vol. 13, N 5.
27. Protection against depletion of stratospheric ozone by chlorofluorocarbons. Washington, D. C. 1975.
28. Robock A. El Chichon cloud cause considerable cooling.— Preprint Univ. of Maryland, 1982.
29. Turco R. P. et al. Analysis of the physical, chemical, optical and hystorical impacts of the 1908 Tunguska meteor fall.— Icarus, 1982, vol. 50, N 1.
30. Turco R. P. et al. Nuclear winter: Global consequences of multiple nuclear explosions.— Science, 1983, vol. 222, N 4630, p. 1283—1291.

Chapter 11

1. Вернадский В. И. Избранные сочинения.— М., 1954, т. 1.— 612 с.
2. Влияние увеличения количества углекислого газа в атмосфере на климат. Материалы советско-американского совещания.— Л.: Гидрометеоиздат, 1982.
3. Израэль Ю. А., Симонов А. И., Цыбань А. В. Исследование загрязнения Тихого океана.— Земля и Вселенная, 1980, № 3.
4. Израэль Ю. А., Цыбань А. В. Проблемы мониторинга экологических последствий загрязнения океана.— Л.: Гидрометеоиздат, 1981.
5. Израэль Ю. А., Цыбань А. В. Экологические основы мониторинга состояния Мирового океана.— В кн.: Комплексный глобальный мониторинг загрязнения окружающей природной среды. Труды II Международного симпозиума. Л.: Гидрометеоиздат, 1982, с. 44—62.
6. Израэль Ю. А., Цыбань А. В. Об ассимиляционной емкости Мирового океана.— ДАН СССР, 1983. т. 272, № 3, с. 702—704.
7. Израэль Ю. А., Цыбань А. В. Экологические последствия загрязнения Мирового океана.— В кн.: Исследования океана. Материалы II Всесоюзного съезда океанологов. М.: Наука, 1984.
8. Израэль Ю. А., Цыбань А. В. Экология и проблемы комплексного глобального мониторинга Мирового океана.— Метеорология и гидрология, 1984, № 8, с. 18—33.
9. Израэль Ю. А. Международный симпозиум «Комплексный глобальный мониторинг Мирового океана».— Метеорология и гидрология, 1984, № 8, с. 5.
10. Исследование экосистемы Берингова моря/Ю. А. Израэль, А. В. Цыбань, Ю. Е. Казаков и др.— Л.: Гидрометеоиздат, 1983.
11. Комплексный глобальный мониторинг Мирового океана. I Международный симпозиум. Тезисы докладов.— М., 1983.
12. Лисицын А. П., Виноградов М. Е. Глобальные закономерности распределения жизни в океане и их отражение в составе донных осадков. Образование и распределение биогенных осадков.— Изв. АН СССР. Сер. геология, 1982, № 4.
13. Мак-Интайр А. Д. Успехи в изучении эффектов загрязнения морской среды.— В кн.: Комплексный глобальный мониторинг загрязнения окружающей среды. Труды Международного симпозиума. Л.: Гидрометеоиздат, 1980, с. 120—130.
14. Морозов Н. П., Петухов С. А. Содержание и распределение тяжелых металлов в компонентах экосистемы Балтийского моря.— В кн.: Исследование экосистемы Балтийского моря. Л.: Гидрометеоиздат, 1981, вып. 1, с. 98—132.
15. Озмидов Р. В. Основные закономерности распространения загрязняющих примесей в Мировом океане.— Метеорология и гидрология, 1984, № 8, с. 51—58.
16. Поликарпов Г. Г., Егоров В. Н. Способность морских экосистем к удалению радиоактивных и химических загрязнений из фотического слоя.— Вестник АН УССР, 1981, № 2.
17. Радиоактивность вод Мирового океана и поведение в океане некоторых осколочных элементов/В. М. Вдовенко и др.—

В кн.: Peaceful uses of atomic energy. Proc. of the Fourth Intern. Conf. Geneva, Publ. UN—IAEA, 1972, vol. 2.
18. Терзиев Ф. С., Симонов А. И. Проблемы мониторинга химических загрязнений Мирового океана.— Метеорология и гидрология, 1984, № 8, с. 42—50.
19. Чумичев В. Б. Стронций-90 в водах Тихого океана в 1964—1966 гг.— Труды ИЭМ, 1974, вып. 3 (42).
20. Цыбань А. В. Бактерионейстон и бактериопланктон шельфовой области Черного моря.— Киев: Наукова думка, 1970.
21. Цыбань А. В. Морская пена как экологическая ниша для бактерий.— Гидробиологический журнал, 1971, № 3.
22. Цыбань А. В. Метод расчета микробной деструкции нефтяных углеводородов.— В кн.: Исследование экосистемы Балтийского моря. Л.: Гидрометеоиздат, 1981, с. 61—68.
23. Цыбань А. В., Теплинская Н. Г. Эколого-физиологические свойства липолитических бактерий в море.— Океанология, 1981, № 5.
24. An assessment of mercury in the environment. Washington, D. C., 1978.
25. Bernhard M., Zattera A. Major pollution in marine environment.— In: Marine Pollution and Waste Disposal. Oxford, New York: Pergamon Press, 1975, p. 195—300.
26. Blumer M., Saas G. Oil pollution persistence and degradation of applied fuel oil.— Science, 1972, vol. 176.
27. Brisou J. The health situation around the Mediterranean.— Ambio, 1977, vol. 6, N 6.
28. Butler J. N. Pelagic tar.— Sci. Amer., 1975, vol. 232, N 6.
29. Chester R., Stoner J. H. The distribution of zinc, nickel, manganese, cadmium copper and iron in some surface water from the World Ocean.— Mag. Chem., 1974, vol. 2, N 17.
30. Cooperative research report on the feasibility of effects monitoring, N 75-ICES, Denmark, 1978.
31. Disposal of radioactive wastes into seas, oceans and surface waters.— In: Proc. Symp. Vienna, IAEA, 1966.
32. Duce R. A., Hoffman J. L. Atmospheric vanadium transport to the ocean.— Atm. Environ., 1976, vol. 10, p. 1989—1996.
33. Elder D. L., Fowler S. W. Polychlorinated biphenyls: Penetration into the deep ocean by zooplankton fecal pellets transport.— Science, 1977, vol. 197, p. 459—461.
34. Fitzgerald W. F. Mercury studies of seawater and rain: geochemical flux transferred. Window and Duce. Toronto, 1976.
35. Fowler S. W., Small L. F. Sinking rates of euphausiid fecal pellets.— Limnology and Oceanography, 1972, vol. 17, p. 293—296.
36. Fowler S. W. Trace elements in zooplankton particulate products.— Nature, 1977, vol. 269, p. 51—53.
37. Global oil pollution. Results o MAPMOPP, IGOS Pilot Project on Marine Pollution (Petroleum Monitoring). Publ. JOC UNESCO. Paris, 1981.
38. Goldberg E. D. Chemical invasion by man.— McCrow Hill Yearbook Science and Technology, 1970.
39. Goldberg E. D. The health of the oceans.— Paris: UNESCO Press, 1976.
40. Harvey G. R. et al. Observations of the distribution of chlorinated hydrocarbons in Atlantic Ocean organisms.— J. Mar. Res., 1974.

41. Ivanov M. V. The global biogeochemical sulphur cycle in some perspectives of the major biogeochemical cycles. SCOPE, 1981.
42. Israel Yu. A., Tsiban A. V. Problems of monitoring the ecological consequences of ocean pollution.—iJ. Oceanogr. Soc. Japan, 1981, vol. 36.
43. Jensen S., Jernelov A. Biological methylation of mercury in aquatic organisms.— Nature, 1969, vol. 223.
44. Jones P. G. W. Review of the distribution of selected trace metals in the water of the North Atlantic.— In: JCES/ACMP Report. Coop. Res., Rep., 1975, vol. 50, Annex 3.
45. Lal D. The oceanic microcosm of particles.— Science, 1977, vol. 198, p. 997—1009.
46. Levy E. M., Walton A. High seas oil pollution: particulate petroleum residues in the North Atlantic.— J. Fish. Res. Bul. Can., 1976, vol. 33, p. 2781—2791.
47. Pesticide Handbook. USA, 1974.
48. Petroleum in the marine environment. Report of U. S. National Academy of Sciences Workshop on Inputs, Fates and the Effects of Petroleum in the Marine Environment. Washington, D. C., 1975.
49. Phtalate ester plasticizers: A new class of marine pollutants/C. S. Giam, H. S. Chan, G. S. Neff, E. L. Atlas.— Science, 1978, vol. 199, p. 419—421.
50. Pilpel N. The natural fate of oil in the Sea.— Endeavor, 1968, vol. 27.
51. Rosenfield A. Recent environmental studies of neoplasms in marine shellfish.— In: Progress in Experimental Tutor Research, 1976, vol. 20 (Tumors in Aquatic Animals).
52. Shuval H. I. Studies on bacterial and viral contamination of the marine environment.— Rev. Int. Oceanogr. Med., 1978, vol. 50, N 4.
53. Simmon V. F. Structural correlation of carcinogenic and mutagenic alkylhedes.— In: Proc. Second FDA Office of Science Summer Symposium. US Naval Academy, 1977.
54. Sorokin Yu. I. Bacterial populations as component of oceanic ecosystem.— Intern. J. on Life in Oceans and Coastal Waters, 1971, vol. 11, N 2.
55. Suess E. Particulate organic carbon flux in the oceans — surface productivity and organic utilization.— Nature, 1980, vol. 288, N 5788, p. 260—263.
56. The chemical composition of a Greenland glacier/ H. V. Weiss, K. Bertine, M. Koide, E. D. Goldberg.— Geochim. Cosmochim. Acta, 1975, vol. 39, N 1, p. 1—11.
57. The World Environment. A Rep. UNEP, 1982.
58. Transport of pollutant lead to the ocean and within ocean ecosystems/C. Patterson, D. Settle, Schauke B., M. Burnett.— In: Marine Pollutant Transfer. Lexington (Massachusetts) D. C. Health and Company, 1976, Ch. 2, p. 23—26.
59. Tsiban A. V. Bacterioneuston and problem of degradation in surface films of organic substances released into the sea.— Progress in Water Technology, 1975, vol. 7, N 3, 4.
60. Tsiban A. V. Scientific approaches to biological monitoring of the Baltic Sea.— Rapp. P-v Reun., Cons. Intern. Explor. Mer., 1980, 179.
61. Walch P. R., Duce R. A., Fasching K. L. Considerations

on the enrichment, sources and flux of arsenic in the troposphere.— J. Geophys. Res., 1979, vol. 84, NC4, p. 1719—1726.

Conclusion

1. Будыко М. И. Эволюция биосферы.— Л.: Гидрометеоиздат, 1984.— 488 с.
2. Вернадский В. И. Несколько слов о ноосфере.— Успехи современной биологии, 1944, т. 18, вып. 2.
3. Всемирная конференция «Климат и человечество». Женева, февраль 1979. Публикация ВМО, 1979.— 406 с.
4. Егоров В. А. и др. Математические модели глобального развития.— Л.: Гидрометеоиздат, 1980.— 192 с.
5. Израэль Ю. А. Охрана природной среды и рациональное природопользование в СССР.— Вопросы философии, 1979, № 6, с. 119—127.
6. Израэль Ю. А. Экологические последствия возможной ядерной войны.— Метеорология и гидрология, 1983, № 10, с. 5—10.
7. Израэль Ю. А. Проблемы охраны природной среды и пути их решения.— Л.: Гидрометеоиздат, 1984.— 48 с.
8. Колбасов О. С. Экология: политика — право.— М.: Наука, 1976.— 230 с.
9. Коммонер Б. Замыкающийся круг.— Л.: Гидрометеоиздат, 1974.
10. Марчук Г. И. Математическое моделирование в проблеме окружающей среды.— М.: Наука, 1982.— 320 с.
11. Материалы Октябрьского (1984 г.) пленума ЦК КПСС.— М.: Политиздат, 1984.
12. Многосторонняя конференция по установлению причин к предотвращению нанесения ущерба лесам и водоемам в результате загрязнения воздуха в Европе. Мюнхен, 24—27 июля 1984 г., резолюция M/II/4 от 27 июля 1984 г.
13. Окружающая среда.— М.: Мысль, 1983.
14. Политика Советского Союза в области охраны окружающей природной среды, Москва, 1974. Заявление советской делегации, представленное в г. Спокане, США, на выставке ЭКСПО-74 23 октября 1974 г.
15. Федоров Е. К. Экологический кризис и социальный прогресс.— Л.: Гидрометеоиздат, 1977.— 176 с.
16. Environmental warfare. A technical, legal and policy appreisal/A. H. Westing (ed.).— London—Philadelphia: SIPRI, Taylor and Francis, 1984.
17. Forrester J. W. World dynamics.— Cambridge, Mass.: Wright-Allen Press Inc., 1971.
18. Holdgate M. Change in world environment during the last ten years and theire emplication for action. Report prepared for Interparlamentary Conference on Environment, Nairoby, 26 November—1 December 1984. IPU, Environment/84/5, Geneva, October 1984.
19. Izrael Yu. World wide environmental issues: atmosphere and oceans. Report prepared for Interparlamentary Conference on Environment, Nairoby, 26 November—1 December 1984. IPU, Environment/84/6, October 1984.
20. Meadows D. L. et al. Dynamics of growth in a finite world.— Cambridge, Mass.: Wright-Allen Press Inc., 1974.

21. Mesarovic M., Pestel E. Mankind at the turning point.— In: Second Rep. to the Club of Rome. N. Y.: E. P. DUTTO and Co., Inc. Readers Digest Press, 1974.
22. The global possible, resources, development and the new century. The Statement of an International Conference, May 2—5, 1984; World Resources Institute, 1984.
23. The global 2000; Antaring the Twenty-First Century. U. S. Government Printing Office, 1980.
24. The World Environment 1972—1982. A Report by UNEP, Ed. by Holdgate et al. Dublin: Tycooby Intern. Publ. Ltd., 1982.

MIX
Papier aus verantwortungsvollen Quellen
Paper from responsible sources
FSC® C105338

If you have any concerns about our products,
you can contact us on
ProductSafety@springernature.com

In case Publisher is established outside the EU,
the EU authorized representative is:
Springer Nature Customer Service Center GmbH
Europaplatz 3, 69115 Heidelberg, Germany

Printed by Libri Plureos GmbH
in Hamburg, Germany